COMPETITION AND EFFICIENCY IN INTERNATIONAL FOOD SUPPLY CHAINS

Why have food crises seemingly become more frequent in recent years, compared to the last few decades? This book examines an array of different issues and distortions that are causing food supply chain dysfunction in many countries, particularly for staple non-perishable foods such as grains, oilseeds, pulses and sugar. It outlines the underlying changes that are currently occurring, which will have an influence on the direction of future food supply chains, and provides some solutions to current food security problems. Based on an analysis from total regulation in the 1950s–60s through to deregulation during the 1980s–90s, as well as post-deregulation, it focuses on liberal trade and deregulation as a more successful solution to creating efficiencies in food supply chains and distribution of storable food commodities. The author highlights a common thread of either farmers using government for vested-interest intervention, or autocratic governments seeking market and supply-chain power.

The book examines the role of government after 70 years of food supply chain intervention. It discusses the role of commercial 'trade' markets and cluster industries and how these can quickly disintegrate when price distortions occur. The author studies both food importing and exporting countries and concludes that commingled commoditization of food has led to increased hoarding, corruption, and dependence on food aid. He argues that a competitive food supply chain that has minimum intervention is more likely to provide future food security. In conclusion the book emphasizes that adequate rewards, competition, and striving for supply chain efficiencies are the essences of sustainable food security.

John Williams is the Executive Director of the Australian Food Security Institute. He has rural science-economics degrees, a master in agribusiness, and a PhD on farmer decision-making psychology from the University of Melbourne, and has studied agricultural food supply chains in 50 countries. His activities include commodity analysis research, journal publications, consultancy and training. He has educated farmers, merchants, end-users, bankers, service industries, and students in agricultural commodity markets, food supply chains, and risk management.

COMPETITION AND EFFICIENCY IN INTERNATIONAL FOOD SUPPLY CHAINS

Improving Food Security

John Williams

earthscan
from Routledge

Routledge
Taylor & Francis Group
LONDON AND NEW YORK

First published 2012 by Routledge

Published 2022 by Routlege
2 Park Square, Milton Park, Abingdon, Oxon OX14 4RN
605 Third Avenue, New York, NY 10017

Routledge is an imprint of the Taylor & Francis Group, an informa business

British Library Cataloguing in Publication Data
A catalogue record for this book is available from the British Library

Library of Congress Cataloging in Publication Data
Williams, John, 1949-
Competition and efficiency in international food supply chains: improving food security / John Williams.
p. cm. – (Earthscan food and agriculture)
Includes bibliographical references and index.
1. Food industry and trade–Security measures. 2. Food supply–Government policy. 3. Food security. 4. Sustainable agriculture. I. Title.
HD9000.5.W474 2012
338.1'9–dc23
2011048756

ISBN: 978-0-415-52071-3 (hbk)
ISBN: 978-0-415-52072-0 (pbk)
ISBN: 978-0-203-12236-5 (ebk)

Typeset in Bembo
by Taylor & Francis Books

CONTENTS

PREFACE

Without the strong beliefs of a number of people, there would not have been the motivation to write such a book on storable food supply. To undertake such topics in the book involved much international travelling and studying detailed history dating back at least 250 years, with more than a basic understanding of psychology, politics, government, agriculture, farm management, economics, ideologies, marketing, supply chain management, operations and logistics, food manufacturing, warehousing, bulk handling and container systems, finance, forward and futures markets, merchant activity, 'trade' markets, storage techniques and hoarding behaviour, consumer behaviour, decision-making psychology, international trade and business, foreign policies, management practices, benchmarking efficiency criteria, organizational management, human resource management, warfare, corruption, food 'aid', contract law, competition law, deregulation and privatization, tendering processes, currency markets, and freight transport involving rail, road, port, and shipping. Any examination of supply chains must involve an enormous multi-disciplinary approach, but this can be quite daunting to an experienced manager or even a government.

Due acknowledgement must be given to Jock Monroe, a NSW wheat farmer whose immutable belief in agrarian socialism to improve his social welfare knows no boundaries, despite little evidence of its success from past experiences. There is also acknowledgement to rural politicians such as Bob Katter and Barnaby Joyce who have unlimited optimism in the ability of government to achieve desired rural social outcomes without any paradoxes, despite much evidence to the contrary from Royal Commissions and parliamentary enquiries into food supply chains over the past 90 years.

Acknowledgement must also be extended to Julian Cribb who is one of the global champions for Armageddon fears over the imminent collapse of international food supply chains, despite the global population recently reaching seven billion at the same time that global farmers have languished in bankruptcy over the past 30 years because of food surpluses and low prices. Those farmers that are not approaching bankruptcy are being supported by farm subsidies that are then bankrupting their governments, despite the advocacy that subsidies are required for 'stable' government. Perhaps it may be governments rather than food supply chains that are more at risk.

Appreciation goes to Peter Reading who in the aftermath of the food price spike and the final deregulation of the Australian wheat industry in 2008 claimed that everything that could ever be written about the commodity food supply chain had already been written. It is expected that this book will stimulate much research by others and generate many more books on a range of topics that have previously remained outside of mainstream thought but nevertheless make a vital contribution to global food supply and security.

Special thanks goes to Ted Kleitsch from PremiumData for support over the last 15 years, and to Elizabeth Firth, a scion of the Burns-Philp family, who provided a useful overview in the early drafts of the book.

The book is especially timely given the current deregulation of the Canadian wheat industry, which will draw to a close a turbulent period of nearly a century in agricultural commodity supply chain intervention that began inauspiciously during the First World War. Disbanding the Canadian Wheat Board has removed one of the remaining vestiges of an era during the 1930s that was so characterized by the domination of an eternal optimistic belief in government ability to achieve socially desirable rural outcomes through supply chain intervention. This book will examine how such philosophies evolved and the damage they provoked.

ACRONYMS

BPI	Bribery Payers Index
CAP	Common Agricultural Policy
cif	commission (cost), insurance and freight – destination port price
CME	Chicago Mercantile Exchange
CPI	Consumer Price Index as well as the Corruption Perception Index
FAO	Food and Agriculture Organization
fob	free-on-board (ship) – origin port price
GCB	Global Corruption Barometer
HIPC	Heavily Indebted Poorest Country debt relief programme
ISO	International Organization for Standardization
JIT	Just-in-time
MENA	Middle East (and) North Africa
OECD	Organization of Economic Cooperation and Development
PL	Public Law in the USA
PVC	Polyvinyl chloride
UAE	United Arab Emirates
UNDP	United Nations Development Programme
USD	United States Dollar
USDA	United States Department of Agriculture
USSR	Former Union of Soviet Socialist Republics

LIST OF FIGURES AND TABLES

Figures

Tables

INTRODUCTION

At the time of Thomas Malthus at the end of the eighteenth century, the global population was approaching one billion, which was mostly supported by the three food bowl regions consisting of Continental Europe that stretched from France to Poland and the Ukraine, the Indus–Ganges river system on the Asian Subcontinent, and the Yangtze–Yellow–Mekong river systems in China and Indo-China. Malthus was concerned that population increased in geometric proportions, while food production increased in arithmetic proportions.

However, the Louisiana Purchase of 1803 and the defeat of Britain in East Louisiana in 1815 opened up the whole of the Mississippi waterway drainage basin that stretched into the southern Canadian prairies as the fourth global food bowl. Britain felt so threatened by potential new food imports from this food bowl region that it immediately introduced the Corn Laws, while many countries in Continental Europe retreated into increased protectionism and isolationism that has largely continued until now.

It only took until 1927 for the global population to reach two billion. The Malthusian fears of population being limited by food shortages were initially very short-lived, but perhaps revived during recent times as the population increased to three billion in 1959, four billion in 1974, five billion in 1987, six billion in 1999, and seven billion in 2011. Current global population growth rates have seemingly levelled at an extra billion people every 12 years, having possibly been steadied by an increased rural shift to urbanization.

The four food bowls of Continental Europe, the Asian Subcontinent, China and Indo-China, and the USA feed most of the 90 per cent of global inhabitants who live above the Equator. Without these four food bowl regions, it is highly unlikely that the global population would have reached current proportions. Whilst this suggests dependence on such regions, it also suggests that food chains have been relatively effective in satisfying the basic needs of much of the global population. However, food insecurity and high death rates have induced higher population growth rates in some countries.

Food supply aggregate effectiveness can mask serious underlying food problems. Research literature has been noticeably deficient in distinguishing between the efficiencies of perishable and non-perishable food supply chains. It is more likely that population growth and nutrition have been sustained more from unregulated but efficient perishable food supply chains than it has from highly regulated and less efficient storable food supply chains.

Perishables such as poultry, eggs, meat, seafood, fruit, vegetables, and fresh dairy products are far more likely to have sustained the large population explosion over the past century than storables such as grains, legumes, pulses, sugar, dried dairy products, and the historically ubiquitous salt. Peripherals such as vegetable oils, coffee, cocoa, and spices lie somewhere in between as semi-storables. Most researchers, commentators, economists and authors have not made distinctions between perishables and storables when investigating food supply chains, which makes this book quite unique. Despite the major role of perishables in the food supply chain, the book will focus exclusively on storables because that is where most price distortion, government intervention, and supply chain problems have occurred through political and vested interest machinations.

Nazi Germany was one of the few governments that attempted through the Reich Food Estate during 1933–36 to intervene and control perishables, with disastrous outcomes. In contrast, most governments have attempted to intervene and control storables, sometimes with dire consequences to the storable food supply chain, such as in Stalin's USSR and Mao's China. These examples will be discussed as case studies in the book.

Despite the importance of perishables in feeding the global population, whenever the price of storable grain rises, supposedly Armageddon is nigh because of food inflation. Blame is proportioned to ethanol, fund speculation, or weather. The paradox is that when grain prices are frequently low, and bankrupt farmers are making decisions either not to plant seed or to hoard, everybody else seems quite content in the status quo. This strange pattern of human behaviour that is interwoven with political ideology, 'equilibrium' delusions, corruption, dependency issues and food 'aid' donations explains one of the motivations for this book.

Whilst perishables might be considered the antithesis to risk management because of the inability to control supply chain risks, except for refrigeration, food security risk for perishables can be low in deregulated countries because of multiple supply sources, substitutability, imports, and high value-to-weight ratios that suit air-freight economies.

In contrast, storables invite risk management in supply chains, but very little storable food chain risk is managed effectively. Paradoxically, government intervention contradicts good risk management practice and is likely to lead to public–private hoarding and speculation. The book will attempt to explain some of the reasons, including commoditization and commingling, traditional bulk handling, and railway-port dependency.

One of the efficiency measures for storable food supply chains is the formation of a forward market to enable forward transactions and risk management to occur. It is only through the establishment of strong 'trade' markets that forward markets and a private warehouse market can be created. The explanation as to how this occurs and the efficiency gained in the supply chain establishes the foundation for the book. There is focus also on how supply chain intervention and related price market distortions can destroy 'trade' markets and hinder the formation of a forward warehouse market. The existence of strong 'trade' markets, forward markets, and private warehouse markets are likely to provide food security for storable products in the supply chain. Food hubs that are associated with industry clustering are more likely to occur under such circumstances.

1

COMMODITY SUPPLY SHOCKS

On the assumption that Continental Europe, the Asian Subcontinent, China and Indo-China, and the USA are so important in feeding most of the global population, it might be expected that food prices would be largely determined by these regions. However, the irony is that global storable food prices are more often supply driven by the marginal producing and exporting countries such as Brazil, Argentina, and Australia in the Southern Hemisphere, and Canada and Western Asia in the Northern Hemisphere. Seasonal weather conditions in these marginal countries frequently determine global prices for storable food.

Although global food demand is increasing due to population increases and income–consumption pattern shifts, demand remains generally far more stable relative to food supply. Therefore, the supply of storable food is often more significant to price movement than food demand, although the latter remains vital.

There are relatively few major exporting countries of storable food. Thailand, India, Vietnam, the USA, and Pakistan are the major exporters of rice. The USA, Canada, Australia, Ukraine, Russia, Kazakhstan, and Argentina are the major exporters of wheat. Brazil is the dominant exporter of sugar followed by Thailand. The main exporters of vegetable oilseeds are the USA, Brazil, and Canada, whereas India, Canada, and Turkey are the biggest exporters of lentils. Thirteen countries are therefore responsible for the majority of global exports of storable food.

Alternatively, African and Middle Eastern countries, as well as Japan, are major global net food importers that are largely dependent on food supply from marginal producing and exporting countries such as South America, Western Asia, Canada, and Australia. These exporting countries are relatively isolated from the main importing countries, thus making them vulnerable to high freight rate variability via crude oil price volatility and currency exchange movement. Freight costs and currency exchange movement are therefore important indirect factors influencing global food supply.

A drought in a marginal producing and exporting country such as eastern Australia could have a disproportionate impact on global wheat prices, as it did in 2006–07, yet Australia exports approximately only 2 per cent of global wheat production and only half of that was affected by the 2006–07 drought. This minute variation in weather-driven production having a short-term divergent impact on global wheat prices is analogous to Lorenz's non-periodic weather theory, whereby a small change could result in a large disturbance before returning to normal.[1] This chaotic price impact is accentuated either through hoarding[2] or government export embargoes in producing countries.[3]

Primary factors driving food prices

Major factors influencing storable food prices include product alignment, weather cycles, other commodities, fund speculation, bio-fuel hoarding, freight costs, and currency movement.

The debate over whether global food surpluses or shortages actually exist and their likely causes generally ignores supply chain alignment. It is frequently assumed that what is agriculturally produced is actually what is being demanded in the food supply chain by food manufacturers, end users and consumers. However, the biggest problem with global storable food production is the surplus of unwanted product, and a shortage of needed product. What is produced may not have any value relevancy within the supply chain either because of product attributes or quality.

Assuming that there is product alignment, another issue is geographical location differences. What may be in surplus in one region may not be aligned to a shortage region either because of inadequate supply chain signals due to price distortions, or because of logistic difficulties and transport costs. Merely aggregating production and consumption data may not accurately reflect regional differences in the specified product availability and logistical capabilities, both of which can drive prices. A country may be incurring food shortages, yet laden ships could be idle offshore.

Weather is a major short-to-medium-term driver of food prices. It can affect agricultural production and prices through drought, floods, temperature extremes, humidity, pests, diseases, and tempest. Often it is a buyer's psychological perception[4] rather than reality that mostly drives food prices. A drought in the Southern Hemisphere might be officially pronounced after three months of dryness, whereas a three-week dry period in a Northern Hemisphere food bowl region has often been referred to as a 'drought'.

The fundamentals of supply offer little guidance as to the relative importance of different global events to food prices, especially those that relate to adverse weather conditions. Floods in the Asian Subcontinent and Indo-China are likely to move international food prices more than a drought in Africa, simply because of the geographic concentration of human population. The fear-driven demand response to a possible food supply shortage is far greater in the four global food bowl regions than it is in the non-food bowl countries that are accustomed to food production marginality.

Many Northern Hemisphere spring crops are planted in late autumn, which then become dormant under snow cover during winter. This never occurs in the Southern Hemisphere. Snow thawing usually provides additional sub-soil moisture in the Northern Hemisphere, whereas the more marginal Southern Hemisphere has to rely totally on natural rainfall for crop production. The agronomic characteristics are therefore entirely different between the two hemispheres, which partially explains the differences in supply importance relativities and the historical human population disparities.

Some southern agricultural regions of the Northern Hemisphere such as the Asian Subcontinent, eastern half of the Middle East, Southeast Asia, East Asia, and Central America rely almost entirely on the annual occurrence of summer–autumn monsoonal rains. Any delays in receiving this monsoonal rain can be devastating to these regions either because of their high population densities or due to their total dependence on its occurrence, such as the Arabian Peninsula.

A confluence of adverse weather conditions can drive prices much more than singular weather events. The 2008 price spikes for grains and oilseeds occurred after the confluence of the 2006/07 drought in eastern Australia, low monsoon rainfall on the Asian Subcontinent and Indo-China during 2007/08, winter thawing and dormancy-breaking problems in Russia and the Ukraine during early 2008, and wet boggy conditions in the US Corn Belt causing critical planting delays to corn and soybeans in April/May 2008. It was a combination of droughts, floods, rain, and snow that contributed to the 2008 price spikes.

Ironically, there was a huge global commodity food surplus in 2006, with doubts from the United Nations even as late as October 2006 that agricultural management could be effectual in restricting sufficient production to cause agricultural commodity prices to increase to alleviate much global farmer poverty.[5] After the perception of global food shortages in early 2008, international food surpluses reoccurred as early as August 2008, which provides some evidence that there may be a very quick response by farmers to prices in deregulated supply chains.

The rise in food prices in 2007–08 was also influenced by the fall in the USD Index, because much of the international food trade is transacted in US dollars. Other influences included the combined impact of rising crude oil prices and bio-fuel demand,[6] whilst US fund managers switched to buying commodity futures in 2007–08 due to the falling equity markets.[7]

Whereas gold may be an inverse barometer to economic expectations, crude oil can be a general direct barometer for major commodity trends, irrespective of the bio-fuel substitute factor. As crude oil prices increased during the 1970s and again in the 2000s, many food prices either rose or became more volatile. When crude oil prices fell in August 2008, the price declines were shared by many food and non-food commodities.

Trade theory suggests that those countries that have trade surpluses have higher domestic consumption, which drives inflation that ultimately erodes the international competitive advantage of the exporting country. This largely was the experience during the 1980s in Japan, which was forced to reduce domestic interest rates to zero to encourage surplus capital outflow from the country.[8] As well, some trade surpluses in China were shifted into the US until 2008.

Much of this Chinese and Japanese capital inflow was destined to fund some of the US government debt through Treasury bond purchases,[9] as well as providing foreign direct capital to offset US trade deficits in the balance of payments. It also dampened US interest rates. However, because of the need to provide higher investment returns, some of this capital inflow was also invested speculatively after 2003, first on equities and then in commodity markets.[10]

Futures exchanges made special provision for these speculative capital funds by classifying them as being 'hedged', thereby allowing them greater volume exposure than the normal speculative trader. A capital fund that had a speculative exposure on the equity market could not be 'hedged' simply by speculating on commodity markets,[11] because the equity and commodity markets were not offsetting, which became only too apparent during 2008.[12] With crude oil rising after November 2001 until July 2008, the spill-over effect on storable food prices was inevitable,[13] particularly when combined with both hoarding and food export bans.

The unique forward contract characteristic of the futures market meant that this speculative commodity bubble was not sustainable for any length of time. Nearby forward contracts that were held by the US funds needed to be exited or rolled over in the short term. It was this exiting of futures market positions from mid-2008 onwards that led to sharp declines in all the major commodities. Crude oil fell from US$146 a barrel in July 2008 to US$36 a barrel in December 2008, while the storable food price spikes were all very short lived.

There is still much debate over the role of bio-fuels in causing the food price spikes of 2008.[14] If global capital fund speculation and hoarding are ignored, then the bio-fuel industry might be perceived to have influenced the food price spikes of 2008, as it did in 1980 when raw sugar prices reached US$0.44 a pound in October 1980. However, there were many prerequisites already in place for corn to spike in 2008, such as wet weather at US planting time, rising bio-fuel and energy prices that were worsened by 2007 hurricanes affecting Gulf of Mexico oil rigs, and promises of increased agricultural subsidies during the 2008 US presidential elections.

High corn prices were unsustainable when the corn-to-ethanol price ratio seriously reduced US manufacturing profit margins in 2007–08. Manufacturing financial losses resulted in much bankruptcy and less demand for corn, with many manufacturers only surviving through government subsidies,[15] import tariff protection, and mandated fuel mixes.[16] Without this protection, the US corn based ethanol industry would have required low corn prices for survival, but this would then require large farm subsidies to prevent farmers going bankrupt. Both the government and taxpayers are caught in a continual debt trap.

Currency movements impact on international food prices but there are trade-offs. In theory, when the domestic currency rises against another currency, domestic imports become cheaper while exports become dearer. Alternatively, when the domestic currency falls against another currency, domestic imports become dearer while exports become cheaper. However, many international food transactions are still priced in US dollars. Therefore, as the USD Index falls, the demand price for international food generally increases. Any rise in the local domestic (home) currency rate that adversely affects exports is likely to be partially offset by the increased international food demand price caused by the fall in the USD Index.

Spot currency rate movements can be affected by physical imports and exports, payment and receipt of international services, or by global capital movements that are influenced by domestic and foreign interest rate relativities. In contrast, the forward currency market depends entirely on these domestic and foreign interest rate relativities. This affects forward food buying and selling prices.

Hoarding is the withdrawal of supply from the supply chain, specifically undertaken in an attempt to benefit from higher prices. There can be public hoarding by government and its agencies, or private hoarding by farmers and merchants. Improved storage bunker technology has increased the propensity to privately hoard.[17]

Whenever adverse weather impacts on global food supplies, the effect is often made worse by hoarders. Hoarding can aggravate either real or perceived food shortages, as did occur in 2008, which tends to cause panic buying and further global food price rises in food import-dependent countries such as in the Middle East and Africa.[18]

Hoarding can have the opposite effect when new production becomes available. Dumping old crop stock onto the market at a time of increased new crop production can have the effect of clogging the supply chain for long periods similar to the 'bull-whip effect',[19] prolonging surplus conditions and depressing prices, which can force farmers to seek alternate agricultural enterprises.

Secondary factors affecting commodity food prices

Other influences on food prices include time lags, pests and diseases, natural disasters, civil unrest, short-term demand inelasticity, information, trade barriers, and supply inflexibility.

Prolonged time lags and possible production cycles tend to be more associated with animal production because of lengthy gestation periods. This contrasts to the rather continuous production of crops because of transitions in the Northern and Southern Hemispheres. As well, grains, sugar, and pulses can be stored for long periods in hoarding and warehousing, with the subsequent supply smoothing out low and high prices. If price spikes occur for these crops, they are likely to be short lived because of the more immediate response mechanism.

The shorter the management response time and production time lag, the more likely there will be just a quick price spike during peak shortage conditions, followed by increased supply and falling prices with no identifiable price cycle. The longer the management response time and production time lag, the more likely there will be a prolonged period of high prices, followed by slowly increasing supply and falling prices, with a greater possibility of a cycle.[20] But these management response times cannot be viewed in isolation from hoarding decisions.

Diseases in perishable food products tend to have a far greater impact on global production and human health, such as the German outbreak of *Escherichia coli* in June 2011, relative to storable food products. Seemingly major storable crop pest damage through a locust plague may affect only a small percentage of the global food supply chain, particularly when there are product substitutes. Such damage may be perceived insignificant when compared to an outbreak of avian influenza that has a direct human impact at higher levels of the perishable food supply chain, and which can consequentially decimate both bird and animal-feed industries.[21]

Some natural disasters that seriously affect food prices include tempest related floods, and earthquake related tidal waves and tsunamis. The impact on price may be very short term as there can be trade-offs in the medium to longer term regarding production and price, and much depends on alternate production localities, product substitutes, and import–export policies.

The 2010 floods in Pakistan and China may have caused food prices to increase in the short term, but in the medium term it signalled the return of the annual monsoon seasons to the Subcontinent and East Asia, thereby ending the low rainfall period. Such events could result in price spikes followed by prolonged periods of low prices. Both the natural disaster damage plus the prolonged period of subsequent low prices can seriously affect the viability of domestic farmers.

Causes and impact of civil unrest on the food supply chain need to be considered case-specific. However, some of the worst global food shortages and civil unrest have paradoxically occurred in global food bowl countries, which may suggest it was government intervention and subsequent distortions that caused the food shortages and rioting.

Elasticity of demand is extremely important in food supply chains[22] because it reflects consumer behavioural response to price changes as well as the amount of product substitutability. Inelastic demand is characterized by a food product with few substitutes, and indicates that end users or consumers will be more willing to pay a higher price during times of scarcity. An example is Middle Eastern demand for basmati rice from Pakistan.

Alternatively, elastic demand is characterized by a food product with many substitutes, and indicates that end users or consumers will be less willing to pay a higher price during times of scarcity. An example is an end user buying cane sugar when there is a range of substitutes such as artificial sweeteners, beet sugar, and corn syrup available.[23]

An essential food product with less substitutes and inelastic demand would be more likely to spike during supply shortages, while a less essential food product with more substitutes and elastic demand would be less likely to spike during supply shortages. Short-term hoarding success is more likely when inelastic demand occurs with few substitutes.

Supply chain information is rarely accurate, but this has little to do with asymmetry of information, exclusivity, or supply chain power. Data collection methods are inadequate in most countries, satellite imagery has proven somewhat ineffectual to date in measuring crop yields, and statistically sampled survey data responses rarely reflect the final outcome regardless of whether they were deliberately misleading or merely guesswork. Time lags involved with releasing 'old' information can destroy any meaningful outcomes, sometimes there is lack of administrative transparency, bias is difficult to minimize, manipulation can occur through vested interests and corruption, while rumours and misinformation abound either deliberately or through carelessness. Seasonal field inspections by independent voluntary 'scouts' using annual relativities on a sample of key farms still remain the best crop intelligence method in the USA.

Privatization with fewer government stock holdings and increased private hoarding has resulted in a decline of official world food stock levels whilst unofficial stocks have risen. Perceptions of food shortages may not be substantiated when more accurate private storages and warehousing information is included.

If food prices reflect the net decision making of billions of consumers and millions of farmers, then it is unrealistic to believe that accurate information is available or possible. Food supply chains can be a mirror image of many human behavioural attributes that reflect human psychology, self-preservation, and vested interest. It is that uncertainty that creates product markets between buyers and sellers.

Government data theoretically relies on regular supply chain intermediary integrity and accuracy, but some intermediaries simply do not know the information required because of complex accounting, just-in-time logistics, and administrative procedures, yet they are required to regularly supply information by filling out government forms. Farmers are never annually surveyed as a whole, unless some emergency occurs that requires data collection for government relief payment. On-farm storage or hoardings are never annually surveyed from an industry perspective, and individual circumstances can quickly change. Governments rarely disclose the stocks they are holding for public institutional catering, yet they represent the biggest food buyer in all countries.

Governments can sometimes manipulate prices and supply chains because of their import–export size. A government could release 'official' domestic drought information that drives up global prices, and then sell their real surpluses internationally at higher prices. Alternatively, the release of 'official' domestic surplus information drives down global prices and then the government buys at lower prices to offset their real shortages. What is 'official' must always be treated with circumspection.

Inaccurate information can be provided by either end users or farmers in food supply chains because of self-preservation. If farmers release accurate information in surplus supply chains, prices fall and they make financial losses. If end users release accurate information in shortage supply chains, prices rise and they make financial losses. There is no financial incentive in competitive food supply chains to release accurate information. This can make it very difficult for product accumulators, but that is what makes them specialists. If specific product accumulation was easy, there would be intense competition with few skills required.

Regardless of the accuracy of the data, the timeliness of release of official data is important. Quite often, the official data that is released has most likely already been factored into market prices. If the market has previously over-factored information, there can be market price corrections that are opposite to the expected market movement from the official data release. Alternatively, the market has to readjust when inaccuracies of the official data become known. This really challenges the usefulness of 'official' data releases when they are frequently out-of-date and cause unnecessary price volatility that suits only the market speculators who pre-empt what the market response may be in game planning the 'official' data release. The upside is that it undermines insider trading.

Trade barriers can have a serious impact on food prices by reducing global demand price in times of surpluses through import trade embargo and restrictive policies, or by increasing global demand price in times of shortages through export trade embargoes and bans,[24] particularly when several major exporting countries ban exports simultaneously. Rice export bans were implemented during 2008 by India, Thailand, Vietnam, Egypt, Bangladesh, and Myanmar,[25] which immediately led to panic buying amongst importing countries such as in the Middle East and Africa, and drove prices even higher. However, these export bans suppressed price in the home country, thus jeopardizing future domestic production of rice in the exporting countries because farmers were denied the occasional windfall price gains that are required to sustain their financial viability through debt reduction and higher investment.

Fundamental theory of supply suggests that farmers will increase production in response to high prices. However, high prices may not result in an immediate or medium-term response to increase farm production. There might be an inability to shift from one farming enterprise to another because of rigidity of machinery and equipment in only performing certain specific farming tasks.[26] Decision-making psychology of farmers is complex, with factors such as crop rotations, weed and insect control, weather, finance, existing debt, experience, and knowledge being important considerations.[27] Price relativities with other farm enterprises and the need for further finance have to be considered. Farmers may have insufficient motivation because of government directives on land they do not own.

The importance of price

During many decades until 2008, global farmers were in a commodity poverty trap,[28] which can be defined as falling real output prices combined with rising input prices. Global food insecurity existed because farm prices were not sufficient to provide longer-term farm profitability, yet global leaders generally ignored this. When the singular price spike occurred in 2008, much blame was apportioned to demand and peripheral supply factors while the perennial farmer poverty dilemma was either ignored or subsumed under farm subsidy policies. Merely increasing farm subsidies does not remove the underlying problems of surplus production and low prices; however, it does increase monetary and fiscal instability through higher interest rate pressures and government debt.

Figure 1.1 indicates annualized US soft wheat prices during 1979–2011 adjusted for US inflation. During the 2008 'food crisis', daily unadjusted wheat prices momentarily reached 10 per cent above the inflation-adjusted wheat price, before falling dramatically below. In 2011, annualized wheat prices were more than 50 per cent below where they should have been if they had been adjustments for inflation. The annual Consumer Price Index (CPI) inflation as expressed through the adjusted wheat price in Figure 1.1 is also indicative of the rise in farm input costs.

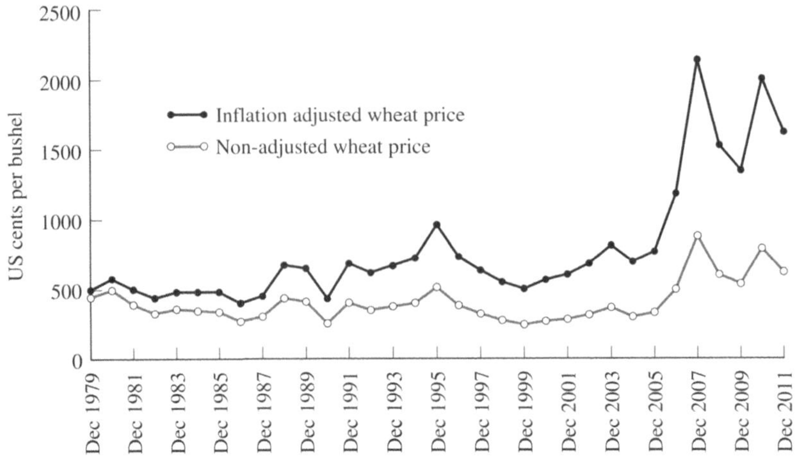

Source: CME wheat futures prices and US Bureau of Labor Statistics, 2011

FIGURE 1.1 US annual averaged soft wheat prices – inflation adjusted: 1979–2011

Whenever there is a price–cost squeeze for farmers, food production is unsustainable. In contrast, farmers generally will be motivated to increase food production and make necessary farm investments whenever sufficient profitability is apparent. For many farmers, 2008 was merely an opportunity to repay some farm debt.

As an example of farmer poverty in developed countries, many farmers in non-food bowl countries such as Canada, Australia, Argentina, and Western Asia often only grow wheat because it fits in with crop rotation programmes, or else they speculate in the hope that it might eventually be profitable.[29] Co-development through joint ownership with merchants, investors and trading banks offered many Western Australian wheat farmers the only alternative during the first half of 2011 to plant wheat, because one-third of them were financially unviable at prices of US$300 per tonne due to marginal production and high domestic currency exchange rate burdens. Price risk management is important to these farmers, but only when short-term price spikes or volatility offer profitable opportunities.

The global food bowl countries have argued that farm subsidies are required to offset this storable food price–cost squeeze to ensure food security.[30] Alternatively, it has been strongly argued by global non-food bowl countries that these farm subsidies are the cause of prolonged low food prices for global farmers.[31] Farm subsidies tend to encourage over-production, which depresses prices, which then requires further subsidies. This explains why the Common Agricultural Policy expenditure represents one-third of the EU government budget,[32] and why increasing expenditures are required to implement the US Farm Bill policies. Ironically, higher prices would generally increase farm incomes and alleviate the necessity to subsidize farmers. However, politicians usually prefer government debt in lieu of the more unpalatable weaning farmers off subsidy dependency with possible higher inflation.

The major difference between the subsidy policies in the USA and EU is that the former results in on-farm hoarding with cash-flow support, while the latter has historically relied heavily on subsidy intervention that also resulted in government hoarding, albeit through licensed warehousing. Whilst the removal of trade barriers might be more important than farm subsidies from a trade and gross domestic product growth perspective,[33] particularly if government debt is low, it is the association of farm subsidies with surplus production and low prices that directly suppresses the motivation to produce in many non-subsidized countries and this distorts production decision making.

If farm subsidies are the cause for depressed prices, hoarding and increased food insecurity in many countries, including food bowl regions, then more emphasis needs to be placed on the removal of these farm subsidies,[34] just as much as the removal of trade-distorting barriers.[35] Reducing government debt due to lower subsidy payments might be traded off with slightly higher food price inflation, but this may be quite small relative to the effects on overall inflation of recent crude oil price rises. Permitting imports would dampen food price inflation, but this would require a huge government mindset change, particularly in global food bowl countries.

The UN Food and Agricultural Organization publicly acknowledging the long-term commodity poverty trap for farmers in 2006,[36] yet since 2008 it has devoted inordinate resources to the Global Information and Early Warning Systems (GIEWS), the Food Security and Early Warning Network (FENIX), and the National Food Policy Capacity Strengthening Programme (NFPCSP), which are designed to monitor and dampen global food prices with economic 'stability' objectives. Both the EU and USA have become large sponsors of such programmes. Given the continuing long-term commodity poverty trap for farmers, such support for any price suppression can paradoxically lead to future global food insecurity, with many farmers exiting unviable production, particularly in poverty-stricken countries.

Price spike theory

International food prices usually respond either to real, expected, or rumoured supply shortages in one or more of the major producing and exporting countries. Major price rises or volatility can therefore occur as a response to small changes in supply availability, similar to Lorenz's butterfly effect in chaos theory,[37] rather than reflecting major global production changes. Any small supply problem then becomes compounded by hoarding[38] and export bans.[39] Under such circumstances, food price spikes are therefore characterized by being very short term, particularly when either private or public hoarded commodity is subsequently dumped quickly onto food supply chains.

Figure 1.2 indicates that the 1993/94, 2003/04, and 2008 price spikes for unhulled rice, as indicated by the Chicago annual futures price, were very short lived. In contrast, a gradual shortage such as occurred in 1995–97 might be more likely to be followed by a prolonged surplus as seemingly occured during 1998–2002.

Figure 1.2 indicates that there was no prolonging of rice prices at the upper level in 2008, which indicates that no serious global food shortage occurred. Prices rose to

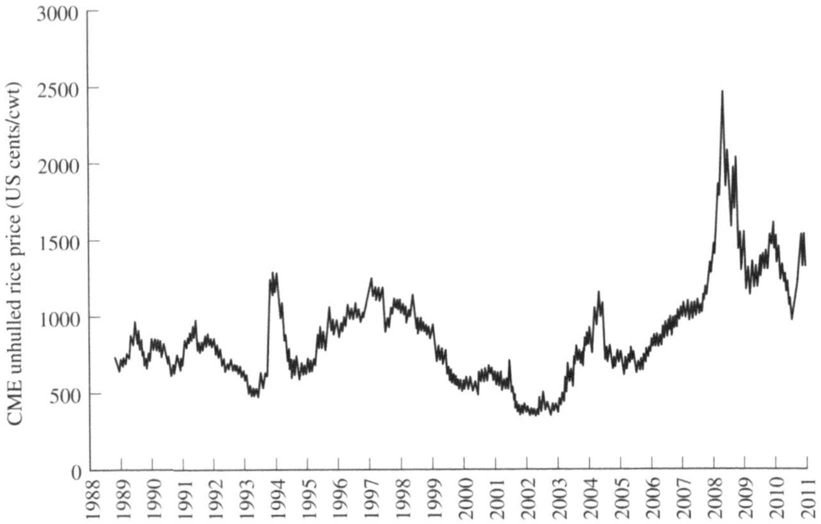

Source: CME prices and Premium Data

FIGURE 1.2 Chicago annual spot futures prices for unhulled rice: 1988–2011

a level where suppliers and ultimately hoarders gradually overwhelmed the buyers, which caused prices to collapse and the removal of export bans.

All storable food commodity prices have a much higher frequency of occurrence at the lower price range,[40] particularly over the past 30 years. This suggests that surpluses exist for the majority of time, which can be explained through high subsidies and other protective policies in global food bowl countries, and that any real or expected shortages are relatively short lived. If price spikes reflect shortages in only one or two of the major exporting countries that are more likely to be marginal producing, then the dominant low-price subsidy effect in food bowl countries will cause the quick return to the commodity poverty trap for farmers.

Figure 1.3 indicates the skewness of international wheat prices, as indicated by the Chicago daily futures price for soft wheat for the period 1979–2011. At US$3 per bushel, there is an approximate 80 per cent probability that prices could go higher, whereas at US$8 per bushel, there is less than a 5 per cent probability that prices could go higher. The median price over the period was US$3.56 per bushel, while the average price was US$3.95 per bushel.

Price spikes need to be viewed in perspective. The price spikes for storable foods in 2007/08 were much less in real terms than the price spike of 1974.[41] However, similar to the 1974 spike, the 2007/08 'price shock' was evident across many commodities, and was temporary.

Storable food price skewness suggests that any high food prices should revert quickly back to the low-price norm, thus supporting the theory of Mandelbrot.[42] Statisticians might refer to this as an 'outlier effect' whereas mathematicians might refer to it as an 'auto-correcting' factor toward an 'equilibrium', but reality suggests

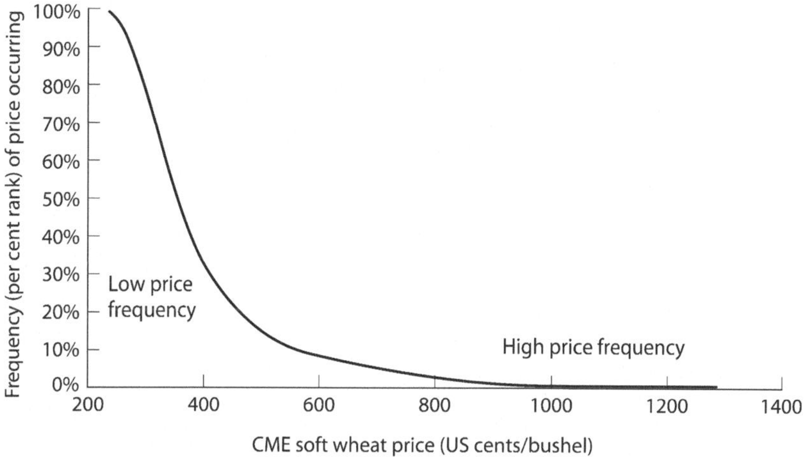

Source: John Williams (2011)

FIGURE 1.3 Skewness of Chicago daily soft wheat futures prices: 1979–2011

that a quick return to low prices is very dependent on the response of hoarders, the removal of export bans, the renewed impact of farm subsidies, and a return to more normal weather patterns. This results in a higher predictability for low prices and a low predictability for high prices.[43]

Notes

1 Lorenz, Edward N., 1963, *Deterministic Nonperiodic Flow*, Journal of Atmospheric Sciences, Massachusetts Institute of Technology, 20, 130–141.
2 Carter, Colin A., and Cesar L. Revoredo Giha, 2007, *The Working Curve and Commodity Storage Under Backwardation*, American Journal of Agricultural Economics, 89:4, 864–872.
3 Piesse, Jenifer and Colin Thirtle, 2009, *Three Bubbles and a Panic: An Explanatory Review of Recent Food Commodity Price Events*, Food Policy, 34:2, 119–129.
4 Ellis, Scott C., Henry, Raymond M. and Jeff Shockley, 2010, *Buyer Perceptions of Supply Disruption Risk: A Behavioural View and Empirical Assessment*, Journal of Operations Management, 28, 34–46.
5 Food and Agriculture Organization of the United Nations, 2006, *Can Supply Management Halt the Decline in Agricultural Commodity Prices?* FAO Commodity Market Briefs No. 1, October, Rome.
6 Tokgoz, Simla, Elobeid, Amani, Fabiosa, Jacinto, Hayes, Dermot J., Babcock, Bruce A., Yu, Tun-Hsiang, Dong, Fengxia, and Chad E. Hart, 2008, *Bottlenecks, Drought, and Oil Price Spikes: Impact on U.S. Ethanol and Agricultural Sectors,* Review of Agricultural Economics, 30:4, 604–622.
7 Haigh, M., Hranaiova, J. and J. Overdahl, 2007, *Hedge Funds, Volatility, and Liquidity Provision in Energy Futures Markets,* Journal of Alternate Investments, 9:4, 10–38.
8 Krugman, Paul, 1998, *Its Baaaaack: Japan's Slump and the Return of the Liquidity Trap*, Brookings Papers on Economic Activity, 2, 137–205.
9 US Department of Treasury, 2011, *Major Foreign Holders of Treasury Securities*, Washington, DC.

10 Haigh, M., Hranaiova, J., and J. Overdahl, 2007, *Hedge Funds, Volatility, and Liquidity Provision in Energy Futures Markets,* Journal of Alternate Investments, 9:4, 10–38.
11 Biggs, Barton, 2006, *Hedge Hogging,* John Wiley and Sons, New York.
12 Yang, J., Balyeat, B.R., and D.J. Leatham, 2005, *Futures Trading Activity and Commodity Cash Price Volatility,* Journal of Business Finance Accounting, 32:1-2, 297–323.
13 Pace, Noemi, Seal, Andrew, and Anthony Costello, 2008, *Has Financial Speculation in Food Commodity Markets Increased Food Prices?,* Lancet, 371, 1648–1650.
14 Perrin, R.K., 2008, *Ethanol Production and Food Prices: a Preliminary Assessment,* working paper 03/08, Agricultural Economics Department, University of Nebraska, Lincoln, May.
15 Gallagher, P., Schamel, G., Shapouri, H., and H. Brubaker, 2006, *The International Competitiveness of the US Corn-Ethanol Industry: A Comparison with Sugar:Ethanol Processing in Brazil,* Agribusiness, 22:1, 109–134.
16 Xiaodong, Du, J. Hayes, and Mindy L. Mallory, 2009, *A Welfare Analysis of the U.S. Ethanol Subsidy,* Applied Economic Perspectives and Policy, 31:4, 669–676.
17 This will be discussed in detail in Chapter 7.
18 Yemtsov, Ruslan, 2008, *The Food Crisis: Global Perspectives and Impact on MENA, Fiscal and Poverty Impact,* Middle East and North Africa Poverty Reduction and Economic Management Department, Washington, DC: The World Bank.
19 Lee, H.L., Padmanabhan, V. and S. Whang, 1997, *The Bull-whip Effect in Supply Chains,* Sloan Management Review, 38:3, 93–102.
20 Meadows, Dennis L., 1970, *Dynamics of Commodity Production Cycles,* Wright-Allen Press: Cambridge, MA.
21 Hammoudi, Abdelhakim, Hoffmann, Ruben, and Yves Surry, 2009, *Food Safety Standards and Agri-food Supply Chains: an Introductory Overview,* European Review of Agricultural Economics, 36:4, 469–478.
22 Panagariya, Arvind, Mishra, Deepak Kumar and Shekhar Shah, 1996, *Demand Elasticities in International Trade: Are They Really Low?* World Bank Policy Research Working Paper 1712, Washington, DC.
23 This assumes that flocking and other product differentiation characteristics are not an issue in comparison.
24 OECD, Annual Reports, *Agricultural Policies in OECD Countries: Monitoring and Evaluation,* Paris.
25 Mitra, Siddhartha, and Tim Josling, 2009, *Agricultural Export Restrictions: Welfare Implications and Trade Disciplines,* International Food and Agricultural Trade Policy Council, IPC Position Paper, Agricultural and Rural Development Policy Series, Washington, DC, January.
26 Shih-Hsun Hsu and Ching-Cheng Chang, 1990, *An Adjustment-Cost Rationalization of Asset Fixity Theory,* American Journal of Agricultural Economics, 72:2, 298–308.
27 Williams, John, 2009, *Decision-Making in Australian Wheat Marketing and Price Risk Management,* School of Land and Environment, University of Melbourne.
28 Food and Agriculture Organization of the United Nations, 2006, *Can Supply Management Halt the Decline in Agricultural Commodity Prices?* FAO Commodity Market Briefs No. 1, October, Rome.
29 Peterson, Willis L., 1979, *International Farm Prices and the Social Cost of Cheap Food Policies,* American Journal of Agricultural Economics, 61:1, 12–21.
30 Brenna, D. Ellison, Jayson L. Lusk, and Brian C. Briggeman, 2010, *Taxpayer Beliefs about Farm Income and Preferences for Farm Policy,* Applied Economic Perspectives and Policy, 32:2, 338–354.
31 Wise, Timothy A., 2005, *Identifying the Real Winners from U.S. Agricultural Policies,* Global Development and Environment Institute, Working Paper 05-07.
32 EU annual budget papers.
33 Anderson, Kym, Martin, Will, and Ernesto Valenzuela, 2006, *The Relative Importance of Global Agricultural Subsidies and Market Access,* World Trade Review, Washington, DC, 5:3, 357–376.

34 Mayrand, K.S. Dionne, Paquin, M. and I. Pageot-Lebel, 2003, *The Economic and Environmental Impacts of Agricultural Subsidies: An Assessment of the 2002 US Farm Bill and Doha Round*, Unisfera International Centre, Montreal.

35 Hertel, T., and R. Keeney, 2006, *What is at Stake: The Relative Importance of Import Barriers, Export Subsidies, and Domestic Support*, in A. Anderson, and W. Martin, (Eds), *Agricultural Trade Reform and the Doha Development Agenda*, Washington, DC: World Bank, 37–62.

36 Food and Agriculture Organization of the United Nations, 2006, *Can Supply Management Halt the Decline in Agricultural Commodity Prices?* FAO Commodity Market Briefs No. 1, October, Rome.

37 Lorenz, Edward N., 1963, *Deterministic Nonperiodic Flow*, Journal of Atmospheric Sciences, Massachusetts Institute of Technology, 20, 130–141.

38 Bradsher, K. and A. Martin, 2008, *Hoarding Nations Drive Food Costs Ever Higher*, The New York Times, 30 June.

39 Mitra, S. and T. Josling, 2009, *Agricultural Export Restrictions: Welfare Implications and Trade Disciplines*, International Food and Agricultural Trade Policy Council, IPC Position Paper – Agricultural and Rural Development Policy Series, January.

40 Hudson, Richard L. and Benoit B. Mandelbrot, 2004, *The (Mis)Behaviour of Markets: A Fractal View of Risk, Ruin, and Reward*, New York: Basic Books.

41 Cooper, R.N. and R.Z. Lawrence, 1975, *The 1972–75 Commodity Boom*, Brookings Papers on Economic Activity, 3, 671–723.

42 Hudson, Richard L., and Benoit B. Mandelbrot, 2004, *The (Mis)Behaviour of Markets: A Fractal View of Risk, Ruin, and Reward*, New York: Basic Books.

43 Kahneman, Daniel, and A. Tversky, 1973, *On the Psychology of Prediction*, Psychological Review, 80, 237–251.

Further readings on commodity supply shocks

Bacon, L.B. and F.C. Schloemer, 1940, *World Trade in Agricultural Products – Its Growth; Its Crisis; and the New Trade Policies*, Rome: International Institute of Agriculture.

Chatrath, Arjun, Adrangi, Bahram and Kanwalroop Kathy Dhanda, 2002, *Are Commodity Prices Chaotic?* Agricultural Economics, 27, 123–137.

Deaton, A. and G. Laroque, 1992, *On the Behaviour of Commodity Prices*, Review of Economic Studies, 59:1, 1–23.

Duyne, Carl Van, 1982, *Food Prices, Expectations, and Inflation*, American Journal of Agricultural Economics, 64:3, 419–430.

Ghosh, Jayati, 2010, *The Unnatural Coupling: Food and Global Finance*, Journal of Agrarian Change, 10:1, 72–86.

Gilbert, C.L. and C.W. Morgan, 2010, *Food Price Volatility*, Royal Society, B:365, 3023–3034.

Haigh, M., Hranaiova, J., and J. Overdahl, 2007, *Hedge Funds, Volatility, and Liquidity Provision in Energy Futures Markets*, Journal of Alternative Investments, 9:4, 10–38.

Kindleberger, Charles P., 1996, *Manias, Panics, and Crashes*, New York: John Wiley and Sons.

Lang, Tim and Michael Heasman, 2004, *Food Wars*, Earthscan, Abingdon.

Meinig, Donald W., 1962, *On the Margins of the Good Earth, 1869–1884*, Chicago: Rand.

Olson, Mancur, 1988, *The Productivity Slowdown, The Oil Shocks, and the Real Cycle*, The Journal of Economic Perspectives, 2:4, 43–69.

Oosterveer, Peter, and David Sonnenfeld, 2011, *Food, Globalization and Sustainability*, Earthscan, Abingdon.

Pace, Noemi, Andrew Seal and Anthony Costello, 2008, *Has Financial Speculation in Food Commodity Markets Increased Food Prices?* Lancet, No. 371, 1648–1650.

Sarris, Alexander and David Hallam, Eds., 2006, *Agricultural Commodity Markets and Trade – New Approaches to Analysing Market Structure and Stability*, FAO, Rome, and Edward Elgar, Cheltenham.

Tonts, Matthew and M.A.B. Siddique, 2011, *Globalisation, Agriculture, and Development*, Edward Elgar Publishing, Cheltenham.

2

SUPPLY CHAIN DISTORTIONS

Intervention, subsidies, protection

Food prices reflect the combined decision making of billions of consumers and millions of farmers in multitudinous global food supply chains. Commodity food buyers and sellers in such markets are price takers. The question is why governments would seek to change price when they are price takers and have no control over the very behavioural mechanism they seek to change.

Market prices cannot 'fail'.[1] Just because consumer–farmer decisions do not always conform to government monetary and fiscal objectives or some egocentric social 'norm' does not translate into failure. Quite often the failure lies in an inability of government to accept the outcomes of billions of consumer and farmer decision-making behaviours. Whether governments could or should attempt to influence such decision-making behaviour is quite another issue.

There can be temporary market–price distortions in deregulated supply chains until arbitrage or competition attempts to correct them. However, more permanent market–price distortions such as illiquid markets are likely to be caused by supply chain intervention or a lack of competition, both of which are more likely to be influenced by government decision making. It is very patronizing to decision makers when the behaviour of billions of global consumers and farmers is claimed to be in a continual state of 'failure' and that this 'justifies' even more government intervention.

Adam Smith recognized that the exchange value of products was reflected through price,[2] while Ludwig von Mises indicated that without a market, there was no functional price system.[3] If governments alter price, they distort the exchange value of products and services. Controlling a market will destroy any functionality of a pricing system. Without value signals through a pricing mechanism, even the smallest individual decision can be an error. The culmination of individual wrong decisions can be catastrophic and eventually destroy governments, as if there was fate and a historical inevitability awaiting such government decision making and empire building that had previously eroded the responsibility of its citizens.[4]

FIGURE 2.1 Luxuriant rice paddies in the Korean demilitarized zone with a wooden pill-box, yet within 100km to the north, famine is occurring
Source: John Williams

North Korea is a contemporary example of everything that can go wrong in the food supply chain through government intervention and distortions.[5] Theoretically, North Korea should be a land with plentiful food (Figure 2.1), yet millions of its people regularly die of starvation.[6] Price signals that are required for efficient food supply chains are either destroyed or distorted by government actions that jeopardize food security and maximize the possibility of protracted queues and regular famine for its citizens.

Price signal distortions affect both functionality and efficiency in food supply chains.[7] Unless clear price signals are transparent at all stages of the supply chain, wrong decisions will cause dysfunction and inefficiencies within the supply chain and can ultimately destroy it. Any objectives of meeting consumer demand are unlikely to be met without such feedback being expressed through price.

The rise of many empires has appeared synonymously with price controls throughout much of history.[8] Centralized government depended on control over economic variables such as price because constant prices facilitated more efficient tax collection. Price variation often meant variations in taxation revenues, which jeopardized public administration and political stability. There was a perception that price controls led to political stability, despite the incongruity of the inference. The paradox was that by destroying the foundation of efficient food supply chains, this sowed many of the seeds of the ultimate destruction of such governments. Empires rose and fell in an inevitable deterministic cycle of historical events that mostly were the result of monetary and fiscal irresponsibility. Wrong supply chain decisions were made,[9] which contributed to the empire being vulnerable to attack by competitors and anarchists.[10]

Early centralized governments were mostly driven by monetary and fiscal policies, with much consideration given to food production self-sufficiency and stable prices. As long as consumers were fed regularly and paid tax, they generally could be ignored.[11] If there was any uncertainty over future food supplies, then storable food could be requisitioned and stockpiled during times of surpluses.[12] This ability to withhold publicly hoarded supply had the added benefit of maintaining higher prices and taxation revenue.

However, food price controls and associated taxes have had a long history of encouraging private hoarding, as well as causing chaos, dislocation, artificial shortages, imprisonment, and even revolution.[13] One of the more public revolts against a government-licensed monopoly that established food prices and imposed tax was the Boston Tea Party revolt against the British East India Company in 1773. This led to the American War of Independence that culminated in 1776, which was the year that Adam Smith published his *The Wealth of Nations*.[14] By the end of the next decade (1789), French consumers were revolting against government food price controls over salt[15] and bread,[16] as well as taxes associated with North American war debts.

The reconsolidation of supply chain intervention

There were major benefits from the consumer-driven French Revolution in terms of food supply chain reform, contestable price competition in food supply chains, and freer trade. The post-revolution period was characterized by the French Liberal School of Jean-Baptiste Say,[17] Bastiat, de Tracy, and de Molinari. However, the movement had been completely eroded by the mid-nineteenth century by French government policies to return to pre-revolution centralised government and price controls,[18] which necessitated the erection of increasing levels of external trade protection.

Consumer value theory in Britain had been consolidated by the Utilitarian movement of Bentham and James Mill during the first half of the nineteenth century. However, Britain had earlier retreated from any freer trade sentiment with the imposition of the UK Corn Laws in 1815, and these were not repealed until 1846 under pressure from the Anti-Corn Law League and Richard Cobden. The repeal of the Corn Laws was perceived as a major free trade victory and a move towards greater deregulation.

However, the battleline had been drawn between consumers who were defended by free traders, Whigs and Liberals, and producers who were generally defended by trade protectionists in an unlikely pairing of Conservatives and Chartists. In the wake of Cobden's visit to France, Germany, Russia, Spain, and Italy in 1847, the 1848 revolutions occurred in many European countries demanding 'freedom' reforms that centred on deregulation and consumerism. Paradoxically, Conservative protectionists and Chartist socialists perceived victory in these riots.

Those who favoured the protection of production responded quickly against the rising free trade and consumer reform sentiment. Engels, who was a son of a German textile manufacturer, together with German economist Karl Marx, had conceived a new production-based protectionist model by early 1848. Consumers would be replaced with 'workers' who would then demand increased trade protectionism to protect their income, at the same time being controlled by conservative production interests who were increasingly being allowed to eliminate private competition. It was an 'economic evolutionary' process that 'inexorably' and 'inevitably' led to centralized control. This unlikely coupling between trade protectionists, monopoly capitalists, interventionists, Chartists, and trade unionists was to dominate global economics for the next 120 years as socialist policy. Any notion of consumer satisfaction (utility) through consumption could be conveniently ignored.

However, it took the socialists an extraordinarily lengthy 80 years to convert the theory into decision-making formulations. Even Stalin allowed some market operations to continue in the USSR food supply chain until 1929 because there was a vacuum of alternate ideas, including Bogdanov's tectology. After all, 'workers' were also consumers, an admission that theorists often overlooked.

In the wake of the introduction of the Smoot-Hawley tariff protection legislation in USA, Taylor in 1929 reiterated the ideals of the production-driven 'socialist state' as those in which the control of the whole apparatus of production and the guidance of all productive operations is in the hands of the 'state'.[19] He then proclaimed that consumers were to be given directive 'authorisations', but no rights or real supply chain function except eating when food was available. As if on cue awaiting some leadership, Stalin then began his food supply chain interventions in 1929, whereas the German food chain interventionists had to wait another four years until the Reich Food Estate had been established.

As trade protection increasingly distorted food supply chains, major trade protection retaliation was regularly occurring by late nineteenth century in many European countries. Britain strongly resisted trade protectionism until 1914, possibly because it was an island-based economy supported by a vast international supply chain network. However, trade protectionism was rife in those Continental European countries that did not have ready access to ports and shipping. A major war therefore became highly probable and its inevitability lay in the envy of the freer trade policies of Britain[20] and the positive consequences of those policies, and the retaliatory protectionist and foreign policies implemented by an axis of Continental European governments after 1850.[21]

Intervention fallacies

Barone was an Italian mathematician and economist who collaborated with Pareto and the German Lausanne School of mathematical economics to enhance the theoretical Walrusian supply–demand equilibrium model into 'trial and error' decision making. His mathematical model in 1908 advocated 'collectivist' centralized government monopoly production controls with price determination based on 'maximum collective welfare'.[22] By using such ambiguous utopian welfare terms, both the definition and method of accurate calculation of these terms could be conveniently ignored through the dubious 'equilibrium' linkage of Pareto 'efficiency' with marginal cost of production.

Either consumers were conveniently ignored in this model, or it was presumed that consumers would achieve an unmeasurable 'maximum welfare' through a price determined by a producer's marginal cost of production, regardless of the producer's efficiency or forehand knowledge of production conditions, and regardless of actual consumer demand.

The application of equilibrium to efficiency in socio-economic decision making using Pareto's 'principle' presupposed that no one could be made better off without making someone else worse off.[23] In physics and engineering, both inputs and outputs were measurable concepts. This contrasted with the unmeasurable theoretical concepts of human behavioural 'efficiency' and 'welfare'. Even Barone conceded that not all losers from production-driven policies, who in reality would be consumers, could be compensated, and that such an 'equilibrium' efficiency could not be determined in advance but rather from historical data records.[24] This reduced the concept to an academic process, rather than a forward decision-making mechanism.

Pareto 'efficiency' does not necessarily result in a 'socially desirable' distribution of resources, and it makes no statement about equality or societal well-being.[25] It assumes that all these terms are objectively measurable, which they are not, and that there are no major differences in human behavioural decision making of millions of citizens, whether they be consumers or producers.

Under trade protection, there was an inexorable slide from private monopoly capitalists to public monopoly authorities[26] as inefficiency and financial losses through a lack of competition overwhelmed the system. However, the awkward dilemma of what government monopolies and public utilities should actually produce in a world devoid of competition and recognition of consumers was not addressed until Taylor, the 'father of modern socialism', issued his directives in 1929. Under centralized controlled governments, citizens would be 'assured' income through the government buying their labour, and the government would 'authorise' them to spend that income on commodities produced by the government.[27]

Somehow, this 'procedure would virtually authorize the citizen to dictate just what commodities the economic authorities of the state should produce'.[28] The problem was that monopoly government production was already occurring without any mechanism to respect consumer demand. Ensuring that the 'peculiarities of tastes and needs characteristic of each individual would not be sacrificed' proved to be impractical when standardization was the centrally controlled operational mode of government monopolies.[29]

Under centrally controlled governments, consumers were only considered in what they were 'obliged' to pay, not what they were prepared to pay.[30] Thus prices were perceived merely as enforced selling prices that would be set by 'economic authorities at a point which fully covered the cost of producing the commodity'. There was no mention of the uncertainties associated with agricultural production, inefficiencies associated with centrally controlled production, cost inflation, the rigid requirement for high-trade protection from imports, and the continual problem of trade deficits and loss of currency reserves.

Problems of production surpluses and deficits occurring under centrally controlled government supervision were to be apparently solved by Barone's 'trial and error' of 'a series of hypothetical solutions' until one proved to be successful.[31] The imputation was that 'wrongs' could be righted, which would lead to an equilibrium state that was defined as the 'equality between the cost of production and the demand price of the buyer', who presumably was not the consumer. It implied that if the selling price of the government monopoly matched the marginal cost of production, then 'equilibrium' would be achieved, which would remove any possibility of production surpluses or deficits occurring.[32]

Knowing the buyer's demanded price then became an illusion and a perceived irrelevance because the selling price was fixed based on marginal cost of production. If actual demand exceeded anticipated supply, there was then the problem that with increasing production and sales above where marginal cost equalled marginal revenue, net profit would decrease although that was not a prime concern for socialist governments. Consumer prices could always be raised and inflation ignored.

Marginal cost–price theory suggests that profit is maximized when marginal revenue equals marginal cost. But this theory is deeply flawed in industries such as agriculture. The theory assumes that there is control over production, which in the case of agricultural crops there is no control over yield or quality. A limited linear relationship between fertilizer and yield might be possible under controlled research conditions, but it is extremely difficult to obtain a consistent linear relationship under field conditions even with precision agriculture. Many factors determine marginal output other than fertilizer input. If marginal output cannot be determined accurately in advance, then marginal revenue cannot be determined, and even that assumes that input and output prices can be predicted or artificially fixed.

Problems with nationalized industries became obvious during the 1930s, but much blame was still attributable to the private sector. There were economists who defended the importance of market-driven prices through a Neoclassical approach, which was a hybrid of both consumerism and production-driven protectionism. Lerner in 1934 opposed price controls, referred to the market price mechanism as 'one of the most valuable instruments of modern society',[33] and described private enterprise as a 'safeguard of the freedom of the individual'.[34] If a monopoly price exceeded its marginal cost of production, Lerner claimed that this was a measure of the market power of the monopoly. Much later, Lerner was to make a major contribution to consumer sovereignty,[35] as did von Mises.[36]

Consumer advocates such as Lerner, a Russian émigré working with Friedrich Hayek in London, as well as Ludwig von Mises in Geneva, drew much ire from German economists. Having had to move from Vienna to Geneva in 1934, and then onto New York in 1940, Ludwig von Mises had previously stated in 1922 that 'even exceptionally gifted men of high character cannot solve the problems created by socialist control of industry'.[37] Students from the German Student Association at 34 German universities burnt von Mises's books in bonfires between April–June 1933 whilst being publicly addressed by economic professors.[38]

Lange countered the Neoclassical opposition in 1936 by advocating central planning boards[39] with the authority to fix prices through the Barone-Taylor method of 'trial and error', which was just after the formation of the second Canadian Wheat

Board in 1935. Instead of relying on supply and demand to determine price, prices would be 'adjusted' based on perceived surpluses or shortages. The problem was that anticipating, forcing, or corrupting these price 'adjustments' could directly lead to food shortages and distort production decisions through private hoarding.[40] Stalin, Molotov, and the young Khrushchev conveniently solved the Kulak farmer hoarding problems through the gulags and the Nagant pistol.[41]

Despite the great failures of government intervention in the food supply chain by both Stalin[42] and Hitler[43] during the 1930s, the Second World War entrenched governments as public monopolists[44] and large institutional catering buyers in the food chain. Many agricultural commodity boards and state buying-selling authorities were established in the late 1930s and early 1940s based on the Lange model. The Australian Wheat Board was established in September 1939 under wartime legislation.

Post-war supply chain intervention

The Structuralists[45] arose during the 1940s, focusing on the German-based Cobweb theory,[46] despite such equilibrium theories being totally discredited in Germany during 1933–39.[47] They rejected human freedom and choice of both consumers and farmers, which may have been understandable under wartime pressures but questionable during the Cold War. Structuralism reaffirmed the dominant role of autocratic government in supply chain activities via statutory intervention,[48] despite all of its distortion and efficiency failures. The name 'Structuralism' suggested that 'orderliness' could be humanly achieved to replace dynamic individual decision psychology, but the end result was more akin to concrete, bricks, and mortar, as well as stagnancy in supply chains.

Concepts of government planning and 'guiding' of the economy was supported by the introduction of computer modelling during the 1950s in supposedly 'scientific social engineering'.[49] Ideals of reaching 'optimality' overwhelmed the reality of ever achieving it, even if it could have been properly defined without unreal assumptions in suspect algorithms. It ignored whether optimality could ever be achieved for both producers and consumers simultaneously in a dynamic supply chain environment.

Many 'benevolent' governments believed that they could actually determine the 'optimum' allocation of resources, thereby 'maximizing national welfare'.[50] 'Optimality' theory assumed that the supply chain was not the most appropriate mechanism to achieve 'optimum' allocation of resources and prices. At a macro level, 'optimality' is impossible to define except in theoretical generalities, while at a micro level it is rather meaningless to consumers and farmers who actually make daily decisions in the food chain. Maximizing government welfare may be entirely different to maximizing farmer and consumer welfare, even if all these concepts could be defined and accurately measured by aggregating individuality.

Many optimalists perceived that market prices led to some social 'perverseness', a concept that is still ill-defined, particularly when commodity markets consist of the culmination of decision making by global consumers and producers. Price distortions from government controls are somehow justified for achieving 'social objectives',[51] without any defining support as to how that is achieved or of the paradoxes involved.

By the 1960s, interventionists were still arguing that deviations from 'optimum' prices were causing resource allocation failure to maximize 'value', which generally referred to an undefined 'social value' rather than the more definable consumer value. Samuelson linked thermodynamic mathematics to economics in a neo-Keynesian linear programming approach,[52] forever seeking to find the 'right' price for efficient allocation of resources. The general consensus still was that markets were inefficient, as were supply chains presumably.

However, intervention by government to correct minor market 'failure' usually resulted in major supply chain failure. Examples during this 1950s–60s era included price control policies, subsidies, import–export controls, and quota schemes that either resulted in disruptions to food supply chains, distortions to consumer behaviour, major funding required by taxpayers, or serious loss of international competitiveness. This was quite apart from bureaucratic inefficiency, mismanagement, and often corruption thereby created under the banner of achieving 'optimality' in resource allocation but which mostly provided only welfare for the privileged few, usually at taxpayers' expense. The Common Agricultural Policy of the EU epitomized the government intervention of that period.[53]

Impact of supply chain intervention

Many governments in the 1950s and 1960s forcibly removed competition and transformed supply chains into grossly inefficient and poorly functioning non-contestable stagnations.[54] Statutory monopolies in exporting countries and monopsonies in importing countries were the usual non-competitive intervention organizations in international supply chains.[55] Any consumer-based organizations that might have driven supply chain efficiency through price signals were either subsumed by statutory authorities, or simply made illegal.

These statutory organizations excluded most traditional supply chain intermediaries such as domestic merchants who had important supply chain responsibilities. This vacuum forced most of the functions of the supply chain back onto a statutory bureaucracy that was ill-equipped to undertake such operations. Such intervention generally resulted in supply chain dysfunction. Under wheat market intervention in Australia, price 'signals that reflected market changes were masked or distorted', which led to 'an inappropriate basis for making investment decisions'.[56]

The core problem with these supply chain intervention policies was the lack of alignment of production output with end-user and consumer specifications.[57] This explained how minor market 'failure' quickly turned into major supply chain failure, with either large quantities of unwanted old-crop commodity being carried over into new crop years and often being recorded as 'reserve-buffer' stocks, or surpluses in one region existing at the same time as deficits and famines in another. The associated price distortions were not conducive to sustainable profitable agricultural food production or to the sustainability of the supply chain. This frequently accelerated farmers leaving their land, while end-users increasingly shifted operations overseas. There is no theoretical physical equilibrium or optimum social outcome under such circumstances, except a rigor mortis horizontal position when citizens actually die of famine in a price-distorted country.[58]

Intervention induced inefficiency because it had no mechanism to reward or motivate change.[59] The increasing lack of modernization was exacerbated by investment disincentives, an inability to actually manage change, and general inefficiencies associated with a lack of competitive pressure and contestability.[60] Blame could be conveniently passed to other countries or to freer trade, or attributed to the lack of 'aid', but usually combined with increased demand for domestic trade barriers.[61]

Vested interests who sought market power because of 'imbalances' in specific industry sectors generally perceived governments as the vehicle for pursuing such power regardless of the subsequent market-price distortions.[62] The outcome frequently was higher price inflation due to higher supply chain costs, import restrictions, lack of consumer choice, and increased risk of fiscal debt with taxpayers being forced to pay for government interventionist bailouts.[63]

Domestic self-sufficiency in food production became an ideal rather than a reality because of trade antagonism with many other countries.[64] The more a country distorted its supply chains and sought refuge under trade protectionism, the more isolated it became from international trade and all the associated growth benefits.[65] Any link between trade protection retaliation and both world wars was dismissed under much Cold War rhetoric.

Whereas consumer-based economics was partitioned off as marketing theory in the 1970s, micro-economic theory still remains predominantly production based even today. The lack of integration of production with consumption continues in micro-economic theory, spills over into agricultural commodity selling, and confuses many food-related issues. It explains why many agricultural market and food supply chain distortions still exist, and why much abstract unrealistic micro-economic theory abounds,[66] which jeopardizes food security.[67]

Case study – North Korea

The significance and vulnerability of Korea lies in its strategic geographical position between Russia, Japan, and China. Its mineral wealth of iron ore, copper, fluorite, and coal in the mountains of the northeast have been in demand since the nineteenth century. Japan annexed the country in 1905 after their naval victory with small torpedo boats against the Russian Pacific Fleet battleships near Port Arthur (Dalian) in the previous year. The USSR controlled the north during 1945–48, after which the control of North Korea was relinquished to the Korean Workers' Party.

North Korea has a government-planned command economy that is highly protected from external trade, thus encapsulating an isolationist policy. After the collapse of the USSR, China has provided North Korea with some assistance, particularly the sale of crude oil at discounted prices.

Food rationing has occurred for much of the time since 1948, despite the UN Food and Agriculture Organisation estimating in 2005 that North Korea was the world's tenth largest producer of fresh fruit and nineteenth for apple production. It still has 37 per cent of its people involved in agriculture.

Since 1961, agricultural production on government-owned land has been managed by work teams under the Taean work system, which establishes management supervision, but lacks any essentials for worker motivation, reward incentives, and initiatives for change. Major periodic famines throughout the country have resulted in small private markets existing in the major cities and in areas close to the Chinese border, which suggests that some food supplies are being smuggled over the border from China to bypass the official rationing system.

There was a major famine period during 1995–97, which caused the deaths of up to a million people a year. UNICEF found in 2006 that 37 per cent of people were chronically malnourished, with some children being severely malnourished and adults being underweight. One-third of pregnant women were malnourished and anaemic. There were major famines in 2005, flood-induced food shortages in 2007, followed by more periods of famine.[68]

Despite much domestic famine, some agricultural production is sold and bartered internationally to repay debts, purchase military hardware, or for money-laundering purposes. Most international food aid is diverted to the military and governing elite. By 2010, currency revaluations had eliminated most of the savings for millions of North Koreans. This crippled small privately operated food markets. More famine and deaths are expected to continue under the current land ownership policies and government interventionist schemes. This truly represents complete market failure when compared to competitive market successes.[69]

Monetary policy distortions

Statistical analysis on World Bank data from 29 developing countries found that price distortions had strong deterrent effects on economic growth, while competition was the driving force for growth.[70] The findings suggested that governments needed to ensure that they did not crowd-out the market mechanism and destroy the economic incentive system. Such distortive policies gave the wrong signal to producers and consumers, which ultimately led to a lower income growth rate and unemployment.

Much government intervention has been supposedly justified for reasons of economic stability. However, the primary reasons for 'stability' lie more in political survival through monetary policies such as maintaining low inflation in food prices rather than benefiting the food supply chain. The worst-case perceived threat for many governments is drought or shortages forcing up food prices, causing higher inflation and interest rates, which then increases civil unrest and jeopardizes political stability.

Volatility in production and prices provides opportunities for profit at different times for buyers and sellers. In contrast, stability in food prices only exists when prolonged surpluses drive low prices for long periods, which completely favours the buyer over the seller.[71] If governments want stability through continual surpluses with low prices, then there is a necessity for government to assume responsibility over production because it is not financially viable for individual farmers to continue farming. This was Stalin's dilemma in 1929.[72] The alternatives are to subsidize farmers or to increasingly rely on imports, which were policies that Stalin could not afford with severe budgetary and fiscal constraints after international demand for Georgian oil declined.

Whereas Stalin commenced the nationalization and collectivization of farmers in 1929 to achieve price stability, other governments such as Nazi Germany bullied farmers from 1933 onwards to achieve the same outcome.[73] Some government policies such as the EU Common Agricultural Policy accumulated commodity food from individual farmers through stockpiling policies whilst paying farmers a guaranteed fixed price despite increasing fiscal debt. Other government policies such as the US Farm Bill subsidized farmers without a fixed price, knowing that this induced on-farm hoarding.[74] Some governments in importing countries accumulate 'buffer' or reserve stocks, either by competing directly with private importers such as in China, or alternatively through a government import monopoly such as in India. Achieving price stability is the over-riding objective in all of these policies despite their origin in food bowl countries.

All such government intervention policies cause market-price distortions, which either destroy or decrease the efficiency of food supply chains. Those organizations that manage the distortions and inefficiency better are more likely to survive, grow, and impose significant barriers for new entrants and competition.

Food surpluses depress prices, which suppresses inflation and the need for interest rate rises. However, low food prices may mask alternate monetary expansion and inflationary policies in other industry sectors, which could favour the expansion of industrial manufacturing but discriminate against the food supply chain. In the longer term, food production is likely to decrease because of low prices,[75] and thenceforth any shortages can place upward pressure on inflation.

Food supply chain intermediaries could be affected by low profitability and investment disincentives, while taxpayers could incur the purchase and storage cost of buffer stocks and associated deterioration, as well as opportunity costs on public asset values if global prices subsequently fall. Consumers could suffer through a lack of choice and poor quality products, as well as paying by being taxpayers.

Undervaluing and under-pricing agricultural output are often based on many false assumptions.[76] These include agricultural prices not being very responsive to changes in supply, that any higher prices would mostly favour larger farmers, that higher food prices would increase inflation and impact the greatest on poorer people, that higher farm input prices from an expansionary industrial sector would not seriously jeopardize agricultural production, and that rapid growth through industrialization required the transfer of income from agriculture to industry. Implementing such pricing policies with low agricultural incentives and inadequate investment in food supply chains led to famine and poverty by the 1960s in many African countries.[77]

Fiscal policy distortions

Fiscal policy distortions can occur through direct taxes on farm inputs and outputs, or a value-adding tax on goods and services for food exports, through mandatory government procurement or acquisition at compulsory low prices, through subsidizing the sale of agricultural products, or through bounties paid to primary industries such as fertilizer companies in the expectation that this will lower farm input prices.

Governments in producing countries such as the EU, which purchase surplus commodity stocks in price stabilization schemes, and governments in importing countries such as the Middle East, which purchase commodity buffer stocks, usually both involve fiscal outlays funded by taxpayers. The usual fiscal justification is that the taxpayer is also a food consumer who will benefit from stable prices, albeit with lower choice and quality. However, without government acquisition demand and with cheaper imports, consumers frequently may have paid even lower prices.

Fiscal problems begin when costs increase through inefficiency. Supply chain bottlenecks can rapidly increase supply costs and taxpayer liability. At worst, importing governments are forced to buy at peak commodity prices to meet short-term shortages,[78] thus jeopardizing other fiscal expenditure policies.

The availability of fiscal surpluses or oil revenues is no surety of food supply.[79] Food shortages can occur in importing countries despite the existence of government food stocks, as did occur in 2008 with distribution failures. Official food 'reserves' may be more aligned with government institutional catering and not satisfy those private consumers who are most in need because of misalignment within the supply chains.

Import purchasing governments are acutely embarrassed when they buy commodity food stocks at record global prices only to find prices subsequently collapsed. Government officials are then forced either to sell at huge losses, as did many Middle Eastern governments during late 2008, or keep consumer prices high, as did many African governments in 2008 because of fiscal necessity.

There is a disincentive for private investment in food supply chains and public investment in infrastructure when government debt forces up interest rates through excessive debt-bond issues.[80] High interest rates act as a disincentive for private capital investment because of the higher borrowing cost and the larger rates of return required for comparable investment benchmarking. As well, government debt and subsequent borrowing can 'crowd out' both private capital investment as well as adequate public infrastructure investment.

Exchange rate policy distortions

The problem with a domestic currency fixed against an international currency exchange benchmark is that foreign currencies are continually moving relative to the fixed domestic currency, and this may seriously distort local prices, destroy capital flows, and hinder trade and domestic investment. Whenever central reserve banks are forced to review fixed currency exchange rates, expectations of outcomes leads to speculation, which usually makes them self-fulfilling prophecies.

During the 91-year period from 1880 to 1971 when the price of gold was generally fixed at a gold-parity value and usually linked to currency conversion either through the Gold Standard or gold convertibility via the US dollar, many currency exchange rates were fixed. This caused domestic industry and supply chain distortions, which led to decreased efficiency and increased trade protection, and thus to regular retaliation.[81] Under this fixed currency regime, there were global depressions in 1891–93 that followed the McKinley US Tariff increase of 1890, and in 1930–33 that followed the introduction of the Smoot-Hawley US Tariff legislation in 1929. As well, there were two world wars. Gold mining collapsed while food wars increased. The 'law of one price' for gold prevailed, which resulted in economic and trade chaos.

One alternative was to have an 'adjustable-peg' system, whereby currency exchange rates are allowed to move within a limited band. This method was used by the Chinese government during 2007–2010 in retaliation for the increasing trade barriers and industry subsidies in the US.[82] When the USD Index rose, the Yuan/USD was allowed to float; when the USD Index fell, the Yuan/USD was pegged to the USD Index. The US government attempted to force China to remove this pegging system without offering any trade concessions.[83]

Currency exchange rates affect the relationship between domestic prices and foreign prices. Under flexible deregulated currency policies, falling domestic currency usually promotes domestic exports and improves the trade balance in the balance of payments. This generally removes the urgency to attract foreign capital inflows to offset any trade deficit, which means interest rates can be lower. Lower interest rates can encourage investment, industry, growth, and employment. However, the trade-off is imports with lower quantities and higher prices, which may add inflationary pressures to the domestic economy, and this may need to be dampened with higher interest rates. In balanced economies, interest rate pressures can tend to equalize each other.[84]

Alternatively, a rising domestic currency usually discourages domestic exports and worsens the trade balance in the balance of payments. There generally is a need to attract foreign capital inflows to offset any trade deficit, which may mean that interest rates must be increased. Higher interest rates may discourage investment, industry, growth, and employment. However, the trade-off is imports with higher quantities and lower prices, which may decrease inflationary pressures in the domestic economy, and this can lessen the need to increase interest rates.[85]

The risks with floating exchange rates can be hedged through currency risk management,[86] but this needs to be done in a forward market. Forward currency exchange rates depend on the relative difference between the domestic interest rate and the foreign interest rate. If domestic interest rates are higher than foreign interest rates, there will be a discount in the forward currency market; if domestic interest rates are lower than foreign interest rates, there will be a premium in the forward currency market. Therefore, interest rate manipulations through monetary policies impact on exchange rates, and have an immediate effect on forward food export–import prices, as well as forward currency hedging.

Labour cost distortions

The relationship between the amount of labour regulations and unemployment has been inadequately researched because of the many variables and market dynamism.[87] Some labour laws do increase efficiency through fewer injuries and absenteeism, and an improved work environment, thereby lowering costs. But labour efficiency should be generally driven by competition and not through increased labour laws or trade protection. It is erroneous to claim that governments should be the sole safeguard against bad working environments when the mortality record of Chinese government coal mines is one of the highest in the world.

The dire labour conditions in Britain as observed by Marx and Engels in the 1840s existed after 30 years of high trade barriers under successive Conservative governments, but the paradoxical Marxian remedy involved even more trade and industry protection. Causal factors for labour problems were therefore greatly enhanced in Continental Europe as it attempted to retaliate against competition from North America, particularly that from clipper ships that allowed US new-crop wheat into Europe in front of their own labour-intensive harvest.

Bismarck's answer to the new North American economic and political union in 1865 was to emulate it through a United States of Europe, whose concept had a similar north–south divide as in the US, but was even more divisive within Europe. At the same time, the Reichstag was passing 'cradle-to-grave' socialist, protectionist, and isolationist legislation spurred on by rising pan-Germanic philosophy and sentiment.[88] A succession of Prussian– and German-instigated wars in 1864, 1866, and 1870 was followed by major economic recessions in the 1870s and depression in the 1890s, which finally culminated in a global war. The ultimate solution to much of the Continental labour surpluses was therefore to be the shedding of blood on battlefields, earlier celebrated by Johann Strauss through such idolatry music as the (Field Marshal) Radetzky March in Verona.

If poor working environments lead to higher costs, then the competitive environment may need improving rather than draconian regulatory measures that merely lead to decreased international competitiveness and persistent high unemployment. Alternatively, improved working environments might lead to higher costs and decreased international competitiveness. It could be that without existing government support, the industry or organization does not have the comparative advantage to survive in a competitive environment. Without this advantage and competitiveness, it is questionable as to why this industry or organization should continue in a protective environment.[89]

Labour laws are most likely to increase costs to the food supply chain if they decrease the ability of labour to offer services, restrict the available time for work, or obstruct labour mobility by removing the ability to move between different employment and regions. Such detrimental labour laws can be reinforced through taxation disincentives that discourage a change or shift in employment. However, there might be occasions when taxation provides offsetting benefits, such as incentives to move to other regions where employment opportunities exist. There can be regional isolation problems involved with some food supply chains that may need some offsetting benefits to attract labour, but these need to be factored into initial capital investment and optimal site–locality decision making.

The 'law of one-price' and purchasing power parities have been rekindled in marketing theory and labour rate negotiations.[90] These suggest that the price of the same product or service such as McDonald's hamburgers should be similar throughout the world and that 'optimum' currency exchange and labour rates should reflect this. But this relates to a highly differentiated branded product in a very specific price-making supply chain. One organisation may attempt a global pricing strategy for its products and services, but that does not make it a 'law' for other differentiated products and services or for undifferentiated commodities under price-taking circumstances. There was no evidence of the 'law of one price' for wheat in the 1980s, even when commodity boards were generally benchmarking their price from US wheat prices.[91]

The 'law of one price' contradicts the Vernon and Heckscher-Ohlin trade theories, does not account for comparative differences in inputs such as labour availability, skills, and costs, ignores competitive differences in management ability and style in deregulated supply chains, and disregards dynamic flexible currency exchange rates. It favours labour unions and organizations that want the rest of the world's labour costs to reflect their domestic work rates, so that they can achieve a competitive advantage through a 'level playing field'. It assumes that when this global 'level playing field' is achieved that somehow domestic labour will automatically achieve competitiveness and industry–organization survival.

The serious problem with this 'law of one price' theory is that trade is dependent on arbitrage opportunities, which would not exist in a 'one-price' world. This trade-arbitrage dependency was the essence of the trade theories of Adam Smith and David Ricardo. If trade was taken out of global economic activity, this would reduce growth and investment to that of third-world country status with a return to sustenance living and huge under-employment. The monetary multiplier effect, or the velocity of monetary transactions, would act in reverse under such circumstances.

Relative price–cost differences between countries drive competitive trade, growth, investment, and employment. As well, consumers want maximum choice of products and services, which causes considerable consternation for processors and manufacturers who desire fewer product lines to minimize costs.

Taylor's international socialist pricing policy of sameness and similarity,[92] with a lack of choice for consumers and close synergies between management and labour, is synonymous to the 'one-price' theory. The paradox is that socialists who are consumers but not manufacturers support manufacturers in their striving for a uniform product but which is anti-consumer. They are global in consumer outlook but parochial in production action. This is particularly the case when inflexible domestic labour laws protect their jobs in a global trading environment, or when subsidies protect farmers.

The problem with inflexible labour laws and rising labour costs is that they seriously impact on the food supply chain and distort domestic prices relative to imports. Higher domestic supply chain costs increase consumer prices unless there is a shift to cheaper import supply chains, whilst inflationary pressures influence monetary policies through interest rates and trade flows.

Increasing labour costs decrease international competitiveness of domestic industries, which can jeopardize these industries unless there are compensating increased trade barriers. The cycle repeats itself. It can be argued that increased labour laws and costs during the 1870s in Europe forced countries into currency exchange fixing through the Gold Standard, which then led to supply chain distortions, trade protection and retaliatory policies, which then led to economic recessions and depressions as well as two world wars. Inflexible labour laws are the antithesis of freer trade.[93] Marx's inevitability under such labour laws and trade protection led to very predictable chaos and possible revolutionary conditions. Global freer trade cannot usually occur without deregulating labour laws.[94]

Industry and trade protection

Government rationale for implementing protection for domestic industry include a lack of comparative advantage in resources, services, labour, land, and capital; uncompetitiveness in industry structure or management; vague policies to solve monetary and fiscal problems; or spill-over effects from fixed currency exchange policies.[95] Industry protection usually results in higher consumer prices, a lack of product choice, and increased costs in food supply chains.[96]

If a particular country does not have comparative advantages in resources, services, labour, land, and capital, then it should import rather than attempt to produce domestically.[97] This was the essence of absolute trade advantage by Adam Smith just before the French Revolution, and relative trade advantage of David Ricardo just after the French Revolution. Alternatively, if the country has sufficient comparative advantages and needs to improve its competitive advantage, then introducing more competition would be a better solution than industry protection policies.

Monetary and fiscal problems are generally the result of proliferate government departments and usually not from industries in a competitive deregulated environment. The question is why should industry protection occur,[98] or increased taxation be incurred, as a result of government inefficiency or prodigality. Trade protection most likely would destroy the competitiveness of industry and diminish the tax-revenue base.

Temporary price distortions

There are many different types of temporary distortions that occur regularly in food supply chains, especially with its biological and weather-dependent characteristics as well as time lags. But given some liquidity and with transaction cost flexibility, temporary market price distortions should be quickly removed through arbitrage, provided that the gain exceeds the cost. However, distortions to food prices due to alternate product uses such as bio-fuels may be more permanent, at least until alternate fuel sources are developed.[99]

Notes

1 McKenzie, D.W., 2002, *The Market Failure Myth*, Ludwig von Mises Institute, August.
2 Smith, Adam, 1776, republished in 2003, *The Wealth of Nations*, New York: Bantam-Dell Random House, 45–46.
3 von Mises, Ludwig, 1951, reprinted in 2009, *Socialism: An Economic and Sociological Analysis*, Auburn, AL: The Ludwig von Mises Institute, Part II:1, Chapter 1.
4 Parsons, Timothy, 2010, *The Rule of Empires: Those That Built Them, Those Who Endured Them, and Why They Always Fall*, Oxford: Oxford University Press.
5 Dong, Yong-Sueng, 2011, *Behind North Korea's Plea for Emergency Food Aid*, Samsung Economic Research Institute, Seoul, July, 127–130.
6 Demick, Barbara, 2010, *Nothing to Envy – Love, Life, and Death in North Korea*, New York: Random House.
7 Rapsomanikis, George, Hallam, David and Piero Conforti, 2006, *Market Integration and Price Transmission in Selected Food and Cash Crop Markets of Developing Countries*, Ch. 8 in Alexander Sarris and David Hallam, Eds., 2006, *Agricultural Commodity Markets and Trade – New Approaches to Analysing Market Structure and Stability*, Rome: FAO and Edward Elgar, Cheltenham.
8 Wickham, Chris, 2009, *The Inheritance of Rome*, London: Allen Lane/Penguin, Ch. 15.
9 Groenewegen, Peter D., 1983, *Turgot's Place in the History of Economic Thought: A Bicentenary Estimate*, History of Political Economy, 115, 611–615.
10 Olsen, Mancur, 1993, *Dictatorship, Democracy, and Development*, American Political Science Review, 87:3, 567–576.
11 See Juvenal's Satire VI: Lines 347–348. 'Who will guard the guards themselves' has been generally interpreted as a general apathy by officialdom towards its citizens, which can lead to tyrannical dictatorship.
12 Kaplan, Steven, 1984, *Provisioning Paris: Merchants and Millers in the Grain and Flour Trade During the Eighteenth Century*, Ithaca, NY: Cornell University Press.
13 Groenewegen, Peter D., 1983, *Turgot's Place in the History of Economic Thought: A Bicentenary Estimate*, History of Political Economy, 115, 611–615.
14 Smith, Adam, 1776, republished in 2003, *The Wealth of Nations*, New York: Bantam-Dell Random House.
15 Kurlansky, Mark, 2003, *Salt – A World History*, London: Random House, p. 233.
16 Miller, Judith A., 1998, *Mastering the Market: the State and the Grain Trade in Northern France, 1700–1860*, New York: Cambridge University Press.
17 Hollander, Samuel, 2005, *Jean-Baptiste Say and the Classical Canon in Economics: the British Connection in French Classicism*, London: Routledge.
18 Miller, Judith A., 1998, *Mastering the Market: the State and the Grain Trade in Northern France, 1700–1860*, New York: Cambridge University Press.
19 Taylor, Fred M., 1929, *The Guidance of Production in a Socialist State*, The American Economic Review, 19:1, 1–8.
20 Trentmann, Frank, 2008, *Free Trade Nation: Commerce, Consumption, and Civil Society in Modern Britain*, Oxford: Oxford University Press.
21 Hopkirk, Peter, 1995, *On Secret Service East of Constantinople*, Oxford: Oxford University Press.
22 Barone, Enrico, 1908, trans. 1935, *The Ministry of Production in the Collectivist State*, in F.A. Hayek, ed., 1935, reprinted in *Collectivist Economic Planning*, Ludwig von Mises Institute, Auburn, AL, pp. 245–290.
23 Barr, N., 2004, *Economics of the Welfare State,* New York: Oxford University Press.

24 Barone, Enrico, 1908, trans. 1935, *The Ministry of Production in the Collectivist State*, in F.A. Hayek, ed., 1935, reprinted in *Collectivist Economic Planning*, Ludwig von Mises Institute, pp. 245–290.
25 Sen, Amartya., 1993, *Markets and Freedom: Achievements and Limitations of the Market Mechanism in Promoting Individual Freedoms*, Oxford Economic Papers, 45:4, 519–541.
26 Noel, Alain, 1987, *Accumulation, Regulation, and Social Change: an Essay on French Political Economy*, International Organization, 41:2, 303–333.
27 Taylor, Fred M., 1929, *The Guidance of Production in a Socialist State*, The American Economic Review, 19:1, 1–2.
28 Taylor, Fred M., 1929, *The Guidance of Production in a Socialist State*, The American Economic Review, 19:1, 1.
29 Taylor, Fred M., 1929, *The Guidance of Production in a Socialist State*, The American Economic Review, 19:1, 5.
30 Taylor, Fred M., 1929, *The Guidance of Production in a Socialist State*, The American Economic Review, 19:1, 3.
31 Taylor, Fred M., 1929, *The Guidance of Production in a Socialist State*, The American Economic Review, 19:1, 6.
32 Taylor, Fred M., 1929, *The Guidance of Production in a Socialist State*, The American Economic Review, 19:1, 8.
33 Lerner, A.P., 1934, *The Concept of Monopoly and the Measurement of Monopoly Power*, Review of Economic Studies, 1, 157–175.
34 Scitovsky, Tibor, 1984, *Lerner's Contributions to Economics*, Journal of Economic Literature, 22:4, 1547–1571.
35 Lerner, A.P., 1972, *The Economics and Politics of Consumer Sovereignty*, American Economic Review, 62, 258–266.
36 von Mises, Ludwig, 1949, *Human Action: A Treatise on Economics*, New Haven, CT: Yale University Press.
37 von Mises, Ludwig, 1922, *Socialism: An Economic and Sociological Analysis*, reprinted in 2009, Aubrun, AL: The Ludwig von Mises Institute, p. 212.
38 US Holocaust Memorial Museum, Washington, DC.
39 Lange, Oskar R., 1936, *On the Economic Theory of Socialism*, Part 1, Review of Economic Studies, 4:1, 53–71.
40 Hoarding will be discussed in Chapter 7.
41 See case study in Chapter 6.
42 Montefiore, Simon Sebag, 2004, *Stalin: The Court of the Red Tsar*, London: Phoenix-Orion, Ch. 4.
43 Evans, Richard J., 2006, *The Third Reich in Power*, London: Penguin, Ch. 4.
44 Keynes, J.M., 1938, *The Policy of Government Storage of Food-Stuffs and Raw Materials*, The Economic Journal, 48:191, 449–460.
45 Myers, Robert J., Richard J. Sexton, and William G. Tomek, 2010, *A Century of Research on Agricultural Markets*, American Journal of Agricultural Economics, 91:2, 376–402.
46 Schultz, Henry, 1930, *Der Sinn der statistischen Nachfragen, Heft 10, Veröffentlichungen der Frankfurter Gesellschaft für Konjunkturforschung*, Bonn: Kurt Schroeder Verlag.
47 The Cobweb theory will be examined more fully in Chapter 7, together with a case study on the German Reich Food Estate.
48 Roudinesco, Elizabeth, 2008, *Philosophy in Turbulent Times*, New York: Columbia University Press.
49 Agarwala, Ramgopal, 1984, *Price Distortions and Growth*, Finance and Development, 21:1, 33.
50 Agarwala, Ramgopal, 1984, *Price Distortions and Growth*, Finance and Development, 21:1, 33–37.
51 Agarwala, Ramgopal, 1984, *Price Distortions and Growth*, Finance and Development, 21:1, 33.
52 Samuelson, Paul, Dorfman, Robert, and Robert M. Solow, 1958, *Linear Programming and Economic Analysis*, New York: McGraw-Hill.
53 Hill, Berkeley, 2011, *Understanding the Common Agricultural Policy*, Abingdon, Earthscan.

54 Whitwell, Greg and Diane Sydenham, 1991, *A Shared Harvest – The Australian Wheat Industry 1939–89*, Melbourne: Macmillan Education Australia, Ch. 10.
55 Clark, J.S., 1995, *Single Desk Selling by the Canadian Wheat Board: Does it Have an Impact?* Canadian Journal of Agricultural Economics, 43, 225–236.
56 McColl, Jim, 1988, *Royal Commission into Grain Storage, Handling, and Transport*, Canberra: Australian Government Publishing Service, p. xxv.
57 Dong, Fengxia, Marsh, Thomas L., and Kyle W. Stiegert, 2006, *State Trading Enterprises in a Differentiated Product Environment: The Case of Global Malting Barley Markets*, American Journal of Agricultural Economics, 88:1, 90–103.
58 Demick, Barbara, 2010, *Nothing to Envy – Love, Life, and Death in North Korea*, New York: Random House.
59 von Mises, Ludwig, 1922, *Socialism: An Economic and Sociological Analysis*, reprinted in 2009 by the Ludwig von Mises Institute, Auburn, AL, p. 211.
60 Mulgan, Geoff, 1995, *Democratic Dismissal, Competition, and Contestability Among the Quangos*, Oxford Review of Economic Policy, 10:3, 51–60.
61 Hillman, Arye L., 1977, *The Case for Terminal Protection for Declining Industries*, Southern Economic Journal, 44:1, 155–160.
62 Lloyd, P.J., 1982, *Protection for Whom?* Economic Papers – The Economic Society of Australia, 1:2, 1–17.
63 Centre for International Economics, 1990, *Free Lunches: How Much Do They Really Cost and Who Pays?*, National Centre for Development Studies, Australian National University, Canberra, November.
64 Stoeckel, A.B., Vincent, D., and S. Cuthbertson, Eds., 1989, *Macroeconomic Consequences of Farm Support Policies*, Durham, NC: Duke University Press.
65 Centre for International Economics, 1989, *What's Wrong with Australia's Export Performance?*, National Centre for Development Studies, Australian National University, Canberra.
66 Mandler, M., 1999, *Dilemmas in Economic Theory: Persistent Foundation Problems of Microeconomics*, Oxford: Oxford University Press.
67 Demick, Barbara, 2010, *Nothing to Envy – Love, Life, and Death in North Korea*, New York: Random House.
68 Demick, Barbara, 2010, *Nothing to Envy – Love, Life, and Death in North Korea*, New York: Random House.
69 McKenzie, D.W., 2002, *The Market Failure Myth*, Ludwig von Mises Institute, Auburn, AL, August.
70 Singh, R.D., 1992, *Government-induced Price Distortions and Growth: Evidence from Twenty-nine Developing Countries*, Public Choice, 73:1, 83–99.
71 Figure 1.3 in Chapter 1.
72 Montefiore, Simon Sebag, 2004, *Stalin: The Court of the Red Tsar*, London: Phoenix-Orion, Ch 3.
73 Evans, Richard J., 2006, *The Third Reich in Power*, London: Penguin, Ch. 4.
74 Hoarding will be discussed in Chapter 7.
75 Peterson, Willis L., 1979, *International Farm Prices and the Social Cost of Cheap Food Policies*, American Journal of Agricultural Economics, 61:1, 12–21.
76 Agarwala, Ramgopal, 1984, *Price Distortions and Growth*, Finance and Development, 21:1, 33–37.
77 Agarwala, Ramgopal, 1984, *Price Distortions and Growth*, Finance and Development, 21:1, 34.
78 Piesse, Jenifer and Colin Thirtle, 2009, *Three Bubbles and a Panic: An Explanatory Review of Recent Food Commodity Price Events*, Food Policy, 34:2, 119–129.
79 van Rheenen, Teunis, Clemens Breisinger, Catherine Aragon, Suresh Babu, and Khalid El Harizi, 2010, *The Global Food Crisis: How Arab Countries Can Emerge More Food Secure*, paper presented at the International Conference, 'Food Security in the Arab Countries: New Challenges and Opportunities in the Context of Global Food Price Volatility', College of Agriculture and Marine Sciences, Sultan Qaboos University, Muscat, 2–4 March.

80 Sturzenegger, Frederico, and Jeromin Zettelmeyer, 2006, *Debt Defaults and Lessons from a Decade of Crises*, Cambridge, MA: MIT Press.
81 Eichengreen, B., Ed., 1985, *The Gold Standard in Theory and in History*, London: Methuen.
82 Thorbecke, Willem, 2006, *How Would an Appreciation of the Renminbi Affect the US Trade Deficit with China?*, Topics in Macroeconomics, 6:3, Article 3.
83 Sharma, Shalendra D., 2011, 'US Debt, Deficit, and the Falling Greenback: Does it mean currency wars and an end to the Dollar's reign?' *SERI Quarterly*, Samsung Economic Research Institute, Seoul, July, 67–78.
84 Blejer, Mario I., and Marko Skreb, 1999, *Balance of Payments, Exchange Rates, and Competitiveness in Transition Economies*, Boston, MA: Kluwer Academic Publishers
85 Sarno, Lucio, and Mark P. Taylor, Eds., 2002, *New developments in Exchange Rate Economics*, Volumes 1 and 2, Edward Elgar, Cheltenham.
86 Ephraim, Clark, 2004, *Arbitrage, Hedging, and Speculation: the Foreign Exchange Market*, London: Praeger.
87 Clark, Kim B., Summers, Lawrence H., Charles C. Holt., Robert E. Hall and Martin N. Baily, 1979, *Labor Market Dynamics and Unemployment: A Reconsideration*, Brookings Papers on Economic Activity, 1, 13–72.
88 Gerwath, Robert, 2007, *The Bismarck Myth: Weimar Germany and the Legacy of the Iron Chancellor*, New York: Oxford University Press.
89 Papageorgiou, D., Choksi, A.M., and M. Michaely, 1990, *Liberalizing Foreign Trade in Developing Countries: The Lessons of Experience*, Washington, DC: The World Bank.
90 Rao, D.S. Prasada, 2010, *Purchasing Power Parities of Currencies*, Brisbane: University of Queensland Press.
91 Ardeni, Pier Giorgio, 1989, *Does the Law of One Price Hold for Commodity Prices?*, American Journal of Agricultural Economics, 71:3, 661–669.
92 Taylor, Fred M., 1929, *The Guidance of Production in a Socialist State*, The American Economic Review, 19:1, 1–8.
93 Zekulich, Michael, 1997, *The Grain Journey – The History of The Grain Pool of WA*, Perth: The Grain Pool of Western Australia, pp. 5–7.
94 Florence, P. Sargant, 2003, *Economics of Fatigue and Unrest and the Efficiency of Labour in English and American Industry*, London: Routledge.
95 Bentil, J. Kodwo, 1985, *Attempts to Liberalize International Trade in Agriculture and the Problem of the External Aspects of the Common Agricultural Policy of the European Economic Community*, Case Western Reserve Journal of International Law, 17:3, 357–388.
96 Anderson, Kym, Martin, Will, and Dominique van der Mensbrugghe, 2006, *Distortions to World Trade: Impacts on Agricultural Markets and Farm Incomes*, Applied Economic Perspectives and Policy, 28:2, 168–194.
97 Irwin, Douglas A., 1993, *Free Trade and Protection in Nineteenth-Century Britain and France Revisited: A Comment on Nye*, The Journal of Economic History, 53:1, 146–152.
98 Lloyd, P.J., 1982, *Protection for Whom?* Economic Papers - The Economic Society of Australia, 1:2, 1–17.
99 Tokgoz, Simla, Elobeid, Amani, Fabiosa, Jacinto, Hayes, Dermot J., Babcock, Bruce A., Yu, Tun-Hsiang, Dong, Fengxia, and Chad E. Hart, 2008, *Bottlenecks, Drought, and Oil Price Spikes: Impact on U.S. Ethanol and Agricultural Sectors,* Review of Agricultural Economics, 30:4, 604–622.

Further readings on price distortion effects

Abouchar, Alan, 1977, *The Socialist Price Mechanism*, Duke University Press, Durham, NC.
Bale, Malcolm D., and Bruce L. Greenshields, 1978, *Japanese Agricultural Distortions and Their Welfare Value*, American Journal of Agricultural Economics, 60:1, 59–73.
Bale, Malcolm D., and Ernst Lutz, 1981, *Price Distortions in Agriculture and Their Effects: An International Comparison*, American Journal of Agricultural Economics, 61:1, 8–22.

Callahan, Colleen M., McDonald, Judith A., and Anthony P. O'Brien, 1994, *Who Voted for Smoot-Hawley?*, Journal of Economic History, 54:3, 683–684.
Ciezadlo, Annia, 2011, *Let Them Eat Bread: How Food Subsidies Prevent (and Provoke) Revolutions in the Middle East*, Foreign Affairs, Council of Foreign Relations, March 23.
Draper, Peter, Philip Alves and Razeen Sally, 2009, *The Political Economy of Trade Reform in Emerging Markets, Crisis or Opportunity*, Edward Elgar, Cheltenham.
Dunn, Robert M., 1970, *Flexible Exchange Rates and Oligopoly Pricing: A Study of Canadian Markets*, The Journal of Political Economy, 78:1, 140–151.
FAO, 1946, *Proposals for a World Food Board and World Food Survey*, United Nations, Rome.
——1956, *Functions of a World Food Reserve – Scope and limitations*, Commodity Policy Studies, No. 10, Rome.
Francois, M. Jacques, 2007, *Even Commodities have Customers*, Harvard Business Review, 85:5, 110.
Greaves, Bettina Bien, 1998, *Interventionism: An Economic Analysis by Ludwig von Mises*, Foundation for Economic Education, Irvington-on-Hudson, New York.
Gupta, Bishnupriya, 2001, *The International Tea Cartel during the Great Depression, 1929–1933*, Journal of Economic History, 61:1, 144–159.
Hilton, Matthew, 2009, *Prosperity For All – Consumer Activism in an Era of Globalization*, Cornell University Press, Ithaca, NY.
Hughes, D., 2002, *Grocery Retailing in Europe and Emerging Routes to the Customer*, Eurochoices 1, 12–16.
Lloyd, T.A., S. McCorriston, C.W. Morgan, and A.J. Rayner, 2003, *Food Scares, Relative Prices, and Market Power*, Discussion Papers In Economics, 18/3, School of Economics, University of Nottingham.
McCorriston, Steve, and Donald MacLaren, 2005, *The Trade Distorting Effect of State Trading Enterprises in Importing Countries*, European Economic Review, 49, 1693–1715.
——2007, *Do State Trading Exporters Distort Trade?* European Economic Review, 51, 225–246.
Miranda, Mario J., and Peter G. Helmberger, 1988, *The Effects of Commodity Price Stabilization Programs*, The American Economic Review, 78:1, 46–58.
Schuh, G. Edward, 1975, *General Economic Policy as a Constraint to the Development of Agriculture*, in *Agricultural Policy: A Limiting Factor in the Development Process*, Inter-American Development Bank, Washington, DC.
Schultz, T.W., Ed., 1978, *Distortions of Agricultural Incentives*, Indiana University Press, Bloomington.
Sharma, Pradeep, K., 1997, *Foodgrain Economy in India: Government Intervention in Rice and Wheat Markets*, Shipra Publications, Delhi.
Soyfer, Valery N., 1994, *Lysenko and the Tragedy of Soviet Science*, Rutgers University Press, New Brunswick, NJ.
Townsend, Robert M., 1977, *The Eventual Failure of Price Fixing Schemes*, Journal of Economic Theory, 14:1, 190–199.
Trentmann, Frank, 2008, *Free Trade Nation. Commerce, Consumption, and Civil Society in Modern Britain*, Oxford University Press, Oxford.
Trotter, Wilfred, 1915, *Instincts of the Herd in Peace and War*, Macmillan, London.
Ulen, Thomas S., 1982, *The Regulation of Grain Ware-housing and its Economic Effects: The Competitive Position of Chicago in the 1870s and 1880s*, Agricultural History, 56:1, 194–210.
World Bank, 1995, *Bureaucrats in Business: The Economics and Politics of Government Ownership*, Oxford University Press, New York.
Zerbe, Richard O., 1982, *The Origin and Effect of Grain Trade Regulations in the Late Nineteenth Century*, Agricultural History, 56:1, 172–193.

3

ROLE OF THE PRIVATE SECTOR, PRICE, COMPETITION, AND GOVERNMENT

There are definitive roles for governments in food supply chains,[1] but maximum limits need to be enforced. Alternatively, governments need to promote competition to ensure proper functioning of the price mechanism. Synergies need to be established without role and function encroachment occurring.

John Locke argued for governmental separation of powers as early as the seventeenth century, and intimated that such separation of power might only occur through revolution and not devolution.[2] It was the height of European empire building with centralized power structures. Locke's anticipated revolution came within the next 100 years.

A confluence of non-interventionist ideas occurred in 1776, with the American War of Independence, Adam Smith publishing his inquiry into *The Wealth of Nations* in Britain, and Turgot's deregulatory food chain policies in France.[3] All three events seriously questioned the authoritative extent of government and licensed monopoly agencies, and sought defined limits of government power and taxation with constitutional balances.

French debt over their involvement in the US War of Independence and increases in the dreaded French gabelle salt tax led to the French Revolution in 1789.[4] The predictions of both Locke and Turgot became a reality with rioting against government bread pricing and shortages.[5] More broader anti-intervention sentiment was expressed in the freer market philosophies of David Ricardo, Jeremy Bentham, James Mill, Jean-Baptiste Say, and John Stuart Mill during the first half of the nineteenth century.

If consumers make the ultimate decisions over what is purchased and eaten,[6] it is rational to assume that they should have the dominant control over the food supply chain, with lesser control by farmers, food manufacturers, and government. It is the pull in the supply chain rather than the push that ultimately determines the functionality of food supply chains. Consumers provide the guiding 'invisible hand'[7] as to what, where, when, and how products are demanded in the supply chain.

Government decision making is a poor substitute for multitudinous daily consumer decisions.[8] Any autocratic enforcement or lack of choice in supply chains will incite consumer resentment, which will have spill-over effects on investment, employment, and economic growth.[9] The uprisings in the Middle East and North Africa during 2011 can be attributable to reaction against such autocratic government intervention,[10] with consumers wanting more control over decision making.

Except for brief periods during the nineteenth century, discussion on competition and deregulatory policies in the food supply chain has only been of recent origin. Frank Knight discussed competition in 1923,[11] but there was a long gap before Lowell Hill attempted in 1982 to clearly elucidate the role of government in agri-food supply chains,[12] and established guidelines for competitive agricultural markets in 1995.[13]

Problems with government supply chain intervention

Price distortions, costs, and resource misallocation can occur if governments arbitrarily intervene in food supply chains.[14] Efficiency is jeopardized, supply chain signals become warped, while decision making becomes difficult despite having public control and responsibility.

The goals of government would not be expected to align with the decisions of millions of consumers, assuming that the latter could be identified and measured in a dynamic domestic–global environment. Public decisions must be made in a vacuum of consumer thought, with assumptions based on historical consumer patterns that are an ill-guide to future consumption patterns.[15]

Governments increasingly assumed control over monetary policy during the twentieth century. Control over commodity prices theoretically were justified through macroeconomic goal priorities of inflation, employment, interest rates, currency exchange rates, wages, economic growth, or political security. This is despite commodity prices rarely causing inflation,[16] which was more likely to be caused by profligate public spending, increasing debt, printing money, or as a result of war and anarchy.[17]

Normal functioning food supply chains and external trade would be expected to decrease inflation as imports balance domestic production shortfalls. In contrast, public intervention into food supply chains combined with trade protection can quickly cause price inflation, as occurred in Germany in the 1930s.[18] To justify more government intervention on the grounds that previous government price controls failed does not engender confidence in the ability of governments to manage microeconomics.[19]

Economists and governments rarely consider human psychology when making decisions, on the assumption that human response is supposedly always 'rational'.[20] However, what is rational to one human individual is probably irrational to another individual, and what may be considered rational today may be considered irrational tomorrow, perhaps after more learning and experience.[21] Public policy decision making cannot be based on the unrealistic assumption of constant 'rationality'.

Governments generally fear deflation, as consumers will delay purchasing decisions because prices will be lower tomorrow, thus lowering economic growth. Inflation is usually favoured because consumers will make purchasing decisions today because prices will be higher tomorrow, thus improving economic growth. If this analogy is understood by government, then the same psychological understanding is required for price.

When governments raise price, there is an incentive for hoarders to withdraw supply in the anticipation that the government will raise prices again in the near future.[22] Food shortages and famine are most likely to occur when governments frequently raise price. Theory would suggest that farmers and hoarders will usually sell when the government lowers price, thus alleviating shortages. That response might occur for merchant hoarders, but it rarely happens with farmers, particularly if they perceive an increased squeeze on their profit margin and sustainability. They are more likely to hoard storable commodities until the government again raises price. The government is in a no-win situation with price controls. Simple human psychology is frequently lost in government decision making, as has been the case in India.

Because intervening governments can be monopsonies by being a single buyer with many sellers, and/or monopolies by being a single seller with many buyers, government prices are generally not based on market criteria but are more likely to reflect the intrinsic supply chain power and opportunism of public authorities.[23] Private competition is likely to be 'crowded out' regardless of whether governments act as a single buyer or seller, or act through licensed oligopoly agents whereby a few organisations have significant market power.

'Crowding out' private competition can result in inefficiency, higher costs, a loss of international competitiveness, increasing dependency on trade barriers and trade isolation, dysfunctional supply chains, and the destruction of 'trade' markets.[24] No adequate benchmarking of government organizations can occur until there is supply chain contestability,[25] as monopoly 'rent' can only be gauged in hindsight after some competition has occurred. Without adequate price-reward signals in the supply chain, domestic merchants will quickly disappear leaving the government with the predicament of managing a limited functioning supply chain.

Governments are bureaucracies with hierarchal structures and divisions that command standardization of procedures to facilitate efficient task functionality.[26] Even decisions based on uniformity can result in complexity under such organizational structures. Price fixing therefore is often reduced to one product for each product-line category, such as one type of bread. Variations are reduced to minimize complexity, despite the multitudinous choice demanded by consumers. Homogeneity replaces heterogeneity under government monopoly bureaucracy. Consumers therefore suffer with such product–price standardization, with negative spill-over impacts on investment, employment, and economic growth.

There is a poor record from governments and their agencies in spot price forecasting. Staff would not be working in mundane jobs if they were good forecasters. Even when global central banks had cornered much of the gold market by 2000, and had some oligopolistic power over gold pricing, they then sold much of their gold reserves at the bottom of market in 2001, and then brought gold back at record high prices during 2011 to rebuild their reserves. If oligopolistic central banks get it wrong, this does not engender much confidence in the ability of government food agencies to get it right. Buying at high prices and worsening inflation, and then selling at low prices and prolonging the misery of farmers, has been very symptomatic of government intervention into food supply chains, especially during 2007–08.

With such a dismal record in the spot commodity market, the construction of a localized forward market has predictably eluded all governments. There is no precedent of a government having successfully established a forward market. All the evidence suggests that no forward market is likely when government supply chain intervention occurs. Yet a forward market is essential to reward supply chain intermediaries to 'carry' inventory that is vital for food security, and to offset forward price risk through hedging.[27]

Government or statutory monopolies are often advocated for reasons of economies of scale, with unit cost falling as quantity rises. However, this theory assumes that these monopolies have efficiency measures to reduce costs. It is very difficult to implement cost efficiencies without competition or contestability, particularly with monopolized operations such as food manufacturing, railways, bulk handling, ports, and shipping.[28]

Fear of government takeover of resources in countries such as Zimbabwe stifles confidence, law enforcement, and investment in food supply chains. Some governments purport to encourage foreign investment and private initiative, only to have secretive secondary takeover and control motives for the future.[29] These uncertainties deter commitment by potential private investors in local food supply chains in many countries.

Change management and flexibility are rarely attributes of bureaucracy, with large errors occurring in being reactive or proactive in decision making.[30] Governments are ill-equipped to have the timely flexibility to adequately and efficiently respond in a dynamic value-changing supply chain environment.

Role of the private sector in food security

In contrast to the problems of government food chain intervention, there are many functions of private intermediaries in the food supply chain that cannot be easily replicated by government or its agencies.

Price is a monetary mechanism that reflects perceived individual value in a transfer of a product or service. Supply chain price derives from a momentary consensus agreement between a buyer and seller. Regardless of the background and motivation of the buyer and seller, any momentary consensus agreement must of necessity be private, even in futures and commodity exchange transactions. Private organizations are better equipped to ascertain all the private behavioural psychology involved with decision making that motivates buyers to buy and sellers to sell, and to ascertain the willingness and ability of each buyer and seller to actually transact. Private organizations are better equipped to measure perceived individual value and its myriad of changes and dynamic transformations.

The mechanism for price discovery is not the simple inevitable 'equilibrium' process portrayed in microeconomic theory. Consumers usually want the best quality for the lowest price, whereas farmers often want the best price for lowest quality. Perceptions of quality and value differ widely, even among consumer and farmer groups. There are frequent stand-offs between the two parties, as represented by unrealistic bid and offer prices in market depth. A separate market price must ultimately be agreed upon to enable each single transaction to eventually occur.

Despite shortcomings of non-convergence[31] and basis risk, futures markets provide public transparency for both spot prices and forward prices. However, there cannot be any forward price discovery mechanism without strong local 'trade' markets. Having been prior established in 'trade' markets, forward prices can then only become publicly transparent through a liquid (high-volume transacted) futures market.

Food is characterized by seasonal production in a limited number of months each year. Non-perishable commodities can be stored to enable better utilization throughout the year. This contrasts with perishable commodities such as meat and dairy products, which incur price disincentives to discourage storage.

Under conditions of uncertainty, end users might undertake their own storage of non-perishable commodities, by buying inventory stocks early. However, under conditions of more certainty, end users usually push this storage–carry function back onto accumulation merchants in the supply chain.

In return for just-in-time delivery from accumulation merchants, end users must be willing to pay a carry charge to merchants for warehousing.[32] The accumulation of specific product, storage, receival of carry payment, quality control, and other aspects of maintaining food reserves are therefore the function of domestic merchants through 'trade' markets.[33] This contrasts with government storage of non-aligned product that is usually unwanted by the 'trade' market because of a lack of product specification, quality risk, product risk, supply chain risk, and untimely.

It is the function of farmers and private supply chain intermediaries in the food supply chain to add or create value in response to continually changing consumer–customer demand. Adding value should not be the sole prerogative of the food manufacturer, whilst perceived value adding might not be synonymous with actual consumer perceptions of value.

Value adding from a farmer perspective frequently does not equate to value creation as perceived by the consumer, which really is a major factor jeopardizing food security. What is produced is not always what is wanted. The main function of food security is to synchronize these two quite antagonistic elements in the supply chain. It is not the role of government to intervene in this stand-off or to favour or protect either party. This is the function of the pricing mechanism.

Consumers demand products that are differentiated through style, choice and range, origin, benefits, treatment, characteristics, content, usage, size, packaging, distribution, and convenience. Adequate private supply chain research and new product development programmes are required, with response teams linked to changing consumer tastes and preferences. It is not the function of government to undertake such consumer-based research, nor is it the function of governments and farmers to enforce production-based product classifications. What is required is a competitive supply chain that facilitates both product flows and the dissemination of information, ideas, and changes.

Functions of price

It is not the function of price to compensate farmers or food manufacturers for their cost of production, or to offset the cost of added services, even if the product is highly differentiated and branded. Both sellers and buyers are generally price takers in competing food supply chains that have both consumer choice and substitute products as their major characteristics. If the product attribute is part of the product specification, then price will reflect product specification and not the cost of the product attribute. Unless buyers and consumers perceive extra value that is beyond the product specification, then no extra price can be expected.

There are a number of identifiable functions of price in deregulated competitive food supply chains.[34] These can be generally separated according to buyer and seller attributes, or surplus and shortage conditions.

Knowledge of correct realistic global and domestic prices is essential to make important management and investment decisions. Few long-term business successes can be expected if prices are distorted, which explains why agricultural businesses and supply chains can perform so poorly under government intervention.

Decisions made by farmers and supply chain intermediaries need to have supply chain scrutiny, because there is no alternative for benchmarking performance. When farmers, merchants, and end users make the right decision, they should receive a reward through higher profits, and receive a penalty through lower profit when they make the wrong decision. If this reward mechanism is absent because government intervention distorts price signals, then poor performance and inefficiencies would be expected to occur, because of disincentives and a lack of competitive motivation.

When surpluses exist, this signal needs to be readily transparent to consumers through an efficient price mechanism. If a government-induced price does not provide a signal to consumers to buy, then surpluses will increase and government storage costs will be borne by taxpayers. Consumers will therefore incur an opportunity cost at the same time that they will be forced to pay taxes for government storage.

Both farmers and food manufacturers need to be able to recognize situations when consumers increase their demand. Increased consumer demand must be allowed to result in higher prices, otherwise there will be no incentive for either farmers or food manufacturers to respond. Clear transparency in the supply chain to facilitate quick transfer of consumer signals is essential.

End users need to know where supplies are located when either buying product or making investment and location decisions. Consumers need to know that choice is available. These signals are best transmitted by different prices in different geographical locations. Merchants can then arbitrage by buying where there are surpluses at low prices and selling into shortage areas at high prices, presumably for a financial margin that is greater than the transport cost.

When surpluses accrue, prices need to fall to act as an incentive for end users and consumers to buy and for farmers to decrease production. If government prices are too high, there is little incentive for end users and consumers to respond, and it is more likely to encourage farmers to produce, thereby worsening the surplus. Under such circumstances, quality usually deteriorates, end users such as food manufacturers become disillusioned about investment, consumers suffer from a lack of choice, especially when cheaper imports are usually restricted, and the taxpayer finances an increasing burdensome stockpile.

Prices should rise in regions where demand is increasing and fall in regions where demand is decreasing. A deregulated competitive supply chain acts as an arbitrage signal for merchants to buy in low price areas and sell in high price areas.

Transparent supply chain price information and efficient communication should act as incentives for farmers to produce or not to produce. If demand signals are not transmitted back in the supply chain because of distorted government prices and regulatory controls, then farmers can be encouraged to keep producing when surpluses abound, and restrict production when shortages occur. Governments need alternative monetary, fiscal, and trade policies other than price controls to remedy inflation and maintain political stability, rather than providing wrong signals to farmers.

Food shortages should cause prices to increase, which then induces farmers to produce more. Private hoarding could be a problem in the short term, but increased production should alleviate long-term problems and remove the incentive for private hoarding.

However, any government hoarding overhangs the food supply chain as a constant threat to farmers because they suspect that as soon as they increase production, the government will sell the public stocks and depress domestic prices. Farmers can easily become distracted from what they do best as producers in an attempt to out-manoeuvre government decision makers.

If prices are distorted by government intervention, professional hoarders will either game-plan expected government decisions, or attempt to corruptly influence government decision making. This has occurred in India under price controls,[35] and it suggests that much private domestic hoarding and corrupt activity can exist whenever government intervention in the food supply chain occurs. Rather than rectifying the domestic intervention problem, governments often seek to conveniently blame other countries whenever food shortages and subsequent riots occur, particularly if other countries are perceived to be more developed than their own or if the other countries are food exporters.

Food surpluses should cause prices to decrease, which then induces farmers to produce less. However, farm subsidies encourage farmers to either maintain or increase production, regardless of the amount of surplus. Usually the greater the surplus, the higher the probability of prolonged lower prices, which then requires even more subsidies from government fiscal budgets and ultimately taxpayers. Farmers can become dependent on government subsidies or income support, which encourages further production. Political instability occurs whenever these farm subsidies or income support policies are threatened by government fiscal restraint.

The role of competition

Competition has a major dynamic role[36] in a food supply chain from many different aspects. It can affect choice, efficiency, performance, costs, risk management, the formation of cluster industries, as well as motivation and innovation.

Competition creates choice, which usually ensures that most consumer needs are satisfied,[37] despite their vagueness and fickleness. Governments tend to favour monopolies, bureaucracy, and standardization, whereas competition drives variety, efficiency, and diversity.[38] Focus is shifted from merely satisfying the organization to meeting customer expectations.

Competition is the driver of supply chain efficiency through contestability of functions.[39] Rivalry between competitors, suppliers, innovators, and substitute products, with few entry barriers, is the key success factor contributing to food supply chain efficiency and developing competitive advantage over other regions and countries. Allowing global supermarkets to operate domestically can increase rivalry and efficiency while decreasing the need for government intervention, provided that there is sufficient competition between the supermarkets themselves.[40]

Farmers sometimes may correctly identify excess profits and superfluous intermediaries in the food supply chain. However, government intervention does not remove these supply chain inefficiencies; rather such intervention tends to entrench such inefficiencies. In contrast, it is more competition rather than less that is required to induce efficiencies and reduce supply chain costs using benchmark performance criteria.[41]

Competition is a strong motivator for better management of financial risk. If financial losses do occur in a competitive environment, then this should act as a strong incentive to minimize them in the future. There are many methods to manage supply chain risks under deregulation, but most require the liquidity (volume traded) that derives from private competitive practices.

Supply chain clusters[42] are interrelated production and service businesses that are geographically concentrated. They may be temporary or have longer-term sustainability, but this depends on international competitiveness, expected profitability, taxation, capital investment, research and development, and monetary and fiscal policies of government. The drivers of cluster industries in a deregulated supply chain may be based on advantages in technology, experience, knowledge, skilled labour, management, capital funding, networks, or the availability of cheap inputs.[43] Key factors that relate to cluster industries include spill-over effects, supply chain integration, shortened value chains, learning curves, knowledge sharing, geographic concentration of buyers, convenience, investment efficiencies, communication, motivation and incentives, high turnover of businesses, infrastructure, joint ventures, high risk, education training facilities, and successful research and development.

Even when monopolies are merely threatened with competition, the theory of contestability[44] suggests that they are still motivated and respond differently compared to monopolies that are not threatened with competition. Competition, or the threat of competition, is a major driver of motivation, innovation, productivity, research, and product–service development. It is more likely that value creation and product diversity will occur under competition than under private or public monopolies.

Without competition, there cannot be benchmarking measures. Performance is a relative measure, which implies that both private and public monopolies and monopsonies have inadequate measurements of perceived performances. Even monopoly profits and rent can only be determined as excessive after contestability or competition has occurred.

Main roles for government in the food supply chain

Despite general criticisms of government involvement in food supply chains, there are defined roles for government in many supply chain facets.[45] The problem is keeping them separate from other functions, and prevent creeping into other roles.

Defence of trade routes and markets from land predators[46] and sea piracy has been one of the prime functions of government since urban clustering developed around fortified geographical locations and ports. The greatest trading nations of the world have historically arisen from those countries with the greatest naval defence. The lingering problem for many governments becomes one of separating private trade from public defence and foreign policy. When public defence and foreign policy affect government monetary and fiscal policy, which also affect private trade, the line separating public from private responsibility is even more difficult to identify, but cannot be ignored.

Governments are buyers in their own right, often with their own food supply chains. Defence personnel need to be properly fed, hospitals need to function with food facilities, prisoners need to be sustained, children need food during schooling hours, and canteens are needed for government staff. The supply logistics coordination behind government requisitioning procurement, and the associated delivery, requires enormous skills and responsibilities.[47] Government institutional catering can be the biggest single food supply chain in most countries.

It is the function of governments to enforce adequate competition and thereby efficiency in food supply chains.[48] However, governments have traditionally been the worst offenders of this principle, introducing or protecting many public–private monopolies, monopsonies, and oligopolies. Merely introducing laws regarding anti-trust or anti-competition legislation is meaningless unless it is enforced and results in more competition. Unless there is some privatization or contestability in institutional catering and supply chains, enforcement of competition laws is very difficult. It usually necessitates the minimization of corruption, combined with an independence of legal decisions.

It is virtually impossible to introduce competition laws when the government is dominant in the food supply chain either through buying power or supply distribution controls. Whenever large public catering contracts exist for institutional needs, governments will continue to dominate food supply chains. Sometimes it is the monopolies created for public institutional catering that spill over into private food supply chains.

Many Asian, African, and Central and South American governments still do not enforce contract law, even though many have enacted the legislation. Under such conditions, international buying and selling becomes extremely opportunistic, with domestic buyers unloading ships only when import prices are rising, and refusing to unload ships when import prices are falling. These contract defaults result in port congestion and chaos, with potential international trade opportunities forever diminishing.

If there is no contract law enforcement, there cannot be any effective forward market for buying and selling, which then limits buyers to primitive scrambling of available spot supplies. Under such rudimentary conditions, food shortages frequently occur, despite laden ships offshore. The forward market is an integral component of importing, exporting, private storage, and food security, but it is very fragile and law dependent.

Whenever non-enforcement of contract law occurs, foreign exporters will frequently demand free-on-board (fob) origin pricing, whereby the domestic importer bears the full cost of freight and insurance. When enforcement of contract law exists, foreign exporters would be more willing to quote a delivered cost, insurance, and freight (cif) destination price, particularly in times of global commodity surpluses and relatively stable shipping freight rates.

Governments need to enforce and protect property rights if adequate food supply chains are to be sustained. Generally, many governments have a poor record on ensuring property rights, even in more developed countries, with farms and businesses being expropriated below market value. Many farmers do not own the land they farm such as in China and North Korea, yet they are expected to be sufficiently motivated to feed the total population and take investment risks. Laws that facilitate private farm ownership can quickly be reversed to accommodate expropriation, such as occurred in Zimbabwe.[49] Pastoral land can be appropriated for unsustainable cropping farms, such as in Australia with its soldier resettlement schemes.[50] Mining access laws still frequently override agricultural land ownership in many economically developed countries. Food security depends on the resilience of farmers to continue with high-risk agricultural pursuits despite such insecure adversities and uncertainties.

There have been many occasions when land 'reform' programmes have not adequately re-compensated the previous legal owner, or when excessive land taxation has jeopardized property rights.[51] Foreign ownership laws can quickly change, thus jeopardizing overseas investment stimuli into local farms. Changes in policies relating to government-owned land, including public leasing, can seriously affect landholders and act as a disincentive for further land investment and development.

Mandating bulk commodity commingling destroys any notion of identifiable property, identity preservation, or traceability. Under existing English law, legal ownership is lost if property cannot be identified as single items, such as individual grains. An Australian High Court decision, *Chapman Brothers versus Verco Brothers* in 1933, removed the farmer's commodity ownership immediately upon delivery into commingled storage, despite it not having been sold. This paved the way for statutory selling boards with ownership or vesting powers.

Family succession generally creates subdivision problems. There is an increasing global tendency for larger efficient-sized land ownership to be subdivided into smaller inefficient-sized landholdings. Government policies must provide flexibility for individuals to accumulate or aggregate larger sized land holdings, but with adequate compensation to the previous small farmers. There is a need for a deregulated, independent, and efficient real estate market, such as that which exists now in Vietnam, which can aggregate farm sizes and remove the need for agrarian revolutions.

The narrow definition of infrastructure includes transport networks, but the wider definition includes provision for industry clustering for learning and skills development, technological transfer, and information systems networking. Global food shortages are not caused merely by local marginal crop failures that are weather related. More often, there are supply chain bottlenecks caused by infrastructural deficiencies and support systems breakdowns, with laden ships frequently offshore waiting to be unloaded. Brazil has the sixth largest global economy in terms of gross domestic product and is extremely dependent on agricultural commodity exports, but it can still take 15 days to load one sugar ship on antiquated conveyor belts exposed to weather conditions.

Many government departments and agencies compete against each other through inter-agency rivalry, which can result in organizational power struggles, obfuscation, and duplication in food supply chains. The problem is exacerbated by lack of relevant training and skills, with frequent promotion not based on ability or competence. There is a major role for government to ensure efficient administration of laws and regulations, without corruption.[52] Unless there is transparency in government transactions, businesses will find alternate investment opportunities overseas.

In a perfectly free-trade world that is driven by consumers and not producers, there would be little need for governments to intervene with international trade barriers and thus cause supply chain distortions. Competition would then drive efficiency, which then drives economic growth and social development.

Given producer-driven protectionist trade barriers imposed by governments, there is a role for more trade-liberal governments to negotiate lower international trade barriers and improvements to global food supply chains. This would then focus attention on meeting consumer demands, increase the production frontier for food and trade, and greatly enhance global food security. A country with food shortages would be quickly satiated by imports from food surplus countries.

Less definitive roles for government in the food supply chain

There are some roles for government in the food supply chain that need to be exercised with caution, otherwise this interference may cause distortions and 'crowd out' private initiative.[53]

A question lies in whether a government can promote an export food business policy without picking winners and losers. It can be argued that if comparative and competitive advantages were already existent for export food businesses, then such private economic investment and development would have already occurred.

Governments can remove infrastructural and bureaucratic impediments, thus lowering costs and increasing efficiencies, which might induce food industries to expand into export ventures. But when governments attempt to encourage export food businesses when there are no comparative or competitive advantages, there is a real danger of food supply chain distortion. This could lead to future industry subsidies and trade protection on the grounds of disadvantages of scale or infant industry support, which would further distort the supply chain.

Many governments have an extremely poor historical record in the development of sustainable agricultural production. Much current government advice is required to minimize historical government mistakes such as previous commodity production–push policies and land subdivision 'reforms', which often have resulted in frequent prolonged surpluses in marginal-producing regions, isolation from main consumption regions, low prices, farmer poverty, small farm holdings, high debt, and farm land degradation.[54] Alternatively, the government may have an important role on production issues such as providing genetic diversity, new crop research, controlling insect and rodent plagues, responding to plant disease outbreaks, and justifiable quarantine restrictions.

The choice of crop varieties should be consumer driven despite agronomic difficulties. A wheat variety might be bred for drought resistance, but have no end-user demand other than for animal feed because of its poor milling characteristics. Fertilizer application should be driven by market grade–price spreads to ensure profitability, rather than mere optimum soil requirements. Independent soil testing laboratories can be more efficient than government providers because of competitive pressures.

Public water irrigation programmes in agriculture can often prove disastrous due to salinity, cost, allocation inefficiencies, environmental issues, and an inability to achieve international competitiveness. The provision of dams is no surety that output will be efficiently produced or competitive advantage achieved. If food production is unsustainable, high in cost, isolated and not aligned with major consumer supply chains, then any public water programme becomes an opportunistic venture with little contribution to food security.

Any alignment of production to government forecasts of weather and climate that are based on historical occurrences has not proven very effective in most countries. Lorenz's chaos theory that challenges weather predictions is a testament to that.[55] Much foreign investment is attracted to marginal agricultural land in good times, only to incur high financial losses in bad times.

Government drought relief and lowering normal bank credit restrictions have frequently prolonged uneconomic farm production in unsuitable marginal regions. The resultant lack of cash flow and farm poverty can lead to land degradation and soil erosion, which often then requires expensive corrective soil-care programmes and land management systems funded by taxpayers.

There may be a role for government in diversified plant breeding and the maintenance of seed stocks that are beyond normal private breeding activities in producing countries. Whilst plant breeding should be the role primarily of private seed companies that can align genetics to both local conditions and end-user needs, loss of genetic vigour over time necessitates diverse breeding programmes to maintain productivity. Such genetic losses could be supplemented by government diversified seed stocks and genetic experimentation. Caution needs to apply to government plant breeding programmes, as evidenced by the Lysenko plant-breeding disaster in the USSR during the 1930s.[56]

If private end users need a certain varietal characteristic, and this variety can be grown effectively and efficiently locally, then the role of government in this varietal production and release process should be minimal through protection of intellectual property rights, unless there are spillover or negative externality effects such as other varietal contamination. There may be a role for government in policing variety releases and distribution.

When sudden agricultural production shocks occur such as the outbreaks of disease or insects, there needs to be adequate prior research and emergency preparation. One of the causes of dependency on government action is the lack of clear roles and responsibilities between the public and private sectors. If the private sector is made responsible for research and emergency preparation, then there may have to be adequate fiscal provisioning in such cases.

Governments have an important role in quarantining pests and diseases, thus preventing negative spillover effects in the food supply chain. This can lead to reliability as a source of 'clean' food, which may consolidate the linkage between exporters and importers in the global food supply chain. However, there is a delicate balance with any trade barrier-protection imposition.

There has been a tendency for bio-terrorism such as the deliberately spoiling or contaminating of food to be included under bio-security. As well, bio-security that normally would focus on quarantine is interchanged with food supply insecurity and national security risk. Referring to food riots as bio-security masks the real problem of initial government intervention causing supply chain distortions and food insecurity.

There have been increasing concerns within society over the treatment of living organisms destined for the food supply chain. The problem with food in society is that consumers want the best quality at the lowest price,[57] which then drives cost efficiencies in production and the supply chains. Quite often, consumers are contradictory in their wants, and it becomes extremely difficult to balance their immediate demands such as high-quality low-priced food at the same time as meeting their social and ethical requirements. Alternatively, consumers may want animal welfare issues minimized, but they are not willing to pay the higher cost-price involved with the required preferential treatment. The role of government in these contradictions and paradoxes in the food supply chain is ill-defined.

Another paradox in the food supply chain is that any animal that is generally well treated in society such as horses, dogs, and cats are creatures that are most often not consumed. This suggests that consumption may be restricted in many countries to only those animals that are perceived not to have preferential treatment. Therefore, animal welfare and consumption habits may be incongruous for many meat consumers.

Food safety, labelling, and traceability issues have mainly arisen from threats of bio-security, acts of bio-terrorism, and frequent outbreaks of unintentional food poisoning that are related to a lack of food hygiene procedures. However, there is an increasing demand by consumers for adequate labelling and traceability for reasons of health concerns, residues and contaminants, containment of allergies, local or foreign loyalties, and novelty through knowing the origin of food.

Consumers generally like the idea of food safety, labelling, and traceability, but usually are not willing to pay for the added cost of buying, segmentation, and presentation. Governments may then be required to enforce the requirement for food safety, labelling, and traceability, but this could jeopardize international competitiveness.

It can be argued that if society eats the wrong types of food, then disease and ailment issues place adverse organizational pressures on a publicly funded hospital system. Governments that allocate public health funding usually have a vested interest in minimizing the number of patients because of fiscal policy constraints, although hospital management, trade unions, and private stakeholder groups might perceive otherwise. Alternatively, it can be argued that public health and nutrition policies shift responsibility from the individual to the government, and undermine private health responsibility.

Government departments can contradict themselves, with agriculture and tourism often promoting the very foods that government nutritionists claim are not conducive for good health. Obesity in developed countries has increased somewhat proportionally to public health warnings on certain food types. As fast as governments attempt to contain one human disease another different disease breakout often occurs. These negativities challenge the role of government in public nutrition.

Governments have a long history of establishing certification grades on behalf of farmers, but which were not aligned with much usefulness to either end users or consumers. Certification must be driven by end users and consumers, not by producers. It is not the role of government to understand food supply chain product segregation and consumer market segmentation.

It can be argued that competition and brand loyalty should drive consumer choice. Under the 'buyer beware' principle, if a consumer does not perceive value after purchasing a product, then the probability of repurchase is extremely low, which quickly causes the exit of such companies and products from the supply chain because no brand loyalty is established. However, problems may lie with 'hidden' value such as organic food that is not readily perceivable, and with free riders such as farmers erroneously claiming to be organic growers. In these cases, governments may have a policing or legal role.

No definitive roles for government in the food supply chain

There are a number of areas that governments should avoid in the food supply chain,[58] otherwise this interference is highly likely to cause serious food supply chain distortions.

Governments are ill-equipped to know all the reasons why consumers accept or reject perceived product value in private food supply chains. Value can only be perceived by consumers, which reflects individual psychology. There is very little role for government in this value adding and creation process, except in facilitating improvement to technology, knowledge, and new genetic varieties.

However, barriers to product value adding and creation can often be due to existing government inefficiencies in infrastructure, administration, quarantine policies, research, and labour laws.[59] Improving government efficiency may therefore be an inducement for increased value adding and creation processes.

Whilst there is an important role for government in ensuring adequate competitive practices in the food supply chain through anti-collusion and anti-monopoly laws, governments should allow competition to ultimately drive supply chain efficiency.[60] Adequate competition provides intrinsic benchmarking performance measures.[61]

There is a special need to ensure transparency with contestability without corrupt practices whenever large government institutional buying occurs with major tender contracts.[62] Strong public–private partnerships are often required to ensure effective delivery of government catering services, but these should be driven with efficiency objectives.

Government subsidies or domestic industry protection can entrench food supply chain inefficiencies, distort price signals, act as disincentives for change and improvement, decrease international competitiveness, and cause potential investment to shift away from the food supply chain.[63] Generally, such subsidies and protection favour some but not others,[64] support the weaker organizations and undermine the strong organizations, whilst politics and often corruption dominate industry affairs.[65]

Favouritism and encouraging politicking through selective protection should not be a role of good governance, rather there should be an emphasis on removing the domestic barriers to achieving international competitiveness of food supply chains. Individual countries cannot expect to have comparative advantages in domestic production of all food products, especially when their agriculture is restricted by geographic latitudes and regional weather conditions. Maximizing choice to the consumer with a balance of local production and imports should be the ultimate goal, not the protection of inefficient domestic industries.

Supply chain distortions can be divided into temporary or permanent. Temporary price distortions could be minimized through facilitating arbitrage activities, and not through supply chain intervention. Whenever more permanent supply chain distortions occur, governments should ensure that there is sufficient competition and infrastructure to minimize any dysfunction. If increasing the size of government institutions leads to greater supply chain distortion, then this might be an argument to decrease the size of government.

Cost inefficiencies in the food supply chain are usually a result of insufficient competition.[66] Government action is required to increase competitive behaviour either by breaking down private cartel operations, removing government monopolies and interventionist policies, or allowing import competition.

However, there are many other factors impacting on food supply chain costs that may not be related to anti-competitiveness, but still come under government responsibility. Inefficient infrastructure, rail-port access policies, and government regulatory policies do impact on costs. Therefore, governments must be held responsible for some aspects of cost efficiency.

Much government data collection and reporting is out of date when it is publicly released. In commodity markets, it is almost a case of public knowledge becoming government knowledge months later, which suggests an inverse information asymmetry. Electronic information transfer generally has widened the gap between public and government knowledge, with farmers now receiving overnight market information that is hours in front of merchants receiving the same information at their office desk.

Information transparency that is required for public organizational efficiency in a non-competitive environment conflicts with information asymmetry that is required for private organizational efficiency in a competitive environment. The Australian government was still releasing monthly commercially sensitive information on private domestic wheat inventories to international competitors, years after supply chain deregulation had occurred.[67]

Government intervention and institutional buying can 'crowd out' private activities and investment.[68] Whilst the benefits of private competition might be recognized, there is the continual possibility of reoccurrence of government interference through regulation–deregulation–reregulation behaviour.[69] There is also the problem of idle public resources seeking some future purpose instead of redundancies, and this might threaten food supply chain efficiency.

The government record in picking winners is historically very low, with selections based on few reliable objective criteria.[70] In choosing 'winners', many governments select favourites through exclusivity, which generally reflects those businesses and families that are the most supportive of the incumbent government, particularly through political donations. Much public corruption is entrenched through governments selecting 'winners' and rejecting 'losers'.[71]

A problem is created when government fiscal policy provides subsidies, tax relief, or trade protection to the real losers. This acts as a disincentive for potential winners to win, and can prolong the life of the real loser, beyond any justifiable reason.

Notes

1 Olsen, Mancur, 1993, *Dictatorship, Democracy, and Development*, American Political Science Review, 87:3, 567–576.
2 Locke, John, 1690, *Two Treatises on Civil Government*, republished by Routledge, London, 1887.
3 Groenewegen, Peter D., 1983, *Turgot's Place in the History of Economic Thought: A Bicentenary Estimate*, History of Political Economy, 115, 611–615.
4 Kurlansky, Mark, 2003, *Salt – A World History*, Random House, London, pp. 233–234.
5 Miller, Judith A., 1998, *Mastering the Market: the State and the Grain Trade in Northern France, 1700–1860*, Cambridge University Press, Cambridge, Ch. 1.
6 von Mises, Ludwig, 1949, *Human Action: A Treatise on Economics*, Auburn, AL, Ch. 15, section 4.
7 A combination of von Mises in his *Human Action: A Treatise on Economics* and Adam Smith in his *The Wealth of Nations*, republished in 2003, Random House, New York.
8 Hilton, Matthew, 2009, *Prosperity for All – Consumer Activism in an Era of Globalization*, Cornell University Press, Ithaca, NY.
9 Ashcraft, Richard, 1986, *Revolutionary Politics and Locke's Two Treatises of Government*, Princeton University Press, Princeton, NJ.
10 Yergin, Daniel and Joseph Stanislaw, 2002, *The Commanding Heights – The Battle for the World Economy*, Simon and Schuster, New York.
11 Knight, Frank H., 1923, *The Ethics of Competition and Other Essays*, republished in 1936, Allen and Unwin, London.
12 Hill, Lowell, Ed., 1982, *Role of Government in a Market Economy*, Iowa State University Press, Ames, IA.
13 Hill, Lowell D. and Karen L. Bender, 1995, *Developing the Regulatory Environment for Competitive Agricultural Markets*, The World Bank Technical Paper 266, Washington, DC.
14 Singh, Ram D., 1992, *Government-induced Price Distortions and Growth: Evidence from Twenty-nine Developing Countries*, Public Choice, 73:1, 83–99.
15 Friedman, Milton, 1993, *Why Government is the Problem*, Hoover Institution Press, Stanford University, Stanford, CA.

16 As indicated by the price skewness in Figure 1.3 in Chapter 1.
17 Kindleberger, Charles P., 1996, *Manias, Panics, and Crashes – A History of Financial Crises*, John Wiley and Sons, New York, Ch. 4.
18 A case study is provided in Chapter 7.
19 Hayek, Friedrich A., with William W. Bartley, Eds., 1988, *The Fatal Conceit: The Errors of Socialism*, University of Chicago Press, Chicago, and Routledge, London.
20 Scitovsky, Tibor, 1976, *The Joyless Economy: An Inquiry into Human Satisfaction and Consumer Dissatisfaction*, Oxford University Press, New York.
21 Radnitzky, Gerard and W.W. Barley, Eds., 1987, *Evolutionary Epistemology, Rationality, and the Sociology of Knowledge*, Open Court, La Salle.
22 This is explained more in Chapter 7.
23 McCorriston, Steve and Donald MacLaren, 2005, *The Trade Distorting Effect of State Trading Enterprises in Importing Countries*, European Economic Review, 49, 1693–1715.
24 'Trade' markets will be discussed in Chapter 10.
25 Baumol, William J. and Andrew E. Burke, 2001, *Competition Policy in Dynamic Markets*, International Journal of Industrial Organization, 19:5, 613–634.
26 Gideonse, Harry D., 1939, *Organized Scarcity and Public Policy: Monopoly and its Implications*, Public Policy Pamphlet 30, University of Chicago Press, Chicago.
27 The benefits of carry will be discussed in Chapter 10.
28 This aspect will be examined in Chapter 5.
29 Hofstatter, Stephan, 2010, *Farmers Furious at 'National Asset' Threat to Land Rights*, Business Day, 16 March.
30 Hayek, Friedrich A., Ed., 1935, *Collectivist Economic Planning: Critical Studies on the Possibilities of Socialism*, Routledge, London.
31 Irwin, Scott H., Garcia, Philip, Good, Darrel L. and Eugene L. Kunda, 2009, *Poor Convergence Performance of CBOT Corn, Soybean and Wheat Futures Contracts: Causes and Solutions*, Marketing and Outlook Research Report 2009–02, University of Illinois, Urbana-Champaign, March.
32 Working, Holbrook, 1949, *The Theory of the Price of Storage*, The American Economic Review, 31, 1254–1262.
33 Carry and the 'trade' markets are discussed in detail in Chapter 10.
34 Williams, John and William Schroder, 1999, *Agricultural Price Risk Management – The Principles of Commodity Trading*, Oxford University Press, Melbourne, pp. 55–56.
35 Kumar, B.V., 2000, *Corruption: An Indian Perspective*, Journal of Money Laundering Control, 3:3, 266–279.
36 Marion, Bruce W., 1997, *Competition and Trade Practice Policies: An Overview*, in L. Tim Wallace and William R. Schroder, Eds., *Government and the Food Industry: Economic and Political Effects of Conflict and Co-operation*, Kluwer Academic Publishers, Norwell, MA, pp. 39–55.
37 Hilmer, Frederick G., 1993, *National Competition Policy: Hilmer Report*, Australian Government Publishing Service, Canberra.
38 Porter, Michael E., 1985, *Competitive Advantage – Creating and Sustaining Superior Performance*, The Free Press – Macmillan, New York.
39 Porter, Michael E., 1998, *Clusters and the New Economics of Competition*, Harvard Business Review, 77, 1–7.
40 Australian Competition and Consumer Commission, 2008, *Report of the ACCC Inquiry into the Competitiveness of Retail Prices for Standard Groceries, July 2008*, ACCC, Canberra, 5 August.
41 Porter, Michael E., 1985, *Competitive Advantage – Creating and Sustaining Superior Performance*, The Free Press – Macmillan, New York.
42 Ryder, Ray and Andrew Fearne, 2003, *Procurement Best Practice in the Food Industry: Supplier Clustering as a Source of Strategic Competitive Advantage*, Supply Chain Management: An International Journal, 8:1, 12–16.
43 Kuah, Adrian T.H., 2002, *Cluster Theory and Practice: Advantages for the Small Business Locating in a Vibrant Cluster*, Journal of Research in Marketing and Entrepreneurship, 4:3, 206–228.

44 Baumol, William J., Panzar, J.C. and Robert D. Willig, 1982, *Contestable Markets and the Theory of Industry Structure*, Harcourt Brace Jovanovich, New York.
45 Rama, Ismo and Sallyann Harvey, 2009, *Market Failure and the Role of Government in the Food Supply Chain,* Policy and Strategy Group, Victorian Department of Primary Industries, Melbourne, June.
46 Olsen, Mancur, 1993, *Dictatorship, Democracy, and Development*, American Political Science Review, 87:3, 567–576.
47 Rimmington, Mike, Carlton-Smith, Jane, and Rebecca Hawkins, 2006, *Corporate Social Responsibility and Sustainable Food Procurement*, British Food Journal, 108:10, 824–837.
48 Industry Commission, 1995, *The Growth and Revenue Implications of Hilmer and Related Reforms*, A Report by the Industry Commission to the Council of Australian Governments, Canberra, March.
49 Chitsike, Francis, 2003, *A Critical Analysis of the Land Reform Programme in Zimbabwe*, paper presented to the 2nd FIG Regional Conference, Marrakech, Morocco, December 2–5.
50 Whitwell, Greg and Diane Sydenham, 1991, *A Shared Harvest – The Australian Wheat Industry 1939–89,* Macmillan Education Australia, Melbourne, Ch. 2–4.
51 FAO, 1984, *Studies on Agrarian Reform and Rural Poverty*, Economic and Social Development Series, No. 27, Rome.
52 Quah, Jon S.T., 1997, *Singapore's Experience in Curbing Corruption*, in Heidenheimer, Arnold J., Johnston, Michael and Victor T. Le Vine, Eds., *Political Corruption: A Handbook*, 4th Ed., Transaction Publishers, New Brunswick, NJ.
53 Rama, Ismo and Sallyann Harvey, 2009, *Market Failure and the Role of Government in the Food Supply Chain*, Policy and Strategy Group, Victorian Department of Primary Industries, Melbourne, June.
54 Whitwell, Greg and Diane Sydenham, 1991, *A Shared Harvest – The Australian Wheat Industry 1939–89,* Macmillan Education Australia, Melbourne, Ch. 2–4.
55 Lorenz, Edward N., 1963, *Deterministic Nonperiodic Flow*, Journal of Atmospheric Sciences, Massachusetts Institute of Technology, 20, 130–141.
56 Soyfer, Valery N., 1994, *Lysenko and the Tragedy of Soviet Science*, Rutgers University Press, New Brunswick, NJ.
57 von Mises, Ludwig, 1949, *Human Action: A Treatise on Economics*, Ch. 15, section 4.
58 Rama, Ismo and Sallyann Harvey, 2009, *Market Failure and the Role of Government in the Food Supply Chain*, Policy and Strategy Group, Victorian Department of Primary Industries, Melbourne, June.
59 Edwards, G., 1997, *Economics, Government, and the Food Industry*, in L. Tim Wallace and William R. Schroder, Eds., *Government and the Food Industry: Economic and Political Effects of Conflict and Co-operation*, Kluwer Academic Publishers, p. 33.
60 Baumol, William J. and Andrew E. Burke, 2001, *Competition Policy in Dynamic Markets*, International Journal of Industrial Organization, 19:5, 613–634.
61 Porter, Michael E., 2003, *The Economic Performance of Regions*, Regional Studies, 37: 6&7, 549–578.
62 Tanzi, Vito, 1994, *Corruption, Governmental Activities, and Markets*, IMF Working Paper 94/99, International Monetary Fund, Washington, DC.
63 Mayrand, Karen S., Dionne, Stéphanie, Paquin, Marc and Isaak Pageot-Lebel, 2003, *The Economic and Environmental Impacts of Agricultural Subsidies: An Assessment of the 2002 US Farm Bill and Doha Round*, Unisfera International Centre, Montreal.
64 Watkins, Kevin, 2003, *Farm Fallacies That Hurt the Poor*, World Bank Development Outreach Special Report: Trade for Development, World Bank, Washington, DC.
65 Bull, Martin J. and James Newell, 2003, *Corruption in Contemporary Politics*, Palgrave Macmillan, New York.
66 Porter, Michael E., 1985, *Competitive Advantage – Creating and Sustaining Superior Performance*, The Free Press – Macmillan, New York.
67 Australian Government Productivity Commission, 2010, *Wheat Export Marketing Arrangements Inquiry Report*, Canberra, 28 October.

68 Rama, Ismo and Sallyann Harvey, 2009, *Market Failure and the Role of Government in the Food Supply Chain*, Policy and Strategy Group, Victorian Department of Primary Industries, Melbourne, June.
69 Christensen, Jorgen Gronnegard, 1989, *Regulation, Deregulation, and Public Bureaucracy*, European Journal of Political Research, 17, 238.
70 Wise, Timothy A., 2005, *Identifying the Real Winners from U.S. Agricultural Policies*, Global Development and Environment Institute, Working Paper 05-07.
71 Tangri, Roger, 1999, *The Politics of Patronage in Africa: Parastatals, Privatization, and Private Enterprise*, Trenton, New York.

Further readings on the role of price and competition

Alsos, Gry Agnete, Sara Carter, Elisabet Ljunggren and Frederike Welter, 2011, *The Handbook of Research on Entrepreneurship in Agriculture and Rural Development*, Edward Elgar, Cheltenham.

Baumol, William J. and Andrew E. Burke, 2001, *Competition Policy in Dynamic Markets*, International Journal of Industrial Organization, 19:5, 613–634.

Bloch, Harry and Peter Kenyon, 2001, *Creating an Internationally Competitive Economy*, Palgrave-Macmillan, Melbourne.

Casson, Mark, 2011, *Markets and Market Institutions: Their Origin and Evolution*, Edward Elgar Publishing, Cheltenham.

Cole, Harold L. and Lee E. Ohanian, 2003, *New Deal Policies and the Persistence of the Great Depression: A General Equilibrium Analysis*, Research Memo, February, Department of Economics, University of California.

Faure, Michael and Xinxhu Zhang, 2011, *Competition and Regulation – Recent Developments in China, the US, and Europe*, Edward Elgar Publishing, Cheltenham.

Gawer, Annabelle, 2009, *Platforms, Markets and Innovation*, Edward Elgar Publishing, Cheltenham.

Griffith, Garry, 2000, *Competition in the Food Marketing Chain*, The Australian Journal of Agricultural and Resource Economics, 44:3, 333–367.

High, Jack, Ed., 2001, *Competition*, Edward Elgar, Cheltenham.

Hill, Lowell, 1985, *Effects of Regulation on Efficiency of Grain Marketing*, Journal of International Law, 17:3, 389–419.

Indian Ministry of Agriculture and the Department of Agriculture, 2002, *Inter-Ministerial Task Force on Agricultural Marketing Reforms*, New Delhi.

Kuchik, Akifumi and Masatsuga Tsuji, 2011, *Industrial Clusters, Upgrading and Innovation in East Asia*, Institute of Developing Economies and Edward Elgar, Cheltenham.

McCorriston, Steve, 2006, 'Imperfect competition and international agricultural commodity markets', Chapter 6 in Sarris, Alexander and David Hallam, Eds., *Agricultural Commodity Markets and Trade – New Approaches to Analysing Market Structure and Stability*, FAO, Rome, and Edward Elgar, Cheltenham.

Motta, Massimo, 2004, *Competition Policy – Theory and Practices*, Cambridge University Press, Cambridge.

Mulgan, Geoff, 1995, *Democratic Dismissal, Competition, and Contestability Among the Quangos*, Oxford Review of Economic Policy, 10:3, 51–60.

Myers, Robert J., Sexton, Richard J., and Williams G. Tomek, 2009, *A Century of Research on Agricultural Markets*, American Journal of Agricultural Economics, 92:2, 376–402.

National Commission on Agricultural Trade and Export Policy, 1986, *New Realities: Toward a Program of Effective Competition*, US Department of Agriculture, Washington, DC.

Rolfstam, Max, 2011, *Public Procurement and Innovation – The Role of Institutions*, Edward Elgar Publishing, Cheltenham.

Stevenson, Lois, 2010, *Private Sector and Enterprise Development: Fostering Growth in the Middle East and North Africa*, International Development Research Centre (Egypt) and Edward Elgar, Cheltenham.

Utton, Michael A., 2011, *Cartels and Economic Collusion*, Edward Elgar Publishing, Cheltenham.

von Mises, Ludwig, 1949, *Human Action: A Treatise on Economics*, reprinted by the Ludwig von Mises Institute as the Pocket Edition in 2010, Auburn, AL.

Wallace, L. Tim and William R. Schroder, Eds., 1997, *Government and the Food Industry: Economic and Political Effects of Conflict and Co-operation*, Kluwer Academic Publishers, Norwell, MA.

4

GOVERNMENT INTERVENTION BY IMPORTING COUNTRIES

Despite the statement by Benjamin Franklin that no nation was ever ruined by trade, there is a big difference between private importing that can enhance trade and government importing that can destroy private trade. As well, supply chain intervention can cause previous self-sufficient or net-exporting countries to quickly revert and become permanent net-importing countries. Myanmar was the biggest global exporter of rice under British administration, exporting a world record of 3.4 million tonnes in 1934, yet it could only export 537,000 tonnes in 2010–11.

If competition and efficiency drive domestic food supply chain performance, the corollary suggests that any factor that restricts domestic competition and efficiency will be detrimental to supply chain performance and can lead to import dependence. This may impact on the balance of trade, and monetary and fiscal policies.

Mao's Great Leap Forward in 1958 paradoxically led to the starving to death of 30 million Chinese between 1959–1961,[1] despite China being a food bowl country.[2] Farmers could still control their production prior to 1958 despite not owning the farmland. Subsequently, they were forced to abandon their private farming pursuits and replace it with commune farming, at a time when much of the farm labour was being forcibly shifted into industrial mining. Agricultural acreages and yields quickly fell. By early 1959, the government had established grandiose urban communal dining halls offering free meals as propaganda, which disguised the fact that one-third of the Chinese provinces were experiencing famine, with tens of millions about to die, despite the huge wastage from these dining halls.[3] These interventionist policies were the start of decreased farm productivity in China and increased dependency on food imports.

African countries have diverse geographic characteristics and weather occurrences, from tropical jungle to arid desert conditions, yet most have become overly reliant on food imports during the past 20–30 years. Most MENA (Middle East and North Africa) countries have sizeable cereal production,[4] despite vast desert interiors. Population growth during the twentieth century that was in response to previously abundant food supplies may be one of the many reasons why African and Middle Eastern countries are now reliant on food imports. There are physical constraints to increased domestic food production such as limited soil types,[5] long supply logistic chains and market isolation, lack of private land ownership, high initial entry costs for farming, inadequate private investment capital, and high risk such as rainfall irregularity that can cause frequent drought.

However, there are many socio-economic factors that can contribute to food import dependency by African countries.[6] These include frequent wars and civil unrest that may reflect cultural tensions and a lack of social cohesion, autocratic family rivalry that leads to government instability, monetary–fiscal policy inadequacies in a high sovereign risk environment, uncertainty over land tenure, food supply chain infrastructural problems, law enforcement issues, a lack of a private innovative culture, and corruption that can be encouraged by food 'aid'.

Many of the factors contributing to African food import dependency do not apply to other major food import dependent countries such as Middle Eastern countries and Japan, which are both characterized by trade barriers and government intervention, with high incomes per capita but with shortages of suitable arable land. One difference is that Japan has a high population density compared to a Middle Eastern country of the same size. Both have some underground water, which is unusually characterized as non-saline. Any conclusion on the reasons for import dependency, apart from direct government interventionist policies, must account for inter-country geographic, geomorphologic, demographic, and in weather and climate differences, rather than mere current income differences.[7]

Current government policies frequently obscure previous mistakes, particularly after successive generations have occurred and intervention is subsequently perceived to be the entrenched historical cultural norm for the country often with quasi-religious sentimentalities. Knowledge of history then becomes important. These current policies may prevent opportunities arising that are required to motivate and encourage risk taking in the food supply chain, which might be the pre-requisite to revert an import-dependent country back to being self-sufficient or a net exporting country.[8]

Historical government intervention in food supply chains sometimes was to imitate other countries, to gain market power for vested interests, or in retaliation to food exporting countries.[9] However, the self-inflicted adverse impact on domestic food supply chains was generally ignored, as was the resulting import dependency and adverse trade balances.

Success in decreasing food import dependency and improving self-sufficiency is unlikely to come from increasing government controls and further regulation. Food self-sufficiency in Holland[10] and south Vietnam was not due to any government intervention, whilst in Japan food export success has mainly arisen in industries such as horticultural hot-house development where the government has had the least intervention.

There are many factors therefore that can historically explain why some countries are net food exporters while others are net food importers, and whilst it is easy to transgress from net exporting to importing under freer trade, it is a very difficult process to reverse net importing to exporting. Apart from the high capital cost factor, the most consistent explanation of why many countries are overly dependent on food imports is because of previous intervention and distortions in their food supply chains that are usually aligned with centrally planned policies, such as those under Mao. This undermined the required incentive and high risk taking necessary to stimulate domestic investment and entrepreneurship.

Import replacement success stories

There are examples whereby previous importing regions have successfully developed domestic industries that replaced the necessity to buy from other regions. However, there are prerequisites and there can be many problems. What was historically done at little cost and high motivation can today be prevented because of high cost and little incentive.

One thousand years ago, Holland was a swampy, tidal, and saline estuary of the Rhine and Ems river confluences. The land was reclaimed by commitment, diligence, innovation, and technology over hundreds of years, until it became a country in its own right.[11] It exists in a region of acutely adverse weather conditions exposed to the North Sea, strong winds, and lacks sunlight for much of the year. Until 1940, Holland was predominantly free trade, with many proclaimed Hanseatic free cities under the old Empire. Despite the small geographical size of the country and population, it is a net exporter of food and is the power house of agribusiness in Europe through its controlled environment techniques in hot-house concentrated agricultural production. The only real rivals for this intense hot-house development are Dà Lat in southern Vietnam (Figure 4.1) and in Japan.

FIGURE 4.1 Rice paddies interspersed with horticultural hot-houses on private land at Dà Lat in southern Vietnam in a food exporting regional hub

Source: John Williams

Similarities exist between Holland, Vietnam, and Japan in terms of size, exposure to the weather, coastline, private land ownership, and historical adversities to agricultural development. Apart from salinity conditions, whereby Holland originally had saline soils whereas Vietnam and Japan did not, the main difference between the countries was that Japan was and still is fiercely protectionist, possibly because it had no major river system linking foreign countries together, while Holland and south Vietnam were predominantly free trade, driven by their geographic location at the mouth of major inter-country river systems abutted by a major ocean. Historically, Japan was quite ready to advance its protectionist state, whereas Holland and south Vietnam were quite ready to defend their free trade status.

Marginal food production in the importing country

There are many problems with attempting to increase local food production in marginal regions within importing countries.[12] This contrasts to corn and soybean crops that have been in continuous rotation for over 150 years in the US Corn Belt with few identifiable unsustainable practices.

Rainfall has a big influence on many soil types, structures, and textures.[13] A general lack of rainfall can be associated with low fertile soils, variable soil structure, and wind erosion. Attempts to improve soils through irrigation with desalinated water for those importing countries near the sea usually prove too costly for sustained agriculture. Desert countries have a distinct disadvantage in food production without underground water because of the high cost of both water[14] and soil treatment. The Dutch turned saline mud into rich arable soil, but that was over many centuries and having the advantages of rainfall, a large continually flowing river system, and a low labour cost.

Saudi Arabia attempted to use crude oil revenue to gain self-sufficiency in some commodity foods at high cost by using desalinated water and subsidizing local production.[15] The economic efficiency in using scarce water resources for low-value agricultural production such as wheat and rice is seriously questionable, particularly when these commodities could be imported and the water diverted to higher value-adding manufacturing industries. If a food importing country is to be innovative, albeit government driven, then it is more likely to be successful by using imported agricultural commodities as bulk fillers for food value creation by its manufacturing industries, and focusing on becoming a regional food exporting hub, similar to Singapore. But that would require supply chain deregulation.

Some food importing countries do not have the long-term stability of government, adequate infrastructure, risk capital, or private innovative culture that are required to undertake technically complex and expensive marginal domestic food production with low rates of long-term investment return at high risk. There are many domestic cultural barriers to such large-scale rural development, apart from the isolation–distance factors, antagonism from existing farmers, and perhaps foreign ownership issues.[16] Alternatively, governments could assist in nurturing a private innovative culture, but be undermined by international 'aid' destroying the integrity of government and the incentive for private innovation.[17]

The existence of food chain oligopolies or public monopolies often prevents private competitors and entrepreneurs who might be more likely to take the risks required[18] to develop marginal domestic food production. Without competition in a strong private food supply chain with low entry–exit barriers, it is unlikely that there will be sufficient motivation and innovation to venture into highly uncertain domestic marginal agricultural production or high-risk food manufacturing. The alternative for governments is to ensure that adequate competition occurs to encourage marginal domestic food production and prevent the dominance of private oligopolies and monopolies from causing negative supply chain motivational and investment spill-over effects.[19]

In a competitive food supply chain environment under freer trade and non-intervention, there is more likelihood of product diversification and greater product range available for consumers. Products are more likely to be sourced from the cheapest international source, thus making food more affordable and encouraging supply chain efficiency. However, a robust domestic private sector that is competitively driven by international efficiency, technology, innovation, and global capital markets is more likely to seek and drive domestic competitive advantages in local agricultural production, food manufacturing, and exporting, and thus is more likely to develop into a food export hub.

It is common for large-scale oligopoly or monopoly ventures in food production and manufacturing to become very dependent on government action, planning, and approval.[20] But it is highly questionable as to why a government would attempt to achieve higher levels of domestic food self-sufficiency or singular exports through such oligopoly or monopoly ventures when private initiative is stifled, fiscal returns are dubious with high taxpayer risk, resources are often inefficiently allocated, and consumer welfare ignored. If imports are subsequently prevented under trade protection, the whole food supply chain becomes distorted when other domestic food manufacturers are often forced to buy inferior-quality higher-priced domestic inputs. The outcome in such countries as North Korea has been frequent abject famine in what should be a land of plenty.

Prudent fiscal policy suggests that any high-risk capital venture into marginal agricultural production or food manufacturing should be incurred by private risk takers. The role of government should then be to create such an entrepreneurial environment with adequate infrastructure and minimize any negative externality impact. Such entrepreneurial and risk-taking encouragement is more likely to keep unemployment rates low compared to any direct supply chain intervention.[21]

The globalization and industrialization of food processing and manufacturing over the last 50 years has resulted in considerable economies of scale with low unit costs because of the large quantity produced, and strong experience and learning curves whereby unit costs decrease over time due to the effect of experience and learning.[22] Hence it is very difficult to establish internationally competitive food processing and manufacturing businesses and new supply chains, particularly in small markets unless it is for a highly perishable product, or the region has a strong comparative or competitive advantage, which allows it to develop an export industry.[23] Local private industry is thus placed at a distinct disadvantage to international suppliers in global food chains, which could be partially due to a legacy of previous government intervention stagnation.

However, there were no domestic economies of scale in Singapore to establish its regional food manufacturing and distribution hub.[24] Its success evolved because of its geographic positioning of a small country surrounded by hundreds of millions of consumers, adequate infrastructure, minimum government intervention, controls over corruption, and the creation of a competitive entrepreneurial environment that encouraged learning, knowledge, risk taking, and technological innovation.

Food hub successes in Singapore might be contrasted to the failure of the United Arab Emirate (UAE) countries to develop into a significant food manufacturing and distribution hub for the Middle East, Northern Africa, and the Asian Subcontinent, thereby endangering future employment for younger generations if ever oil revenues dwindled. The geographical location in the centre of hundreds of millions of consumers would suggest that a food manufacturing and distribution hub could evolve in the UAE region because of the potential for external economies of scale.

There are some current examples of UAE food manufacturing successfully re-exporting such as sugar and vegetable oil refining of imported raw product, but these are still very commodity based with much undifferentiated product in highly interventionist supply chains.[25] Blaming supply chain dilemmas on the lack of local food supplies ignores the success story of Singapore, which is nearly totally reliant on the import of raw food ingredients and was not unduly affected by the 2008 price spikes. Input prices become rather insignificant when real value creation occurs in the food supply chain.[26] There is a big difference between value adding that requires significant risk management and value creation that relies more on supply chain integration (Figure 4.2).

FIGURE 4.2 Oman Flour Mills Company attempting supply chain integration under severe supply chain regulation
Source: John Williams

Food hubs might evolve on their own volition, given adequate external economies of scale and geographical comparative advantages, provided that there are the required non-intervention, competition, and adequate price signal prerequisites in the supply chain. Based on these criteria, a food manufacturing and distribution hub might be possible on the African continent and in Middle Eastern countries. However, the current problem is that supply chain price signal distortions predominate, which is destroying the food chain prerequisite conditions for regional hub development.

Much marginal domestic food production is already occurring in many countries regardless of their economic development status, yet it is failing to find food supply chains. There could be many reasons for product exclusion in supply chains, including its unsuitability to meet the product and quality specification required by end users and consumers. However, the lack of adequate infrastructure frequently is a major isolationist problem and barrier to creating place value for agricultural output.[27]

The provision of adequate infrastructure usually is required to precede any additional agricultural production, otherwise the output effort is mostly wasted. However, improved infrastructure often is the driver of domestic regional food production, without any need for government intervention into production.[28] The extension of railway networks and the development of new farmland during the nineteenth century is a good example.

Even after initial infrastructural investment, many governments shift priorities away from food distribution and maintenance to supposedly more social concerns. Food supply chain infrastructure either disintegrates quickly or does not keep pace with food distribution demand. A developed country can quickly revert to less developed country status without continual adequate infrastructural investment that is timely, efficient, and adequate.[29]

The introduction of supermarkets has changed both the product offering and the direction of transport networks.[30] Supermarkets demand highly differentiated food products, which necessitate either refrigerated or non-refrigerated containers in highly integrative logistical distribution systems.[31] Cities need to be connected to ports, urban regions interconnected with one another, with whole new road–rail–air freight networks synergized. Any country without a strong supermarket-driven supply chain is unlikely to have the incentive to improve infrastructure that could be a prerequisite for the transition to a future food export regional hub.

In many countries, it has been the pressure from supermarkets, and ultimately from consumers, which has forced many governments to respond with better domestic distribution infrastructure for both local producers and manufacturers as well as for importers. The efficiency-driven supermarket supply chains might add the competitive pressures required to stimulate new investment into rural arterial road transport that might encourage high-risk domestic marginal production.

Failures in government import tender purchasing

Import tendering is a passive monopsony purchasing method whereby the buyer only has market power when commodity surpluses exist, but no market power when shortages occur. If the sole purpose of government purchasing tendering is to secure food supplies in times of shortages, then paradoxically the tender-purchase method is the least effective method to achieve food supply chain security. This has important ramifications when governments urgently need to satisfy the demand of large food catering institutions that are dependent but constrained under fiscal budgeting.[32]

Procurement food tenders are a relic of the Cold War era when government-to-government bulk commingled commodity transactions abounded in port-rail centricities. Their usage has been unnecessarily prolonged by an inability to manage change, which has increased the potential for civil unrest and political insecurity. The solution is to allow merchants to enter a deregulated and competitive domestic food supply chain, enabling them to perform their required product accumulation, warehouse and storage, and distribution functions efficiently.

Public officials are ill-equipped to respond with decisions in a timely manner within a bureaucratic hierarchical organization that has inherent change management difficulties. Much commodity food import tender buying occurred under panic at peak global prices early in 2008 in many importing countries, particularly in the Middle East and North Africa. After global commodity prices fell in late 2008, unwanted old crop surplus commodity was dumped onto local supply chains during early 2009 in front of their major Middle East and North African harvest period of March and April.

Taxpayers bore the brunt of the financial loss between the purchase and sale price. The offloading of public old-crop stocks also caused major supply chain disruptions to domestic food manufacturers at a time when they could have imported better quality new crop inventory cheaply if allowed to do so. As well, there can be nothing more disconcerting for a consumer to pay for bread or rice at a high government tender purchased price when current international prices are substantially cheaper.

It also had devastating low price consequences to local farmers who were attempting to maximize the previous high price benefit to offset historical losses. There can be nothing quite as destructive to domestic farmer morale than for a government to be importing large quantities of commodity and accumulating huge domestic surpluses just in front of major selling by domestic farmers, thereby undercutting their price. This explains why countries can regress from being net exporting to net importing very quickly.

Turgot warned of such dire outcomes and consequences in the lead-up to the 1789 French Revolution.[33] The 2011 riots in North Africa and the Middle East merely reflected what Locke was warning several hundred years ago.[34] Many Middle Eastern and North African governments repeated their 2008 procurement mistakes by issuing large grain purchasing tenders in February and March 2011 at peak international grain prices. They were effectively importing huge quantities of grain imports and accumulating large surplus reserves at a time when their own farmers were about to harvest. During their 2011 harvest-selling period (March and June), wheat prices fell by up to 30 per cent. It was bureaucratic mistakes mired in Taylor's trial-by-error psychology.[35]

If commodity food surpluses exist for approximately 85 per cent of the time,[36] the buyer has strong bargaining power compared to the seller's weak position. There can be an illusion that such supply chain power is permanent. When shortages ultimately do occur, there is generally a state of unpreparedness, particularly when hierarchal promotion designates such tasks to unskilled and inexperienced government staff, which can lead to panic buying and overreaction.[37]

Because the import tender method is weakest during commodity shortages and inadequate in demand–supply alignment, any decision delays and delivery time lags associated with procurement and tenders can exacerbate shortfalls for domestic buyers. Government officials frequently overreact by over-ordering, which results in over-supply, which causes the supply chain multiplier or 'bull-whip' effect.[38] This ultimately necessitates domestic dumping, often of low-quality old-crop product, which causes a disruptive shock to local food supply chains, and jeopardizes the shelf-life of manufactured food products.

Dumping that is a result of the 'bull-whip' effect induces serious prolonging of low prices for local farmers, decimates their financial viability, with subsequent pressure for government intervention to assist these same farmers with subsidies. Such buy high price and sell low price dumping can also seriously impact on taxpayers through large government financial losses and debt recoupment.

Procurement tendering usually is based on ad-hoc subjective decision making, with time lags due to group-decision indecisiveness in consensus over quantities, timing, specification, and delivery. Decisions are reactionary and juggled based on historical supply and demand data. Timeliness in purchasing and alignment with meeting specific demand in the domestic food chain is usually non-existent, except when accumulating for institutional catering.

These time lags in tender procurement can seriously distort the domestic market, with large oscillations between acute shortages and low-quality old-crop surpluses. Supposedly transparent international pricing by government officials is really historical global pricing because of serious time lags. Domestic food supply chains cannot operate efficiently with such time lags and product risk, which discourages investment into the food supply chain, hinders innovative practices, and acts as a disincentive for export development.

The import purchasing tender method relies on uncertain foreign supplier networks, which exist when sellers need buyers, often with suppliers colluding to win tender supply contracts, but quickly disappearing when buyers need supply. This relationship uncertainty is not conducive to supply chain integration or food security.

Procurement tendering prevents organizational and operational fusion with international merchants because of the innate ad-hoc decision making, with little continuity and regularity except for product that is destined for government institutional catering. Merely improving tender-procurement skills is not likely to deliver physical product in times of scarcity.

The action of matching current tender purchases to historical demand data cannot be referred to as supply chain alignment, nor can annual arbitrary increments be an efficient decision-making method. Yemen merely 'scales up' each year in its food procurement tenders.[39] This contrasts with deregulated supply chains that are more likely to have supply chain intermediary integration based on change management within synergies and relationships that are consumer driven.

The public import-tenderer has little intimate knowledge of global demand, food locations and stocks, or supply availability because they are not enacting daily international transactions and have no global intelligence network. These are the functions of international merchants, not of government procurement tenderers who are extremely susceptible to misinformation.

Using official historical global stock data to make decisions is spurious because of unknown private hoardings, inadequacy of data collection techniques and timing, and survey response psychological distortions. Vested interests are unlikely to accurately declare their inventory stocks publicly with separated old–new crop information, even if all such stocks and inventory could be identified and surveyed in deregulated international food supply chains.

Public authorities were never designed for complex risk-taking merchant functions such as accumulating product when shortages actually occur, import specialization, integrated supply chain synergies and logistical operations, just-in-time deliveries, marketing, quality assurance, or price–currency risk management. Governments act as pseudo-merchants whenever they purchase by taking legal title in an accumulation function, behave as warehousers through supposed 'buffer' stock policies, and subsequently sell as wholesalers. Public stock policies act more as a shock to the food supply chain rather than any 'buffer' because of the impact on food supply chain operations through their size and lack of synchronization with integrative quality-driven logistics.

There were instances in early 2008 when Middle Eastern flour mills could not buy or import suitable quality spot wheat supplies, irrespective of how much price was raised, much to the chagrin of other importing governments who could not afford to import at these high prices. This had the effect of exacerbating the shortage, because why would a seller or exporter agree to sell today when the price was going to be higher tomorrow.

This is exactly what the speculators and hoarders wanted, particularly when most government tender procurement decisions were synchronized in panic. It also motivated some exporting countries such as India, Bangladesh, Egypt, Myanmar, Vietnam, and Thailand to impose rice export bans, which further increased the grain price spikes. It is difficult to distinguish between hoarding and export bans, particularly when governments control both.

The government import tender method of ad-hoc subjective buying, warehousing, distribution and selling usually results in the public sector generally dominating prices and food supply chain logistics to the exclusion or 'crowding out' of the private sector, which is totally dependent on accurate, transparent, and dynamic price signals. If governments use international prices for their transactional benchmarks and proclaim competitiveness, this is price distortion because it ignores the local supply and demand issues being experienced domestically.

Government buyers may satisfy their institutional catering needs, but it is delusional to believe that they likewise can satisfy private individual consumers. Turgot perceived the inevitable lack of consumer choice, uniformity of product, and frequent shortages that would eventuate from such public actions to the exclusion of private supply chain intermediaries, and that this would inexorably lead to civil unrest.[40]

Unless domestic supply and demand dynamics are allowed to influence price, local spot price discovery cannot occur. Merely forcing international prices onto local food supply chain prices under the guise of 'competitiveness' after tender purchases have occurred will prevent important domestic 'trade' markets developing. If these 'trade' markets are prevented, no forward pricing market can develop that would enable the development of private storage in the domestic food supply chain.[41]

'Trade' markets can only exist with long-term supply contracts between supply chain intermediaries. Such long-term supply contracts are unlikely to exist whenever monopoly import tendering prevents supplier relationships from developing or when there is dependency on government for purchases, sales, and prices. If there is a lack of 'trade' markets and long-term supply contracts, supply chain activity is reduced to opportunistic behaviour rather than risk management, which can lead to rampant speculation.

Merchants incur huge entry barriers into countries with tender import policies because of government monopoly buying, government control over warehousing and storages, market-price distortions, the potential for corruption, and uncertainty over supply chain investment. The normal functions of merchants such as accumulation in integrated supply chains, private storage, and just-in-time delivery are most likely absent, which then requires governments to fill the vacuum.

Problems of public dependency then become entrenched often under the delusion that only the government can provide food 'security'. Import tenders and government pricing cause the public sector to have the dominant influence over food chain decisions, with any remaining private food supply chain operators becoming dependent on government decision making. This makes the public sector vulnerable to manipulation and corruption.[42]

Whenever food shortages occur, tender buying can quickly transgress to tender selling to the highest bidder. Wealthy country buyers can jump the queue to purchase scarce supplies. The capacity to pay higher prices is exactly what the exporter needs to force up prices by withdrawing sales in shortage markets. Propensity to pay drives the motivation not to sell. A mild global shortage will quickly become acute scarcity under such circumstances, as did occur in early 2008 when rich oil-producing food-importing countries raised their food-buying prices. Ironically, this is exactly how oil-producing countries became wealthy from crude oil cartel selling.

Quality problems can occur either in the actual product tender specification offering, non-delivery of the product specification against the import tender, subsequent quality problems in government warehousing, or corruption in the actual import-tender process.[43] This would indicate that public food tendering is an extremely high-risk venture whose stakeholders include domestic end users, consumers, taxpayers, and exporters.

Examples of public tender corruption include government officials accepting lower quality specifications after awarding the tender, which unduly favours the supplier and penalizes the domestic end user.[44] Excessive continuing poor-quality problems with a favoured supplier can be ignored despite the import-tender specifications and despite the domestic supply chain ramifications on consumers and potential exporters. Alternatively, over-buying in the import-tender process by public officials to meet end-of-financial-year expenditure budgets[45] can result in storage problems, selling delays, and deteriorating quality, quite apart from associated product shelf-life issues for end users and consumers. Huge wastage often results, usually at taxpayers' expense.

The import tenderer might accept less than the quantity specified in the tender contract, which unduly favours the awarded supplier.[46] As well, there may be tender-purchase preferences for government institutions,[47] which may distort quantity flows to the private food chain. There can be government internal tenders in the distribution process as well as external tenders to purchase the original supplies. This interdepartmental distribution system can be a precursor to much corrupt activity just to maintain the operation of the system, quite apart from time wasting and institutional–organizational power struggles.

Investment in overseas agricultural production

Much of the foreign direct investment into marginal food production programmes in non-food bowl countries to achieve food 'security' in the home country has ended in failures, particularly in Australia, Africa, and South America. The associated problems include inexperience, poor management, high investment risk, dependency, and possibly more food insecurity.

One classic example was the debacle in Australia and Africa by the British Government Food Corporation[48] under the direction of Lord Beveridge and John Strachey.[49] Just as the Food Corporation's peanut scheme in Tanganyika was being aborted under abject failure to supply Britain with vegetable oil, a joint venture was established in 1948 with the Queensland government (Queensland and British Food Corporation) to 'feed the British population'. The Corporation owned almost 200,000 hectares of brigalow land in Central Queensland by 1949. Apart from the agronomic difficulties in controlling brigalow regrowth, there were successive years of floods, droughts, fires, frost, mice, and wild pigs, all the very reasons why the country had been pastoral land beforehand. Its aggrandizement was extremely short lived, surviving with heavy debt until 1952 when the Queensland government paid out the British government debt in settlement.

Relatively little foreign direct investment has flowed to agricultural production in the Northern Hemisphere where the global food bowls exist, although China has recently invested in agricultural projects in marginal Western Asian countries. In contrast, most of the foreign agricultural investment has occurred in Southern Hemisphere production that is very marginal, higher risk, and seldom sustainable from a production or financial perspective.

Diversification of industry investments in different countries may theoretically spread risk, but there is a risk that the overall threat to profitability and investment returns might actually increase. A diversified high-risk portfolio is not the same as a diversified low-risk portfolio, particularly when risk taking is not adequately rewarded with high return opportunities.

The perception of potentially profitable under-utilized arable land is often illusory. In some countries, agricultural land may be cheap with low cost of water, rural services may not be charged at full cost, labour costs may be low, and the current impoverished occupiers of the land might be removed without adequate compensation. But this belies the real possibility that the land and its occupants might be impoverished for good reason, which might be caused by agronomic deficiencies or induced by government policies. There is the patronizing assumption that foreigners know best, that money will overcome all problems, that good seasons will forever prevail, and that the local people are not motivated as individuals. What is mostly overlooked are the factors that prevent these local people from motivating themselves. Mere investment generally does not remove the obstacles that lead to de-motivation, which ultimately will challenge the new foreign investor.

Middle Eastern and Chinese corporations have recently been directing foreign direct investment into overseas marginal agricultural development, particularly in Africa and Australia, but this time with much local political and social criticism.[50] However, such criticism often ignores the high investment risk characteristics of these ventures and the positive spill-over effects for domestic-host economies, such as attaining access to scarce global capital to develop its agricultural industries, create employment, infrastructure development, and rising land prices although this may make it more difficult for new farmers to buy land.

Foreign direct investment into regional diversification of food production might be justified on plant quarantine reasons, such as disease risk minimization or plant genetic requirements. However, to undertake investment in offshore agricultural production in marginal agricultural regions for financial or 'food security' objectives is frequently undermined by economic and agricultural realities. Having the global capital or mobilizing investment capital for such 'development' is no recipe for successful financial or food supply chain outcomes in marginal producing countries.

Foreign direct investment into overseas agricultural production ventures often presumes that foreigners have more knowledge about agricultural production than the local population and that capital investment is the only factor limiting development.[51] However, it is usual for experienced local farmers to understand the risks of marginal farming better than any newcomer. Seemingly vacant land may be left undeveloped for very good reasons, either because of high risk or previous land degradation. Cycles of good rainfall during a few years frequently mask decades of inadequate rainfall. In marginal producing regions, the better farmers generally enter on low rainfall and exit on good rainfall, whereas the worst farmers and foreigners usually enter on good rainfall and exit on low rainfall.

Foreign direct farming investment is generally aligned with corporate farms, because of the difficulty in negotiating with individual farmers. However, any assumption that corporate farms are more successful than individual family farms is often false.[52] Size of the farm is not the sole prerequisite to sustainable and efficient financial outcomes.

There are many serious issues and risks in achieving effective and efficient outcomes from overseas agricultural production investments from an organizational and operations perspective, regardless of whether management is local or foreign sourced, centralized or decentralized in control, or whether shareholders are public or private. There will always be difficulties in contracting experienced agricultural managers and skilled labour in remote localities.

Corporate management with large organizations are unlikely to be able to make the timely decisions in agronomy, pricing, selling, and investment, compared to efficient family-operated farms that have clear succession planning. Time lags in decision making within hierarchal structures can destroy opportunities.

Because countries currently undertaking foreign direct investment into overseas agricultural production generally have higher levels of income per capita, the perception of 'land grabbing' by the wealthy has been an issue.[53] However, usually it is a government–taxpayer issue rather than a private investment concern because corporate risk is usually far too high and financial returns far too low for private organizations to invest in marginal overseas farmland. Both the Chinese government and the oil-driven Persian Gulf countries have huge trade surpluses that need to be shifted offshore to reduce domestic inflation. The Qatar Investment Authority had $60 billion directed for foreign food and energy investment in 2009.[54]

The argument that the rich countries are taking food away from poor countries such as Africa can be challenged in the host-recipient countries. The African continent was a net food exporter until 1988,[55] a time when many newly formed socialist governments were nationalizing food supply chains. War, civil unrest, appropriation of agricultural farmlands, land 'reform' policies, government intervention, food supply chain distortions, investment disincentives such as the prevention of profit repatriation, and corruption have mostly led to African agricultural and food supply chain demise, rather than any issue of foreign direct investment.

Given that much foreign direct investment into overseas agricultural production is through home government authorities, the fiscal distortions and costs in the home country could be substantial. What could be invested into domestic home-country infrastructure is now being expended on pursuing marginal benefits at high risk in a host country.

It is difficult to measure the net cost impact when government-to-government arrangements occur in overseas agricultural production, because of fiscal audit inadequacies, industry subsidies, opportunity costs, and the impact of 'crowding out' potential private investment benefits in both the home country and the host-recipient country. Any attempt to artificially lower the domestic home purchase price of such foreign production would incur an even greater opportunity cost-loss to the venture, from an activity-based costing perspective. This could decrease the potential trade balance benefits in the host-recipient country, which could lead to retaliation by the host government.

Some producing countries imposed export bans in 2008. If the foreign direct investment in overseas agricultural production had occurred in these countries for the purpose of food security in the home country, then the scheme would have failed to secure supplies for the home country due to these export bans in times of shortages.

If the objective of foreign direct investment in overseas agricultural production is food security in the home country, then the added risks in marginal host countries might further jeopardize food security in the home country. This is especially the case when there is dependency on this particular method to secure domestic food supplies, inadequate coordination in operations management, and a lack of timeliness. There is the political risk that the host country could either change policies or renege on previous commitments. At worst, foreign investment ventures are vulnerable to appropriation, nationalization, confiscation, or retaliation by the host-recipient country with inadequate redress or compensation.

Overseas food supply chain investment

Overseas food chain investment by an importing government could be either through a takeover of an existing company or establishing a new company, such as a food processor or manufacturer. However, any supply chain objective by the importing country to alleviate food insecurity in times of shortages can be quickly thwarted by exporting governments imposing either export bans or higher taxes on food exports.

Sovereignty issues and foreign direct investment laws usually exist in the host country when importing governments undertake foreign direct investment in sensitive food supply chains of exporting countries. There can be many investment restrictions and rejections, while success may only be achieved through compromising original objectives and goals.

There is little synergy of objectives with an importing country attempting to achieve domestic food security while undertaking commercial business practices. Ad-hoc supply diversions for food-security reasons are not generally synergistic with financial objectives such as profitability and adequate returns from the investment. Ad-hoc supply diversions at critical times for reasons of providing food security to the home country could destroy supply chain contracts and long-term relationships, resulting in lost profit, opportunity costs, contract default, and expensive legal proceedings.

Regardless of the type of management, there are still major adaption issues in the host country, such as different cultures, business ethics, and legal proceedings. Quite often, the importing government authority has little control over foreign management and business policies in the host country, which then jeopardizes the initial goals of providing food security to the importing home country.

It is unusual for an importing government to undertake a joint venture with a host government. A government-controlled flour mill in the host country would be unlikely to enter into a joint venture with another foreign government for the purpose of providing food security to the home country. As well, importing governments are unlikely to enter into a joint venture that involves a complete overseas food supply chain. Rather any joint venture is more likely to be confined to a single operation within a complex food supply chain, such as a private rice mill, or an abattoir.

There could be problems in agreeing to share capital costs and expenditure in any joint-venture agreement, particularly when public fiscal arrangements need to synergize with tight private budgetary controls. Operational management synchronization would be expected to be very difficult to achieve because of conflicting objectives.

Goal alignment within a joint venture based on a private partner is unlikely when the goal of the importing country to provide food security in times of shortages contradicts the profit-maximization and shareholder-return goals of the private joint-venture partner. Any demand for an ad-hoc short-term food supply diversion to satisfy one partner would be expected to lead to serious managerial conflict and be perceived as the vested interests of one partner leading to the detriment of the overall joint-venture operation. Many joint ventures are undertaken simply for private operators to gain funding that otherwise would not have been readily available to an individual organization. Goal incompatibilities can quickly undermine any joint-venture relationship.

There is no surety of any management control under most joint-venture agreements, despite the commitment of investment and working capital. Many joint ventures succeed because of power imbalances, whereby the weaker partner loses management control to the stronger partner. Where power is more balanced, it can be very difficult to reach consensus management agreement, particularly when there are divergent business goals.

Government cartel-buying arrangements

Some governments of food-importing countries have advocated a monopsony cartel with discretionary decisions over food purchasing and subsequent selling in the various home countries, so as to reverse the market power of sellers in times of food shortages. The problem is that laws in most countries preclude the formation of cartels in host countries through anti-collusion legislation. It would also contravene World Trade Organization guidelines in that it might encourage serious trade retaliation.

Some countries contradict themselves by allowing the formation of commodity food domestic buying and selling cartel groups through producer cooperatives, single desks, and selling authorities, even though anti-collusion legislation exists. Government policies can be open to vested interest manipulation and potential corruption.

An international buying cartel could attempt to 'corner' a global market that was in short supply. Success would depend on enticing hoarders to sell. However, any buyer collusion in a shortage market is a clear signal of major buyer panic. The more urgent the purchases during shortages, the less likely the hoarder will sell, in the knowledge that prices will rise tomorrow. Both private and public attempts in market 'cornering' have generally ended with huge financial losses for buyers. In contrast, there have been notable hoarding successes for sellers.

Buffer and reserve stocks in the food supply chain

Government reserve stocks are equivalent to public hoarding, albeit for different reasons to the private hoarder. The latter withholds supply to force up price, whereas governments may attempt to stabilize price, but usually this causes dire consequences to the supply chain.[56]

Both Benjamin Graham[57] and Keynes[58] theorized over a granary 'asset' that could smooth out economic production cycles and prices, thus eliminating unemployment and inflation. The theory was extremely weak as to exactly who would want to buy old-crop commingled commodity that was often moth eaten, insect riddled, with uncertain chemical fumigation controls. Keynes presumably supposed that people would suffer such inconveniences under wartime pressures, which they eventually did under adverse circumstances.

There were serious flaws in Graham's in and out 'stability' theory that had been extended by Davies.[59] It assumed that the government could buy at low prices and sell at higher prices, thus generating easy public revenue and supposedly 'stabilizing' the economy through price averaging. However, intake accumulation quantities were largely unknown because of weather-driven biological production, the serious impact caused by dumping unspecified high risk product on a fragile food supply chain was dismissed, and the question as to whether consumers would tolerate or want such product was totally ignored.

The real problem with any government accumulation programme that precedes the formation of a 'buffer' stock is that it stimulates private hoarding. Any government announcement heralding large buying activities by public authorities is a great stimulus to privately hoard. Unless there is a Stalin and an enthusiastic Khrushchev underling, hoarders will proliferate.

The other real problem for the theories of Graham and Keynes was that agricultural commodities are global price takers. It assumed that either the government could influence price or that prices would move favourably. The problem for 'buffer' stock accumulators in 2008 was that they bought at peak market prices early in the year, and were then forced to sell old-crop stock in the latter half of the year when prices had collapsed. Both Graham and Keynes did not account for the frequent high losses incurred by governments and taxpayers in a buy high and sell low price disaster.

Government stock is product limited by quality, commingling, place value, and form value, with chemical fumigant contamination. It is not aligned to the private food supply chain because of high product specification risk, quality risk, with deteriorating shelf-life prospects. The stocks would be expected to incur huge price discounts at disposal time, except under government-induced shortages. Quite often, buyers are expected to pay for bagging commodities from public wholesale depots for transporting on small open trucks. Dependency on such methods for public distribution is becoming archaic for most countries.

Usually the only foods that governments are able to stock for long periods are undifferentiated commingled commodities that are storable. Perishables such as meat, fruit, vegetables, and dairy products that are highly nutritious are excluded, as well as products with high oil content because of rancidity, and those with moisture issues such as rice that can only be stored for less than 12 months. This limits long-term government food storage to low nutritional product that can withstand high chemical fumigation treatment with tolerances for moisture and low stock turnover. Given choice, food manufacturers would completely reject government food stocks as ingredients, as would consumers.

'Buffer' stocks often conflict with the objectives of achieving food security because of their disruptive impact on price signals and supply chain integrative operations. Commodity reserve stocks are the antithesis of an efficient functioning food supply chain.[60] None of the proclaimed benefits of 'buffer' stock schemes meet any final consumer needs[61] or end-user demand specifications, except as starvation rations.

Sometimes there are dubious objectives of 'buffer' stocks to meet monetary targets such as inflation control, despite being a price taker in a global commodity market. Alternatively, it can be used for self-preservation by incumbent autocratic governments concerned over civil disruption, under 'give them bread'[62] policies. There might also be a 'buffer' objective in the balance of payments or in fiscal budgetary policy.

'Buffer' stocks can be a tool for leveraging foreign policy objectives and used as trade-offs in government-to-government transactions. Whenever trade protection increases for a country, the more likelihood commodity reserve stocks will be used in trade and foreign policy trade-offs. Generally, it is only when trade protection decreases under freer trade does the dependency on government reserve stocks decrease.

Long-term commodity storage is the antithesis of consumer-based product quality. Farmers are more likely to sell their better-quality product to the private food chain or black market, and sell only their lower quality product to the government usually as a last resort sale.

It is usual for buffer or reserve stock policies to be the least discriminating regarding product accumulation quality simply because intake targets and government policy directives need to be met, and administration job positions safeguarded. There could be poisonous or obnoxious weed contamination, admixtures of other crops and varieties, soil, stones, sand, and vegetable matter, with moisture levels that could cause fungal outbreaks, hot spots, storage fires, and explosions.

Farmers are unlikely to have quality control procedures when they deliver into government buffer stocks, which can result in deliveries contaminated with weevils, borers, beetles, and moths that can seriously jeopardize the usefulness of the final product. Much buffer stock is sold into supply chains that can disguise the source of product origin, such as animal feed or bulk fillers in food manufacturing.

Government reserve stocks must incur regular fumigation treatment to control insects and moths, sometimes for five years as has occurred in China, before the food stock is released into the supply chain for human consumption. Organophosphorus compounds such as phosphine (previous fumigants included methyl bromide) are still the most effective fumigants against grain weevils, borers, and beetles despite increasing insect-resistance levels, but they require an inordinate amount of protective clothing and breathing masks during fumigation to prevent human poisoning.[63]

Product shelf life can be dramatically shortened when ingredients are sourced from previous government storage stocks. Domestic food quality is thus tarnished, which affects price, choice, brand loyalty, and availability. It could act as a signal for consumers to buy inferior products cheaply, or a signal to switch to a substitute product. This has an enormous impact on private decision making in the food supply chain, and can impact negatively on investment decisions.

The rapid sale and dumping of high-priced reserve 'buffer' stocks in late 2008 resulted in a shock effect on domestic food chains, which distorted domestic price signals, caused bureaucratic chaos amidst financial losses, and undermined domestic farmers who were about to plant their 2008–09 crops. These ad-hoc government policies are not conducive to effective efficient private decision making in production, operations, and supply chain management.

The most effective buffers against the 2008 global grain and oilseed crisis were the sophisticated food chain networks that have proliferated since the introduction of supermarkets and the diverse range of new food products, as evidenced in Singapore. These were then buffered by private storage in food supply chains through numerous 'trade' markets and funded by carry charges,[64] and the ability to globally source perishable and non-perishable food products within 24 hours through air freight. These acted as natural buffers in deregulated food chains.

The greater the domestic buyer dependency on any government 'buffer', the more likely are food queues, government stock rationing, black-market activities, and corruption to occur. Domestic buyers would defer purchases if there was an expectation of 'buffer' stock dumping by government, even when people are starving. Why buy today when prices will be lower tomorrow? Food chain decisions become dependent on expected government action or inaction. Such activity then becomes susceptible to bribery and corruption.

Government reserve stocks can cause distortions in imports, distribution, price, consumers, and re-exports. Existing supply chains become stagnant with erroneous price signals, while food products tend towards uniformity and standardization. When governments force private buyers to take delivery from public depots, there are many issues such as disrupting normal supply chain logistic operations and preventing seamless just-in-time deliveries, as well as decreasing the shelf-life of products.

It is usual for the government to monopolize the right to import under such commodity reserve stock policies, which immediately imposes barriers on domestic importers as well as the entry of merchants and overseas private end users. Food supply chain integration with global markets is stifled, whilst dependency on government action increases.

Government commodity disposal prices rarely are good barometers of spot price discovery. Port-based government prices usually do not reflect domestic supply and demand. Alternative methods to fix government prices can distort import prices, domestic prices, and impact on export profitability. Added costs in private supply chains due to government stockpile policies and disposal generally cannot be passed onto retail prices because of price controls or competition with substitute products.

Distorted price signals act as a disincentive for food supply chain investment, particularly in potential food export industries, and discourage innovation through research and development. This penalizes local private companies, generally with the loss of competitiveness to imported substitute products. Rather than demonstrate a robust capacity to counter food shortages, government reserve stocks decrease the international competitiveness of domestic food supply chain companies, thus increasing the probability of domestic food deficits.

Reserve stocks can be owned directly by government, through semi-government authorities, or via government-licensed private agencies. Governments can mandate minimum private stocks to be held in the food supply chain, but usually require the private operators to bear the cost of storage, product shrinkage and deterioration, and subsequent supply chain losses.

In some countries such as the Persian Gulf states, private-sector companies such as flour mills and vegetable oil refiners are required to maintain minimum flour and edible vegetable oil reserves in private storages for the government. There is little evidence that the existence of these publicly determined private reserve stocks made any contribution in avoiding the 2008 global grain and oilseed panic.[65]

Apart from the financial losses and supply chain disruption not being adequately compensated by government, this mandated private reserve-stock policy can seriously impact on the food supply chain through quality problems, decreased product shelf life, disrupt normal supply chain logistic operations, and prevent just-in-time delivery. The outcome usually is a loss of domestic and international competitiveness combined with investment disincentives.

The situation is exacerbated when competing or substitution foods are allowed to be imported while the domestic private food companies incur higher costs, lower quality, and decreased supply chain efficiency associated with government-mandated minimum private food stocks. This results in the loss of competitiveness compared to those companies that import substitute products, and lowers the potential for developing an export food industry that could ensure domestic surpluses. There is a huge paradox between the objective and outcome.

Privatized agency responsibility of reserve stock policies does not change the supply chain distortions or achieve the desired outcome any better than government control. It is the concept of having the reserve stock overhanging the food supply chain and 'crowding out' the normal private sector functions that is the primary problem. Private administrative efficiency might be better in achieving reserve stock policies, despite the lack of legal ownership, but the distortions in the food supply chain still remain, as do questions relating to patronage, favouritism, and corruption in the agency selection process. The blame for supply chain distortions, higher costs, and any international legal ramifications can be conveniently blamed on the agency.

Alliances with international merchants

If a government appointed international merchants as their private agents, assuming that such arrangements can be made legal and directives made transparent, this does overcome some of the problems associated with government procurement and storage. Internationally procured product could be warehoused either locally or internationally. Forming such integrative alliances is a positive step, and one that would have removed much of the government panic that occurred over imports in 2008.

International merchants are highly specialized in export–import facilitation, complex logistical operations, just-in-time deliveries, quality assurance, market intelligence, and risk management. They have the ability to accumulate even in times of supposed shortages, to deliver even in times of disruptions, to manage the risk of adverse price and currency movements, and to provide quality assurances. However, the appointment of international merchants for the purpose of government accumulation and storage for buffer stocks does not alter the distortions associated with reserve stocks overhanging the domestic food supply chain and 'crowding out' the domestic private sector in their normal supply chain functions.

Regardless of such import alliances and ownership transfer with international merchants, government decisions still need to be made regarding size of the reserve stock, the timeliness of purchases and sales, and distribution policies. Making such timely and relevant decisions in a bureaucratic environment regarding a dynamic commodity market with speculative purchases and sales, and effectively communicating these decisions to international partners, is difficult. However, any mistakes in decision making such as buy high and sell low can conveniently be blamed on the international merchant, which might make the alliance untenable for many international merchants.

Notes

1 Ashton, Basil, Hill, Kenneth, Piazza, Alan and Robin Zeitz, 1984, *Famine in China, 1958–61*, Population Development Review, 10, 613–645.
2 Rhodes, Richard, 1988, *Man-made Death: a Neglected Mortality*, Journal of American Medical Association, 260, 686–687.
3 Chang, G.H. and G.J. Wen, 1997, *Communal Dining and the Chinese Famine of 1958–61*, Economic Development Culture Change, 46, 1–34.
4 The World Bank, 2009, *Improving Food Security in Arab Countries*, Washington, DC.
5 Sanchez, Pedro A., 2002, *Soil Fertility and Hunger in Africa*, Science, 295:5562, 2019–2020.
6 Lyon, Fergus and Gina Porter, 2009, *Market Institutions, Trust, and Norms: Exploring Moral Economies in Nigerian Food Systems,* Cambridge Journal of Economics, 33:5, 903–920.
7 Darity, William and Lewis S. Davis, 2005, *Growth, Trade, and Uneven Development*, Cambridge Journal of Economics, 29:1, 141–170.
8 Feng, Haifa, 1999, *Agricultural Development in the Netherlands: An Analysis of the History of Dutch Agricultural Development and its Importance for China*, Agricultural Economics Research Institute, The Hague, June.
9 Hopkins, Raymond F. and Donald J. Puchala, 1978, *Perspectives on the International Relations of Food*, International Organization, 32:3, 581–616.
10 Jardine, Lisa, 2008, *Going Dutch*, Harper Collins, New York, Ch. 12.
11 Blom, J.C.H. and Emiel Lamberts, 1999, *History of the Low Countries*, translated by James C. Kennedy, Berghahn Books, New York.
12 Dubai School of Government, 2008, *Food Security in the United Arab Emirates*, Policy Brief, 9 October, Dubai.
13 Birkeland, Peter W., 1999, *Soils and Geomorphology*, Oxford University Press, New York.
14 Reed, Michael, 2010, *Economic Theory and Policy Implications for Arab Oil-Rich Countries*, paper presented at the international conference: 'Food Security in the Arab Countries: New Challenges and Opportunities in the Context of Global Food Price Volatility', College of Agriculture and Marine Sciences, Sultan Qaboos University, Muscat, 2–4 March.
15 van Rheenen, Teunis, Breisinger, Clemens, Aragon, Catherine, Babu, Suresh and Khalid El Harizi, 2010, *The Global Food Crisis: How Arab Countries Can Emerge More Food Secure*, paper presented at the international conference: 'Food Security in the Arab Countries:

New Challenges and Opportunities in the Context of Global Food Price Volatility', College of Agriculture and Marine Sciences, Sultan Qaboos University, Muscat, 2–4 March.
16 von Braun, Joachim and Ruth Meinzen-Dick, 2009, *'Land Grabbing' by Foreign Investors in Developing Countries: Risks and Opportunities*, International Food Policy Research Institute, Washington, DC, Policy Brief 13, April.
17 Reusse, Eberhard, 2002, *The Ills of Aid – An Analysis of Third World Development Policies*, The University of Chicago Press, Chicago.
18 Porter, Michael E., 1985, *Competitive Advantage – Creating and Sustaining Superior Performance*, The Free Press – Macmillan, New York.
19 Alston, Julian M., 2002, *Spillovers*, Australian Journal of Agricultural and Resource Economics, 46:3, 315–346.
20 Bunting, Edward, 1952, *Land Development and Large Scale Food Production in East Africa by the (British Government) Overseas Food Corporation*, Economic Botany, 6:1, 55–68.
21 Bloch, Harry and Peter Kenyon, 2001, *Creating an Internationally Competitive Economy*, Palgrave-Macmillan, Sydney.
22 Messinger, Paul R., and Chakravarthi Narasimhan, 1995, *Has Power Shifted in the Grocery Channel?* Marketing Science, 14, 189–223.
23 Humphrey, John, 2007, *The Supermarket Revolution in Developing Countries: Tidal Wave or Tough Competitive Struggle?* Journal of Economic Geography, 7:4, 433–450.
24 Annual Seminar series of 'Singapore as a Food Hub', Singapore Economic Development Board, Singapore.
25 Yemtsov, Ruslan, 2008, *The Food Crisis: Global Perspectives and Impact on MENA, Fiscal and Poverty Impact*, Middle East and North Africa Poverty Reduction and Economic Management Department, The World Bank, Washington, DC.
26 Marion, Bruce W., Meuller, Willard F., Cotterill, Ronald W., Geithman, Frederick E., and John R. Schmelzer, 1979, *The Price and Profit Performance of Leading Food Chains*, American Journal of Agricultural Economics, 61:3, 420–433.
27 The World Bank, 2007, *Connecting to Compete: Trade Logistics in the Global Economy*, Washington, DC.
28 Brooks, Douglas H. and David Hummels, 2009, *Infrastructure's Role in Lowering Asia's Trade Costs*, Edward Elgar, Cheltenham.
29 Sankaran, Jayaram K., Gore, Amanda, and Baz Coldwell, 2005, *The Impact of Road Traffic Congestion on Supply Chains: Insights from Auckland, New Zealand*, International Journal of Logistics Research and Applications, 8:2, 159–180.
30 Kinsey, J.D., Senaur, B., King, R.P., and P.F. Phumpiu, 1996, *Changes in Retail Food Delivery: Signals for Producers, Processors, and Distributors*, Working Paper 96-03, The Retail Food Industry Center, University of Minnesota, St Paul.
31 Beuthe, M., V. Himanen, A. Reggiani, and L. Zamparini, 2003, *Transport Developments and Innovations in an Evolving World*, Springer-Verlag, New York.
32 Rimmington, Mike, Carlton-Smith, Jane, and Rebecca Hawkins, 2006, *Corporate Social Responsibility and Sustainable Food Procurement*, British Food Journal, 108:10, 824–837.
33 Groenewegen, Peter D., 1983, *Turgot's Place in the History of Economic Thought: A Bicentenary Estimate*, History of Political Economy, 115, 611–615.
34 Locke, John, 1690, *Two Treatises on Civil Government*, republished by Routledge, London, 1887.
35 Taylor, Fred M., 1929, *The Guidance of Production in a Socialist State*, The American Economic Review, 19:1, 1–8.
36 Refer to Figure 1.3 in Chapter 1.
37 Aucoin, Peter, 1990, *Administrative Reform in Public Management: Paradigms, Principles, Paradoxes, and Pendulums*, Governance, 3:2, 115–137.
38 Lee, Hau Leung, Padmanabhan, V., and S. Whang, 1997, *The Bull-whip Effect in Supply Chains*, Sloan Management Review, 38:3, 93–102.
39 The World Bank, 1995, *The Wheat Subsidy in Yemen*, Washington, DC.

40 Groenewegen, Peter D., 1983, *Turgot's Place in the History of Economic Thought: A Bicentenary Estimate*, History of Political Economy, 115, 611–615.
41 The importance of 'trade' markets in the formation of forward markets and the provision of private storage and carry will be discussed in depth in Chapter 10.
42 Singh, Balwinder, 1999, *India: Procurement Frauds*, Journal of Financial Crime, 7:2, 183–185.
43 Her Majesty's Treasury, 2005, *2004–2005 Fraud Report: An Analysis of Reported Fraud in Government Departments*, HM Treasury, London.
44 Singh, Balwinder, 1999, *India: Procurement Frauds*, Journal of Financial Crime, 7:2, 183–185.
45 Singh, Balwinder, 1999, *India: Procurement Frauds*, Journal of Financial Crime, 7:2, 183–185.
46 Singh, Balwinder, 1999, *India: Procurement Frauds*, Journal of Financial Crime, 7:2, 183–185.
47 van Weele, Arjan J., 2010, *Purchasing and Supply Chain Management – Analysis, Strategy, Planning and Practice*, Cengage Learning, Andover, Ch. 6.
48 Bunting, Edward, 1952, *Land Development and Large Scale Food Production in East Africa by the (British) Overseas Food Corporation*, Economic Botany, 6:1, 55–68.
49 Lord Beveridge was the secretary of the UK Ministry of Food and author of *Planning under Socialism* (1936) and *Pillar of Security* (1948), and John Strachey was the UK Minister of Food and author of *The Coming Struggle for Power* (1932) and *Why You Should be a Socialist* (1938).
50 Cotula, Lorenzo, Vermeulen, Sonja, Leonard, Rebecca and James Keeley, 2009, *Land Grab or Development Opportunity? Agricultural Investment and International Land Deals in Africa*, FAO, Rome.
51 von Braun, Joachim and Ruth Meinzen-Dick, 2009, *'Land Grabbing' by Foreign Investors in Developing Countries: Risks and Opportunities*, International Food Policy Research Institute, Washington, DC, Policy Brief 13, April.
52 Strange, Marty, 1988, *Family Farming: A New Economic Vision*, University of Nebraska Press, Omaha.
53 Cotula, Lorenzo, Vermeulen, Sonja, Leonard, Rebecca and James Keeley, 2009, *Land Grab or Development Opportunity? Agricultural Investment and International Land Deals in Africa*, FAO, Rome.
54 von Braun, Joachim and Ruth Meinzen-Dick, 2009, *'Land Grabbing' by Foreign Investors in Developing Countries: Risks and Opportunities*, International Food Policy Research Institute, Washington, DC, Policy Brief 13, April.
55 UNCTAD policy statements, 2008–2010.
56 Governments do accumulate some reserve inventory for institutional catering, but this is normally excluded in a 'buffer' scheme.
57 Graham, Benjamin, 1937, *Storage and Storability*, McGraw-Hill, New York, reprinted 1998.
58 Keynes, John M., 1938, *The Policy of Government Storage of Food-Stuffs and Raw Materials*, The Economic Journal, 48:191, 449–460.
59 Davies, Joseph S., 1938, *The Economics of the Ever-Normal Granary*, Journal of Farm Economics, 20:1, 8–21.
60 Makki, Shiva S., Luther G. Tweeten, and Mario J. Miranda, 1996, *Wheat Storage and Trade in an Efficient Global Market*, American Journal of Agricultural Economics, 78, 879–890.
61 von Mises, Ludwig, 1949, *Human Action: A Treatise on Economics*, Ch. 15, section 4.
62 See Juvenal's Satire VI: Lines 347–348. 'Who will guard the guards themselves' has been generally interpreted as a general apathy by officialdom towards its citizens, which can lead to tyrannical dictatorship.
63 Canadian International Grains Institute (CIGI), 1993, *Grains and Oilseeds: Handling, Marketing, Processing*, Winnipeg, Vol. 1, 4th Ed., 309–354.
64 'Trade' markets and carry will be discussed in Chapter 10.
65 Yemtsov, Ruslan, 2008, *The Food Crisis: Global Perspectives and Impact on MENA, Fiscal and Poverty Impact*, Middle East and North Africa Poverty Reduction and Economic Management Department, The World Bank, Washington, DC.

Further readings on import intervention

Hammond, R.J., 1946, *British Food Supplies, 1914–39*, Economic History Review, 16, 1–14.

Hatanaka, Maki, Carmen Bain, and Lawrence Busch, 2006, *Differentiated Standardization, Standardized Differentiation: The Complexity of the Global Agrifood System*, Research in Rural Sociology and Development, 12, 39–68.

Hayenga, Marvin L. 1979, *Risk Management in Imperfect Markets: Commodity Procurement Strategy in the Food Manufacturing Sector*, American Journal of Agricultural Economics, 61:2. 351–357.

Kastner, J.J. and Block, C.L., 2009, *The Global Food Supply Chain*, in John G. Voeller, ed., *Wiley Handbook of Science and Technology for Homeland Security*, John Wiley and Sons, New York.

McCorriston, Steve and Donald MacLaren, 2005, *The Trade Distorting Effect of State Trading Enterprises in Importing Countries*, European Economic Review, 49, 1693–1715.

Ouyang, Yanfeng and Carlos Daganzo, 2006, *Characterization of the Bullwhip Effect in Linear, Time-Invariant Supply Chains: Some Formulae and Tests*, Management Science, 52:10, 1544–1556.

Peterson, W.L., 1979, *International Farm Prices and the Social Cost of Cheap Food Policies*, American Journal of Agricultural Economics, 61:1, 12–21.

Rasiah, Rajah and Johannes Schmidt, 2010, *The New Political Economy of Southeast Asia*, Edward Elgar, Cheltenham.

Ryder, Ray and Andrew Fearne, 2003, *Procurement Best Practice in the Food Industry: Supplier Clustering as a Source of Strategic Competitive Advantage*, Supply Chain Management: An International Journal, 8:1, 12–16.

Sarris, Alexander and Jamie Morrison, 2010, *Food Security in Africa*, FAO and Edward Elgar, Cheltenham.

5

THE RISE AND FALL OF COMMINGLED COMMODITY HANDLING

Traditional bulk handling can be defined as a homogenized handling system for commingled commodity that usually was rail dependent. Initially there was the primary wooden-structured elevator that lifted the commodity into a storage bin. Then came balloon or crib annexes. Concrete silos were a later modification of the elevator method. Much later came flat storage with either round or rectangle warehouses that were usually erected around the initial elevator or silo.

Commingled bulk commodity handling and rural railways were usually financially co-dependent in producing countries. However, it was not until the 1960s that ship designers introduced upper ballast tanks to stabilize bulk ships that facilitated the introduction of the Handysize and Panamax bulk carriers for exports and imports, but by then commingled bulk commodity popularity was at its zenith and was in general decline within food supply chains by the early 1970s.

The notable exception to rail dependency was intensive barge and river transport within the food bowl regions of Europe, the USA, China and Indo-China, and the Asian Subcontinent, as well as in Brazil. Bulk handling gradually replaced bags in this transport system. But this barge transport method mostly was to remain privately owned rather than be government controlled, and be independent of increasingly inefficient government-owned railways.

This contrasted with most rail-dependent countries whereby governments either owned the rail network from the beginning, or were pressured to take ownership of previous privately operated railways once road transport became a major competitor, such as in Canada and Australia. Railways in the US mainly remained with private ownership, albeit often with subsidies, because they crossed many state borders and there was great difficulty in the federal authorities negotiating with individual state authorities for public takeovers of rail networks.

Since the 1850s, many of the handling procedures of agricultural commodities were designed for railway network flow effectiveness, rather than for food supply chain efficiency or value creation from an end-user or consumer-demand perspective. Traditional bulk handling was generally adopted in order to improve rail transport logistical flows at a time when road transport efficiency was rapidly improving. Bulk handling gradually replaced bags in many countries during the late nineteenth to early twentieth century.

One of the many reasons why traditional bulk handling failed was through its dependency on an increasingly inefficient rail network system in most producing countries, which were mostly operated by a government monopoly. Railways had all the disadvantages associated with a monopoly, as well as becoming increasingly uncompetitive with road transport that bypassed both the rail network and the traditional commingled bulk-handling system. Even worse, the traditional bulk-handled product had little synergy with domestic food supply chains in importing countries, and mostly caused port congestion and urban bottlenecks.

The other major problem with commingled bulk handling was its irrelevance to end users in the food value chain.[1] Product differentiation and potential value for the end user was lost through commingling, particularly for grains, oilseeds, pulses, and even cane with sugar flocculation differentiation.

An irony lies in the comparison with perishable food products. These were unsuitable for bulk storage, generally bypassed the whole rail system, rarely coming under the control of governments, always highly product differentiated, mostly under private ownership and individual direction, and nearly always aligned with buyer requirements. This is what mostly fed consumers when government controls over commingled bulk handled storable commodities failed.

Because the rise of traditional commingled bulk handing coincided with the development of new railway networks, new bulk-harvesting inventions, the formation of cooperatives, the rise of international socialism, increasing government intervention, government-to-government transactions, and the rise of the international merchant, it will be important in this chapter to investigate dependency and complementariness between these various occurrences.

General weaknesses in commingled bulk handling

The theoretical advantage of traditional bulk handling was standardization, but this was actually its weakest link in the food supply chain.[2] Traditional bulk handling attempted to align a biologically diverse production system with an equally diverse human consumption–behavioural system through a process of standardization. As such, an efficient food supply–demand system was never going to be achieved through the traditional homogeneous bulk-handling system.[3] Supply chain inefficiency and confused responsibilities were generally going to distort accurate farmer and consumer signals.[4]

When traditional bulk handling was introduced in the nineteenth century, bread making was often no more than homemade unleavened bread or damper. It was then that the commoditized bulk-handling system suited the product. However, there were to be rapid technological improvements in bread making in the late nineteenth and early twentieth centuries in the understanding and usage of yeast cultures, as well as gluten, protein quality, impurities, and dough characteristics. By 1912, Rohwedder had invented the bread-slicing machine in the USA. From a specialized dough-making technological perspective, traditional commingled wheat bulk-handling methods had become obsolete by the First World War.

Inadequate investment capital, innovation stagnation, inefficient operations, disincentives in performance and productivity, and a general lack of motivation were the main drawbacks to statutory bulk-handling authorities.[5] The underlying causes of these problems were the inherent vagueness of ownership and responsibilities of the commodity within the supply chain, and the lack of competitive pressures. Slight 'modification' to the traditional methods of bulk-storage handling was an inadequate response to new innovations and technological advances that were occurring elsewhere in other food manufacturing supply chains.

Inventory was perceived as waste and high risk by many industrial manufacturers as far back as the early 1920s,[6] and this thinking ultimately flowed over to food manufacturers although there were major differences. Food ingredients were biological products produced within two to three months of the year.[7] The new delivery philosophy of just-in-time (JIT) that arose during the 1980s assumed that product accumulators would be more suitable as inventory managers than end users.[8] The cost and previous risk of warehousing inventory was therefore pushed back in the value chain to product accumulators, despite the new supply risks to end users such as product risk and continuity-flow uncertainty.[9]

At the same time that food manufacturers were attempting to introduce more logistics-efficient methods, the government was still pushing the usage of traditional commingled bulk commodity handling to recoup railway debt.[10] Whereas JIT required specific product with value to the accumulator through time, place, form, and service, traditional bulk-handling methods consolidated the accumulation of commingled non-aligned product with little value.[11] Something had to give way if logistics and supply chain flow efficiency were to be achieved. Commingling was incongruous with JIT. It was ironic that by the time that governments had fully mandated the traditional bulk-handling method, it had been made redundant by JIT.

The impact of railways on bulk handling

During the mid-nineteenth century, railways linked many of the new agricultural producing countries with ports, making the supply chain cheaper with less risk for local merchants to buy and sell, or sell and then buy, over large distances. By the 1870s, private railway networks were spreading out over the United States and Canada. However, competition with river barges and inland ships in these countries was intense, which induced motivation for innovation.

The handling of bags onto the increasing number of rail wagons was time and cost inefficient to the private railway companies. These private rail companies then offered free land and special privileges to anyone who had the skills and investment capital to build wooden structures known as country elevators adjacent to the rail lines, which could receive bulk grain, elevate it, gravitate the grain into storage bins and then load onto rail wagons.[12]

Whilst the link between traditional bulk handling and railways is strong, it is also strong with barge transport. In contrast, this linkage tends to be broken for road transport, which has placed enormous competitive pressures on traditional bulk handlers over the past few decades under market deregulation. Road transport has become much more aligned to the food supply chain compared to rail transport because of its ability to meet just-in-time deliveries of smaller lots of exact specification to the end user, which are driven by consumer demand. Rail transport is the antithesis of meeting just-in-time delivery requirements in the food supply chain.

The relationship of bulk handling to bulk harvesting

There is little evidence to suggest that bulk handling caused the invention of mechanical harvesting, with the latter usually following the former after a long time lag. Whilst railways caused the early introduction of bulk-handling structures, farmer adoption of delivering bulk commodity to silos and warehouses was usually delayed for decades because of the poor rural road conditions and the lack of suitable trucks capable of delivering product to the elevator–railway siding. Even when mechanical harvesting was introduced, the harvest bins were mostly emptied straight into bags.

Any similar timing of the rise of bulk handling with mechanical harvesting was therefore largely coincidental. Inventions in mechanical harvesting were more likely to have been driven by the rising labour cost of manual reaping and increasing tendency towards organized labour unions from the 1880s onwards.[13] Associated rising labour costs caused by trade union organization induced farmer motivation for labour-saving devices. The private railway companies were encountering these same issues when they pushed for bulk handling.

The association of bulk handling with cooperatives

Despite the coincidence in the simultaneous rise in both the cooperative movement and commodity bulk handling during the second half of the nineteenth century, the two remained unconnected in many areas even up to half a century. Many producer cooperative pooling arrangements commenced with bag handling and cooperative-owned bagged storages. Therefore there is little evidence that one caused the other, rather they ultimately complemented each other. However, each had their own problems, and any linkages ultimately compounded the supply chain problems.

There was a gradual evolution of cooperative activities into commodity bulk-handling facilities and the associated dependency on the government-controlled rail and traditional bulk-handling network. Inherent problems with farmer-selling cooperatives such as the lack of consumer focus and inadequate consensus decision making were immediately compounded by the inherent bulk-handling difficulties of commingled commoditization and railway dependency.

The problem of non-alignment with the food supply chain through production-push policies was the major obstacle for most selling cooperatives. Many farmer members firmly believed that they could force commoditized product onto a supply chain, through rail systems that had no links to the food supply chain, onto ships that had little alignment with international food supply chains, and expected 'adequate compensation' for doing so.[14]

Because of their product misalignment and alienation from the food supply chain, some of the early producer cooperatives in 1915–17 struggled to survive the First World War.[15] It was in the wake of these commodity selling cooperative failures that early demand for statutory marketing authorities arose. But these public authority entities were not equipped to either recognize the underlying food supply chain misalignment problems, nor were they inclined to change policy because of their vested fiscal interest in railways, ports, and often the traditional bulk-handling system itself.

When the producer cooperatives had to both finance bulk-handling facilities as well as finance payments to farmers when there was no long-term sale contracts, which was due to product accumulation not being aligned with the food supply chain, the provision of bank credit security failed.[16] Farmer selling cooperatives either were forced onto government financial guarantees to maintain survival, or they ceased to exist. Until the 1960s, semi-government authorities in Australia and Canada were often forced to take over the bulk-handling assets of failed cooperatives.[17]

Predisposition for government intervention

For countries such as Canada and Australia, and for most countries in Asia and Africa, the catalyst for much government intervention was a traditional bulk-handling system that was aligned with a public railway system, with commingled product going nowhere in the domestic food supply chain, thereby creating investment disincentives. The rail-port centricity also discouraged investment in the development of private domestic 'trade' markets and cluster industries.

The initial problem with traditional bulk handling was its expensiveness, requiring huge early capital investment for a very long-term uncertain financial return, particularly in marginal producing countries. Neither farmers nor domestic commodity buyers could be expected to finance initial investment into bulk-handling infrastructure, although they would be required to finance its operations. Such infrastructure in producing countries could lay idle whenever drought or production shortfalls occurred, and not be revenue earning. Governments and taxpayers were increasingly perceived to be the conduit for ownership and financing of bulk-handling facilities as well as the railway systems.

The exception to this was the bulk handling associated with private barge transport in Europe, the USA, and Brazil, which was predisposed to smaller operations and the development of strong merchant activity. Brazil was an exporting country that did not have extensive rural-based railway systems, had little early market intervention by government, and was more likely to develop domestic 'trade' markets and cluster industries.

This may suggest that it was not the bulk-handling system per se that necessitated public funding and dependence on government intervention, but rather it was the government-controlled railways that were not linked to the food supply chain, which prevented bulk handlers integrating with forward domestic supply chain transactions. Bulk handlers were restricted to intake, storage, and out-load operations, and unable to participate in further supply chain operations with the final end user because of railway-port restrictions. However, the intrinsic disadvantage of commingling was always to be a major relevancy problem for bulk handlers and elevators in any supply chain.

The impact of traditional bulk handling affected countries differently because of their diverse geographic characteristics.[18] Traditional bulk handling in producing and exporting countries such as Canada and Australia, which were dependent on vast rail networks, usually required large infrastructural investment both at major ports such as terminal elevators, new wharves, and dredging channels, as well as up-country facilities. It was increasingly common for governments of these countries to own the railways in between and control ports through statutory authorities.

Most of the impetus for traditional bulk-handling infrastructure and public ownership arose more from government fiscal stimulus policies and a desire to be 'export focused', rather than any demand from farmers,[19] end users, or global importers. The latter were most reluctant to change to bulk because of inadequate private investment, port-urban congestion, and the poor state of the local road transport. Each successive economic recession and war from the 1870s to the 1960s tended to encourage even greater government fiscal expenditure and statutory involvement in rail-port-based commodity infrastructure, despite increasing economic inefficiency due to the monopoly organizational structures, their propensity for supply chain dysfunction with no alignment with end users, and increasing reliance on government-to-government international transactions. It was a government solution rather than a supply chain or even a farmer-based solution.

Governments in importing countries were forced to intervene when foreign exporters demanded bulk commoditized shipments at a time when the local food supply chains of the importing countries were still oriented towards bags and smaller, highly differentiated lots. There were occasions when the government of the exporting country such as Canada and Australia financed or provided 'aid' for bulk unloading and warehouse facilities for the government of the importing country. However, this import infrastructural financing inconveniently finished at the port, which could be located in a very congested urban area and very unsuitable for subsequent bulk rail or road transport.

This bulk imposition on importing countries immediately created a barrier to their local food supply chains. Local end users in importing countries were increasingly forced to be dependent on a commingled unsuitable uniform product that was a legacy of traditional bulk handling, and be increasingly dependent on government bulk warehouses at the port. In many importing countries, the local food supply end user had to pay for the bagging of the bulk commodity at government port storages.

There were claims by the bulk exporting countries that importing countries were 'undeveloped' because they lacked bulk-handling facilities. This was despite the destruction of food supply chains and 'trade' markets in both the importing and exporting countries, as well as the imposition of government dependency. The paradox was that this same bulk commingled handling system was to cement the 'undeveloped' status of importing countries.

Regardless of whether commodity ownership was government, statutory authority, or co-owned through a cooperative, any pooled product in the producing country needed to be financed, either through a private bank or from government fiscal revenue, so that farmers could receive payment before the commodity was sold. Long-term bank credit security was virtually impossible because the product was not aligned to any end user, which suggested that even bank managers perceived the supply chain dysfunction. This need for advance finance to pay farmers before the product was eventually sold,[20] usually at a heavy discount, epitomized everything that was wrong with the traditional commingled bulk-handling system.

Governments were forced to intervene under farmer pressure to provide either direct finance or guarantee credit security for bank financing of commodity pools. Grains, oilseeds, and pulses had seasonal production based on crop years, but any unsold non-aligned old crop had to be stored into the new crop year. It was therefore financially prudent to pay producers based on crop-year commodity, which necessitated separate pooling when crop-year storage overlapped. Once again, the need for crop pool separation epitomized the problem with selling, because it provided proof that commingled bulk product was often left unsold even after 12 months. Some governments conveniently reclassified old-crop carry over as 'buffer' stocks.

Commodity arrivals in importing countries also needed to be financed because of the incapability and unwillingness of small local end users to immediately purchase large bulk shiploads, particularly when Panamax ships were introduced during the 1960s. These ships climaxed the financial problems inherent in the traditional bulk-handling system. Many importers were going bankrupt because they were being forced to buy imported commingled product that had no local food supply chain alignment or little consumer relevance. It was becoming a government intervention disaster from one end of the commodity chain to the other, which tended to create even more pressure for greater government involvement to solve the mess that had originated with the exporting country through the enforcement of the railway-port commingled bulk-handling system.

There is little evidence that the impact of commingled bulk handling on the food supply chain in both the export and import countries was ever seriously examined. It was not even mentioned in the recent study on significant issues in agricultural markets over the past century.[21] It mostly was driven by logistics and fiscal stimulus policies in a production-driven period, especially during the Cold War, when export pressures and foreign policy directives were paramount in order to counteract currency exchange rigidities.

When production-driven initiatives failed because of misalignment with the food supply chain, governments were forced to further intervene into selling functions in producing countries, or buying functions in importing countries. This merely prolonged the production-driven system, mostly destroyed the whole consumer-driven food supply chain, and corrupted the government of the importing country. 'Aid' from the exporting country merely consolidated the corruption in the importing country.

Merely changing bulk handling and railway ownership from public to private under new privatization policies did not change the production-push mentality or the non-alignment with the domestic and international food supply chains. The problem was not just due to uncompetitive inefficiencies, rather it was a supply chain alienation problem linked with organizational structural misalignment. Privatization, contestability, and competition were real shocks to traditional commingled bulk handlers because their structure and operations had never been aligned to the food supply chain or road transport networks. Alliances between privatized traditional bulk handlers and international merchants or end users have usually failed because of goal incompatibilities with operational and structural dysfunction that was associated with commingled rail-port-centric bulk commodity.

Usually, privatized bulk-handler survival has been dependent on forcing some food supply chain relevance through private takeovers and joint ventures within the existing food supply chain, but noticeably outside the railway network. This supports the evidence that the problem is not with bulk handling per se, but rather the linkage with both commingling and railways that is the real problem.

It could be argued that food supply chain deregulation in both Canada and Australia was only made possible when road infrastructure reached a certain minimum stage. The co-dependency between commingled bulk handled product and railways could then be broken, and segregated product could become integrated with segmented domestic food supply chains through road transport.[22]

Rise and fall of government-to-government transactions

Traditional commingled bulk handling ultimately became dependent on government-to-government transactions because much of it was unsaleable due to the lack of alignment with international private food supply chains and any old crop carry-over had suspect quality. Government dependency increased during the 1950s[23] especially when bulk ships reached a size that was beyond any capacity of private importers to finance and make reliable payment for such shipments.

Private food supply chains that were actually aligned to meeting consumer demand were shattered because of this rising interdependency between traditional commingled bulk handling and government transactions. Strachey-modelled public authorities[24] increasingly were forced to fill the void left by departing private food supply chain intermediaries, at the same time that creeping intervention pushed more citizens into institutional living. Bulk procurement policies, 'buffer' stocks, public food manufacturing, and price controls proliferated. Economic 'development' was perceived in those countries that adopted such policies and systems, despite the many paradoxical outcomes.

The competing Cold War environment between the East–West producing regions made government-to-government sales based on foreign policy trade-offs inevitable,[25] until the Soviet grain shortages of 1972–73 when the USSR was forced to buy grain from the USA. These Soviet grain purchases effectively brought the Cold War to a close.[26] The Australian commingled bulk-handling system collapsed almost as fast as the fall of the Berlin Wall, with domestic wheat deregulation occurring in 1989.

Complementariness with international merchants

Numerous up-country bulk handling and warehousing infrastructures that were dependent on inefficient government owned rail networks seemingly going nowhere slowly from a food supply chain perspective was the antithesis of domestic merchant operations in producing countries. The combined introduction of railways and traditional bulk handling systems and their interdependence quickly caused the demise of the local merchant in many producing and exporting countries by the 1920s. Whenever statutory authority regulations were introduced, domestic merchants were usually mandatorily excluded from domestic transactions.[27]

In Canada, whereas previously there had been major cross-border bagged grain sales from local merchants to US merchants and onto end users,[28] particularly through Cargill in Minnesota via the rail link to St Paul or by steamer and rail to Duluth, this was effectively stopped by the direction of new local Canadian railways and with new traditional bulk-handling methods. Canadian bulk grain then travelled in an East–West direction along the Canadian Pacific Railway or via another rail line to Thunder Bay on Lake Superior, rather than flowing south through local merchants as differentiated product to end users.

Whilst domestic merchants existed, they were the convenient scapegoat[29] for many farmer woes that were mostly caused by attempting to produce in marginal farming regions that were isolated from end users. With the demise of the local merchant through mandated commingled rail-centric bulk-handling systems went many of the merchant functions provided in an efficient food supply chain.[30] Governments were then increasingly pressured to fill the domestic void and provide these same functions when public officials really had no experience or motivation to undertake them.

Merchants who survived this onslaught generally were less dependent on the railways, and had more access to river-ways in the US, Europe, and Brazil, with more exposure to international food supply chains. They were therefore in a better supply chain position to capitalize on the weaknesses in the traditional commingled bulk-handling and railway systems in more interventionist countries.

Government authorities in railway-dependent and traditional bulk-handling countries increasingly encountered dysfunctional logistical commodity non-flows in shattered food chains exacerbated by misaligned product. Various governments were forced to seek merchant advice in damage-control mode. The growth of international merchants prior, during, and after the Second World War was often the result of powerful alliances between an increasing number of statutory buying–selling authorities and rapidly growing international merchants who could assist these statutory authorities.

Cargill arose from small domestic merchant activities via barge and rail in the northern USA, while Continental (Belgium), Louis Dreyfus (France), and Andre (Switzerland) arose from private barge and merchant activities in Europe,[31] Toepfer (Germany) arose from river–canal–port trading in Hamburg, while Bunge (Sweden) had the Scandinavian river barge and Baltic trade. Prior to government intervention into food supply chains, these merchants were small non-descript largely domestic entities.

The advent of the international merchant during the inter-war period facilitated international trade for commingled bulk commodities particularly from failing farmer-selling cooperatives and statutory organizations whose only real function in the supply chain was non-aligned product accumulation. Whenever surpluses could not be disposed under international merchant-selling arrangements, statutory organizations then were forced to rely on government-to-government dumping transactions. Price could always be discounted to dispose of unwanted commodity to an overseas buyer or a trade-off with a foreign government, with the price impact passed back to the farmer.

An international merchant operated quite independently of domestic railways and other local infrastructure investment, both for the exporting and importing country. A local leased office and the chartering of a ship was all that was now required (Figure 5.1). Even merchant risks were reduced to selling-agency risks because the international merchant had no need to take ownership or title. The bulk handler or a statutory authority performed the port-accumulation task, whilst farmers paid all costs to the free-on-board (fob) stage. If governments could not finance all the operations, then the international merchant arranged the financing function, thereby assuming or complementing merchant bank operations.

FIGURE 5.1 Realities in cooperative commodity selling
Source: Stanley Toohey

Other problems with traditional bulk handling

There were many problems with traditional commingled bulk handling, other than its historical dependence on railways and governments, its dysfunction with the food supply chain, and its reliance on international merchants.

The era of early traditional bulk handling coincided with the commoditization of other industries such as heavy industries such as automobiles, trucks, tractors, and machinery.[32] This led global economies down a production-driven pathway at least until the 1950s.[33] It was assumed that customers could be found for whatever commodity was produced.

Traditional commingled bulk handling forced farmers to be commodity producers, characterized as price takers of an undifferentiated product that had little alignment to the food supply chain except as a bulk filler in food manufacturing. Under such conditions, price increases could not be expected to keep up with domestic inflation. Farmers regularly incurred commodity price deflation during times of surpluses, at the same time that input costs kept rising. The end result was a commodity trap, with falling real output prices combined with rising input costs, which provided little incentive for further production.

Traditional bulk handling had no method to preserve product origination or identity segregation, as it was not designed for differentiated product that could be traced back to its origin. Commingling was the antithesis of product segregation. It led to commoditization because large logistical systems rarely are capable of adequately differentiating a product. Logistical efficiency enforced commingling of commodity.[34]

Few commodity grades were able to be segregated by the bulk handler because of logistics restrictions. Commodity grading was therefore logistically driven and largely determined from what farmers actually delivered. Usually the farmer was not overly concerned with limited commodity grades because of on-farm blending, provided that the agronomics and genetic varieties could support such gradings. This all drove the production push towards a more uniform product.

The problem was that a commingled logistics-driven and averaged-graded product did not meet end-user and consumer specifications in an efficient food supply chain. Many governments enforced compulsion to use the traditional bulk-handling commingled system, thus removing any opportunity of aligning production with what was actually needed by end users, and destroying any incentive to differentiate output. It was the failure to address international market segment needs and JIT requirements that ultimately destroyed the foundations of mandatory traditional commingled bulk handling.[35] Potential 'trade' markets went elsewhere.

Many end users demand the matching and sourcing of product specification by district and farmer. This may have technological importance in food manufacturing or in meeting consumer demands. However, statutory selling of a traditional commingled bulk-handled wheat product forced end users to accept a non-differentiated, unsegmented, and standardized product, generally referred to as 'fair average quality', which in Australia lasted until 1974, and which had no 'guide to baking value'.[36] The subsequent manufactured product was mostly mundane, accepted by a few consumers, and tolerated by the remaining consumers mostly because their ignorance, especially when international travel was only in its infancy and consumers could not take delight in what was being offered overseas. With greater international trade and travel, consumers became more discerning by the late 1960s.

The objective of traceability might be disease control or contamination issues, health factors, or simply to know the origin of products from a global perspective.[37] As well, preserving a farmer's identity with a product is becoming increasingly important in tracing products back to their production source. There is little indication as to how traditionally rigid commingled bulk-handling methods can readily adapt to these new consumer demands and be more relevant in a modern food chain environment.

Bulk or bag stacks were equally dependent on hygiene. The mere conversion from bags to unsealed bulk-handling facilities did not solve the hygiene problem; rather it potentially worsened the problem with insect, rodent, and bird infestations, moisture build-up, mould, biochemical changes, weather damage, whilst increasing temperatures could ignite fires or explosions.

Fumigation via organophosphates had been made mandatory for bulk-handling facilities in many countries during the 1950s in an attempt to control weevils, borers, beetles, and moths. Early control successes slowly evaporated through the perennial problem of increasing insect resistance to chemical treatments. Higher doses and more frequent fumigation offered only a short-term solution until another chemical was found.[38]

The structural weakness of traditional bulk handling was that they could not be hermetically sealed. There were alternative methods available, but these required vacuum sealing. The general lack of adoption of new technologies through inadequate investment capability generally prolonged out-dated and ineffective methods of insect control in traditional bulk handling facilities. When a range of new non-chemical alternatives became available, the rigidity of the traditional bulk handling method prevented their adoption.[39]

Whilst chemical treatment was being prolonged in out-dated bulk handling facilities, there was an increasing need for global supervision and control over the use of chemicals in insect fumigation. Organizations such as the Food and Agriculture Organization, World Health Organization, and various medical research councils and institutes, developed exponentially under chemical fumigation used in bulk grain handling facilities.[40] The integrity of these public organizations is required to ensure that the public health and safety of consumers are not compromised.

Even as late as the 1960s, insect infestations were still creating enormous problems for bulk handling both within and between countries. Up to 80 per cent of Australian grain that was shipped to the UK was arriving with insect infestations, which brought the wrath of the London Corn (Grain) Traders Association.[41] The blame could be laid all along the supply chain including the export country, the hygiene of ships, the conditions at ports of call, and during unloading procedures.

Farmers could hardly be expected to carry such blame, but they were expected to pay the cost up to when the commodity title was transferred, which could be at either the origin or destination port. They bore the supply chain risks and responsibilities, and incurred price discounts for damaged commodity. Importers only had responsibilities after title had been transferred.

It is questionable as to why all this responsibility should be the function of farmers, or even end-users. Until the introduction of traditional bulk handling, all these risks and responsibilities lay with merchants. Traditional bulk handling was to seriously distort the function and role of supply chain intermediaries, and cause immense misallocation of human resources and capital.

From a consumer–consumption perspective, food contamination issues through insecticide fumigation residues were publicly known and discussed by UN organizations, but generally ignored because of the need to feed the global population, and the unwillingness to confront the real problem of global food supply chain dysfunction caused by traditional commingled bulk handling. In Australia, there is continual government monitoring of export grains chemical residues through the National Residue Surveys.

Traditional bulk handling was regionally centralized, usually at railheads or major rail sidings. The whole production from a region was forced to unload in one intake centre. As well, seasonal harvesting seriously concentrated intake for only two weeks of the year for grains, oilseeds, and pulses, yet for the other 50 weeks of the year, the bulk intake resources lay idle.

Early traditional bulk handling intake was originally designed for the slow horse and cart, with bags being unloaded into bulk hopper bins by hand. As external road transport improved with bigger, faster trucks, nothing much changed for traditional bulk handling intake design or operations. Each year, as mechanical harvesting got faster and delivery trucks got bigger, the external logistics became more efficient, but the queues at congested traditional bulk handling centres got longer as intake became more inefficient.

Despite the seemingly large structures, traditional bulk handling elevators and silos usually operated with a single-entry weighbridge, single-sampling and inspection stations, and single-bin unloading facilities. The weighing, sampling, inspection, and subsequent documentation, with many administrative inefficiencies, caused immense delays with long queues of fully laden trucks exposed to the external elements, sometimes for days. Queues were more likely to form at crucial harvesting periods when weather conditions might cause the difference between a successful harvest and failure. Farmers then were at the mercy of the bulk handling administrators.

Harvesters could lay idle for 24 hours or longer while trucks waited for unloading at congested traditional bulk handling depots. When rain or storms hit mature crops while trucks waited in queues, production from whole areas was destroyed or downgraded. In Australia, insurance was the responsibility of the farmer, but usually it was prohibitive under the circumstances. When double-vehicle trucks began arriving in intake areas during the 1980s, governments were powerless except to privatize public monopolies and deregulate previous government-controlled supply chains.

If traditional bulk handling caused the demise of domestic merchants, the impact on food supply chain end users was equally devastating. A wedge was driven between the farmer and the end user through mandatory commingled bulk handling, which destroyed existing end users and discouraged the development of local cluster industries (Figure 5.2). Traditional bulk commodity handling regulators did not allow integration with other supply chain intermediaries.[42]

(a)

(b)

FIGURE 5.2 Supply chain deregulation contrasted with supply chain regulation (a) Brazilian supply chain clustering (b) Lonely silo under Australian regulation
Source: John Williams

The starkness of the lonely silo silhouetted against the horizon in Western Canada and Australia under mandatory traditional bulk handling contrasted with the hub of local post-harvest industry complexes and cluster industries in the USA and Brazil under non-regulation over the same period. This contrast was truly perplexing.

What made matters extremely worse was that end users and food manufacturers disappeared from towns in isolated marginal production regions. The Eyre Peninsula of South Australia lost a flour mill and local feed-milling industries after mandatory traditional bulk handling was introduced.[43] The demise of these previous end-user industries in isolated regions and the importance of cluster industries were conveniently forgotten, with retrenched labour often accepting military deployment in regular world wars. Potential investment went elsewhere.

The overall impact caused by the destruction of local end-user cluster industries was to make agricultural output in countries such as Australia and Canada almost entirely dependent on exports. It was a litany of food supply chain disasters, starting with railways only going to ports, which then caused the rise of dysfunctional traditional commingled bulk handling, followed by the destruction of end-users and other intermediaries in the food supply chain, followed by governments being forced into government-to-government transactions to maintain the whole supply chain disaster.

Only since Australian domestic wheat deregulation (1989) has the re-intensification of its animal feed industry occurred to an extent that was comparable to overseas developed countries. Some of this activity might be considered mere coincidence, but again the post-deregulatory cluster industrialization period stood in stark contrast to the period of statutory commingled bulk handling and monopoly selling of grains mostly to export markets.

When commodity food producing countries were forced into export expansion because of the quick demise of local end-user cluster industries under regulation, there was little or no consideration from the perspective of the overseas end user. When importing governments were forced into bulk commingled handling through foreign exporting country pressure, there was usually a quick demise of end-user cluster industries in the importing country. This then resulted in greater government supply chain intervention in importing countries. The current supply chain situation in many importing countries, as well as in many previous export countries, owes its origin to the evolutionary demise of once healthy food supply chains and cluster industries caused by the introduction of traditional commingled bulk handled commodity methods.

Importing food supply chains quickly became not just dependent on government, but sometimes the government was forced to take ownership through nationalization of the whole commodity food supply chain just to keep it operational. However, centralized bureaucracies required logistically based product simplicity and consumer behaviour uniformity, otherwise inefficiency and bottlenecks caused food supply shortages and civil unrest. The alternative was product segregation and recognition of consumer behavioural differences, but this invited complexity and decentralized logistical operations, which bureaucracies were ill-equipped to provide.

When this intervention inevitably failed either through misalignment, mismanagement, or corruption, there were calls for international 'aid'. This has had serious ramifications in many countries in Africa where previously there was general self-sufficiency in food and relatively efficient supply chains, but now there are food shortages.

Under the bagged selling method, the farmer was not directly responsible for subsequent supply chain costs and risks that were borne by domestic merchants, although the local price had to reflect these costs and risks. Merchant competition ensured that supply chain costs could be kept minimal.

When traditional commingled bulk handling was introduced with port-centric railways, the monopoly organizational structure had no competition or contestability to minimize costs and often no cost transparency. Farmers were forced to pay the supply chain costs regardless of the system efficiency. It was often falsely assumed that because bulk commingled handling had cost advantages over the bagged method, there was no need to be concerned over cost control.[44] Even by 1927, farmers were complaining that they increased their efficiency, but the monopoly bulk handler did not improve its efficiency in management, handling, or storage.[45]

As early as 1920, the West Australian Grain Growers' Cooperative Elevators Ltd. had doubts about real savings of the bulk handling method.[46] Major benefits of traditional bulk commingled handling had accrued to the railways through improved economies of scale, with decreasing unit cost as quantities increase, as well as increased cost–time efficiency compared with bags. However, there had been little consideration as to the impact of traditional bulk handling methods on farmers, end users, port operators, shippers, and importers, both from a cost and quality perspective, and product specification from the end-user perspective.[47]

There may have been some bulk handling port infrastructure in importing countries, but this was to no avail when buyers had to transport bulk product through narrow congested streets and over pot-holed dirt roads, and when end users had no bulk unloading facilities. The commingled commodity therefore usually had to be bagged in the bulk handling import terminal, before being stacked onto small trucks.

Labour costs in these importing countries during much of the twentieth century enabled bagging by hand to be cost effective. However, rising labour costs and increasing food cost inflation threatened the traditional port-based bulk-to-bagging handling methods of importing countries, which were increasingly inefficient through decades of either little or haphazard investment.

Roads in importing countries were becoming more suitable for 20-tonne shipping container transport, which many end users were already using for just-in-time output. Ports were becoming more decentralized. The only impediment for expanded shipping container usage and greater cost efficiency was government mandated traditional bulk handling rigidity.

Traditional bulk handling encouraged cost averaging in operations,[48] which usually resulted in the misallocation of resources. 'Inefficient facilities were over-utilized whilst efficient facilities were under-utilized.' Cost averaging resulted in 'substantial cash transfers between farmers, which resulted in some producers subsidizing others'.

Farmers who were near export ports subsidized more distant farmers, whilst some regions subsidized others, all of which was not disclosed to farmers.[49]

Under traditional bulk handling, commodity did not 'flow along least-cost pathways'.[50] It 'prohibited or impeded the use of some services' that might have lowered costs because there was no organization to assume 'responsibility of taking decisions' in the supply chain. Traditional bulk handling usually existed in a vacuum devoid of domestic merchants whose function was to manage the local supply chains.

The traditional bulk handling method did not meet benchmark 'criteria in economic efficiency and cost effectiveness'.[51] Instead it was unnecessarily propped up by government support for many decades, usually by statutory selling authorities, which cemented the systemic inefficiencies and dysfunctions. It is still being propped up in many importing countries, which are now struggling to reverse their current undeveloped status.

If traditional bulk commingled handling grew out of railway efficiency measures, serious reviews should have been undertaken of traditional bulk handling methods when railway inefficiencies rapidly occurred by the 1930s. Being bundled with railway inefficiency and dysfunctionality in the food supply chain, the demise of traditional commingled bulk handling should have occurred in the 1950s.

Despite the first bulk handling facility in Australia being completed in 1918, 40 years after their introduction in North America, up-country bulk handling facilities were rarely fully utilized even by the 1930s, which created problems in public debt repayment. Many farmers wanted a selling method choice without the compulsion of delivering into a selling system that lacked cost transparency, contestability, or benchmarked efficiency.

By 1933, only 50 per cent of the wheat produced in New South Wales was handled in bulk, yet there were mounting public debts relating to bulk handling infrastructure and an inability of farmers to fund these debts.[52] In response to this increasing debt problem, the Australian government used war-emergency powers in September 1939 to take control of the whole domestic wheat supply chain through the Australian Wheat Board. It was debt repayment problems rather than any endeavour to rectify the problems of traditional commingled bulk handling that drove much of the intervention into the food supply chain by statutory authorities. But it was to be wheat farmers and not government who were made responsible for this debt repayment.

Location accessibility was one early problem of traditional bulk handling, but it was compounded by the reluctance or inability of farmers to provide funding for bulk infrastructure. The introduction of statutory selling authorities forced all farmers to pay for the traditional bulk commingled handling infrastructure, regardless of location accessibility and geographical differences. There were to be no 'free riders'. This was despite storage in the food supply chain not being the function of the farmer, and despite rail-centric traditional bulk commingled handling being a monolith imposition on the food supply chain. Because a few farmers were conveniently elected to these statutory authority boards, albeit under much rural politicking, some farmers deluded themselves that these authorities existed for the benefit of farmers. That delusion was often shattered when farmers were levied or received reduced financial returns.

Traditional bulk handling generally transferred supply chain costs and responsibilities from the local merchant to the farmer. There was nobody left but the farmer to pay for traditional bulk handling infrastructure and logistical operations after the demise of the domestic merchant. Supply chain debts did not simply disappear. Sometimes consumers were taxed on flour purchases, particularly when governments assumed responsibility over flour mills. However, both governments and taxpayers generally did not perceive the commodity food supply chain costs to be their responsibility.

Government policy during much of the twentieth century involved controlling inflation, countering balance of payments problems, and attempting to reduce public debt. Having commodity food surpluses ensured that these policies could be achieved by reducing food inflation, creating international financial reserves through exports, and forcing farmers to repay public debt on bulk handling, railways, and port infrastructure. Hence, there were frequent production-push promotions by many governments.

The implication of such government policy was that the bulk handling system would cope with production increases without the necessity of major fiscal investment. Commodity chain problems inevitably occurred either because of huge congestion at existing bulk handling facilities that were under-resourced, or production occurred in regions with few bulk handling facilities because of investment inertia or insufficient time to erect new facilities. The reality was that traditional bulk handling could never meet short-term macroeconomic goals of government because of its design.

The impact of frequent government production-push promotions on farmers' economic viability was devastating. There was no thorough investigation as to whether commodity sales could absorb the subsequent surpluses from such production-push policies. Commodity prices disadvantaged farmers,[53] and agriculture was the one industry sector that continually suffered falling real prices of output as deflation. The end result of induced surpluses was prolonged low prices and poverty amongst farmers. Governments were then required to intervene with price support or income subsidies, particularly when the discipline of returned soldiers after both world wars demanded it through political action. Both the EU and USA responded to surpluses and prolonged low prices by heavily subsidizing agricultural production, which only perpetuated low prices and production disincentives for other farmers.

Some producing countries responded with government 'price stabilization' schemes (1950s) and production quotas (1960s). 'Solutions' were still being sought via government intervention.[54] There was little attempt by governments until the 1970s to find alternate non-interventionist monetary solutions such as financial deregulation of exchange rates and interest rate policies, which ultimately were to cause a profound change in how governments perceived intervention into agricultural commodity supply chains. Some farmers were slow to perceive this change through their delusion that the government was always acting in their best interest.

Under government-mandated traditional bulk handling, there was no competition because of the single controlling statutory authority. Costs were 'higher than they otherwise should have been' because of 'restricted competition', whilst monopolized services were not aligned to meet 'the requirements of users'.[55] The regulated industry organizational structure rigidity associated with traditional bulk handling 'tended to be perpetuated, thus making the system less capable of responding flexibly to changing market circumstances'.

Without competition, there were increasing problems over cost-management performance efficiency and no audit benchmarking opportunities, whilst innovation stagnation caused new investment indecision. There was scarce finance provisioning due to asset insecurity and non-ownership of product. A general lack of motivation amongst traditional bulk handling management staff enforced the status quo, particularly when many of the staff were casuals due to activity being seasonal, while permanent staff could be retrenched during drought periods.[56] There was 'little contestability' of most service providers, which usually resulted in higher administrative and operations costs. It was 'very difficult for bulk handling authorities to create the types of incentive structures found in competitive markets'.

The farmer had no choice other than to deliver commodity to the nearest available up-country bulk handling depot. A local monopoly intake situation occurred first up-country, but this monopoly was then usually extended through the whole commodity logistics chain, via the bulk handling system, railways, exporting port, and finally to another monopoly entity at the importing port.

Similarly, end users in both the domestic and foreign countries were usually forced to buy from a designated railway-port centric depot that had little logistics functionality and product alignment for end usage. Logistics competition and contestability in the supply chain from farmer to port through to the importer were therefore largely non-existent.

The demise of traditional bulk handling

Both the second Canadian Wheat Board (1935) and the Australian Wheat Board (1939) had barely mandated compulsory pooling through their traditional bulk commingled handling systems when serious threats to this bulk handling procedure emerged. There were storage bunkers used at Fremantle in Western Australia during the Second World War,[57] but these were very exposed and rudimentary with tarpaulins. However, grain was bunkered during the Second World War in both Argentina and Uruguay using deep pits lined with bricks or straw and wood, or a soil cement mix, and roofed with bitumen-treated sheeting and often covered with soil.[58]

It was easy for the traditional bulk handling authorities to dismiss such threats initially because of the weaknesses in the hermetic sealing process and in the grain handling procedures. However, it was more difficult for traditional bulk handling authorities to accept that the technological development and innovative processes of alternative bulk handling methods were to quickly overcome the previous problems to an extent where they could pose a real threat to the traditionalist's very existence and surpass them in both efficiency and effectiveness.

Hermetic sealing refers to the creation of airtight spaces that prevent the entry of biological organisms, similar to tin cans. Although hermetic sealing dates from the seventeenth century, it was not used in the traditional method of bulk handling in the nineteenth and early twentieth centuries.

Research into underground storage of food grains in the 1950s[59] continued during the 1960s[60] until the mid-1970s.[61] When the prevailing fuel crisis stimulated new innovations in the petrochemical materials industry during late 1970s, bunker storage reverted from below ground back to above ground.

After several Middle Eastern wars, Israel experimented in 1979 with above-ground levelling with three-sided earth ramps, lined with polyethylene sheeting.[62] After loading the grain in from the fourth side, thick PVC formulated sheeting was used as an over-liner. The liners were overlapped, folded, and buried in a trench to form a hermetic seal. After 15 months' storage of 15,567 tonnes of wheat, there was minimum mould damage (0.06 per cent) while insect damage was only 0.15 per cent.[63]

These exceptionally low damage percentages when compared to traditional bulk handling methods were achieved through no chemical fumigation treatment, which was a cost saving. As well, there was no chemical residue threat and no resistance build-up for insects. Other advantages of hermetic storage bunkers included the low infrastructural capital investment cost, the ease of in-loading and out-loading with less congestion at harvest time provided that there were ample weighbridges and sampling inspection points.

Traditional bulk handlers were increasingly under threat because of the higher capital investment cost, much higher operational expenditures, reliance on expensive chemical fumigants, increasing insect resistance, and chemical risk on foodstuffs. These bulk handlers were continually being plagued by congestion queues that reduced the efficiency and often productivity of local farmers.

There were still many initial problems with above-ground bunkering including handling losses, contamination, hermetic seal breakage, and water seepage through damaged plastic that was often caused by birds, vermin, wild animals, and hail damage. However, new material processes thickened and strengthened the outer plastic over-lining, which could be double or trebled lined.

Aeration equipment added a new dimension to bunkering by being able to dry commodities that were delivered with high moisture, and control the commodity temperature during maintenance procedures. This occurred very cheaply without huge capital investment in expensive aerated storage buildings and without on-going high operational costs that were usually associated with traditional bulk handling methods.

Hermetic bunkering opened up new food supply chains based on organic marketing. It began just when consumers were questioning the necessity for chemically treated foodstuffs, and when scientists were becoming increasingly concerned over insect resistance to previous chemical fumigants.

Hermetic sealing using plastic liners in shipping containers for specific bulk product created a new international supply chain that connected previously hidden global food markets to the outside world. This method required minimum financing, little capital investment, relatively low cost, and encouraged a myriad of new competitors to international food supply chains. The motivation and initiative of domestic merchants with their food supply chain functions and skills was rekindled in deregulated supply chains.

The competitive advantage of international merchants during the 1960s in the chartering of bulk carrier ships was removed in less than 40 years when these large carriers became largely obsolete in the food supply chain. Financing these shipments, quality, and commingling were the biggest destroyers of the Panamax ships for freighting commodity foods. Just-in-time product-specific segregated shipping container trade undermined this bulk commingling and reignited international food supply chains. Traditional international commodity merchants were overtaken by more end-user-driven logistics operators and domestic merchants integrated within local 'trade' markets.

Sealed shipping containers also had the added advantage of delivering product packaging flexibility. There could be hermetic sealing within the lining of a shipping container, or bags could be stacked on pallets and hermetically sealed before lifting into a shipping container. As an added measure, each individual bag could be hermetically sealed. Mechanically palletized bagging induced cost savings. The traditional bulk handling and statutory selling methods never could provide this product packaging flexibility or hygiene control to the final customer when they required fumigation at all stages of the supply chain, and especially when bulk commodity had to be bagged by hand in the importing country.

There have been problems, but these may be classified as normal supply chain risks. Product quality caused by agronomic-harvesting problems is not going to be improved through hermetic containerization. Meeting customer–product specification has to be treated in operations management quite independently from a mere logistics improvement.

Suitable food container availability, sourcing, and back-freighting of empty containers are ongoing global problems. Normal arbitrage between surplus and deficit regions may not occur because the container price differential is frequently smaller than the transport cost. This creates huge container stacks in one country while others seek desperately for such food containers. As well, containerization does not make shipping times and freight scheduling issues disappear.

This all assumes that container ports exist that are accessible by road. The legacy of a century of rail-bulk-centric ports has discriminated against the provision of port containerization and associated infrastructure in many countries. Governments are frequently reluctant to establish separate specialized container ports as national infrastructure, often preferring private investment. This is particularly the case when governments prefer social debt to infrastructure debt, with the outcome generally contributing to lower productivity and efficiency with higher unemployment and social cost.

Mandatory traditional bulk handling acted as a disincentive for farmers to install small-sized sealed on-farm storages for vermin and humidity control. Traditional bulk handling prevented end users from transmitting supply chain signals to farmers about when and how to store on farm. As well, compulsion to use the traditional bulk handling facilities necessitated a cost that was additional to the on-farm storage cost.

The vested interest for farmers in using on-farm hermetically sealed storages came through storing seed. Any problems with seed quality were going to cause an immediate agronomic reaction when planting subsequently occurred. Hence the farmer was simultaneously both a producer and end user with quality performance being paramount. Grain hygiene standards generally did not fall significantly with on-farm storages under supply chain deregulation, as earlier predicted.[64]

There is an increasing market for organic non-fumigated hermetically sealed stored differentiated product, in contrast to unsealed fumigation-stored commingled commodity. As well, quality-certified farmers can sell their product to local merchants or end users, and then have the opportunity to store on-farm on their behalf, receiving storage rent in exchange. The longer-term benefits for farmers investing in meeting minimum supply chain quality assurances for specific end users usually far exceeds the costs.

The real advantage to farmers with hermetically sealed storage is that the 'trade' markets that exist between end users and domestic merchants have now been extended to farmers. Being included in the 'trade' market can result in end users being more willing to compensate farmers for storage costs.[65] Sealed on-farm hermetic storages have opened up new avenues for farmer discourse directly with end users.

Apart from earlier research on underground storages, Argentina was also to lead the development during the 1990s with on-farm silo bags, each with a capacity of between 200–250 tonnes. Specially designed machinery enabled the bags to be loaded directly from the harvester and hermetically sealed. Once again, this is transforming traditional agricultural supply chains, logistics methods, and food supply chains.

The advantages of storage bags include the sealing of an organic product that could be sold directly through the organic food supply chain, full traceability with no commingling, identity preservation, grade and product segmentation can occur, no chemical fumigation residue, or insect resistance issues, quality control as long as the hermetic seal was unbroken, avoidance of the high traditional bulk handling and warehousing costs, avoidance of harvest congestion at bulk handling depots, lower off-peak transport costs, and it eliminated the necessity for high capital investment in on-farm sealed silos, which particularly suited farmers leasing rather than owning land. Only a 12-volt battery is required to form the hermetic seal.

Early problems were similar to those of the hermetically sealed bunkers, particularly the damage caused by birds, vermin, wild animals, and hail to the over-liner. Both vermin control and thickened PVC have minimized these problems, but there may have to be further improvements. There was the added machinery cost both to load and unload the silo bags. Some quality issues have occurred with 'bad odours' in silo bags as well as sealed bunkers, and these problems need to be addressed. But the advantages generally have proven greater than the disadvantages.

On-farm hermetically sealed storages have occurred in an age of telemetry where distant measurement of moisture, humidity, and temperature can occur through metering from remote offices. Both sealed silos and silo bags were to synergize with the new shipping container market. Farmers now have more control over both the product and the supply chain, and more supply chain opportunities than ever occurred under the old traditional bulk commingled handling methods.

Notes

1 Whitwell, Greg and Diane Sydenham, 1991, *A Shared Harvest – The Australian Wheat Industry 1939–89*, Macmillan Education Australia, Melbourne, Ch. 11.
2 International Wheat Council, 1980, *Problems in Grain Handling and Transportation*, London.
3 Whitwell, Greg and Diane Sydenham, 1991, *A Shared Harvest – The Australian Wheat Industry 1939–89*, Macmillan Education Australia, Melbourne, p. 266.
4 McColl, Jim, 1988, *Royal Commission into Grain Storage, Handling, and Transport*, Australian Government Publishing Service, Canberra, p. xxv.
5 Whitwell, Greg and Diane Sydenham, 1991, *A Shared Harvest – The Australian Wheat Industry 1939–89*, Macmillan Education Australia, Melbourne, Ch. 7.
6 Kingsman, Brian G., 1985, *Raw Materials Purchasing: An Operational Research Approach*, Pergamon Press, New York.
7 Heifner, Richard G., 1967, *Managing Seasonal Grain Inventories*, Michigan State University Press, East Lansing.
8 Beuthe, M., Himanen, V., Reggiani, A. and L. Zamparini, Eds., 2003, *Transport Developments and Innovations in an Evolving World*, Springer-Verlag, New York.
9 Karmarkar, Uday, 1989, *Getting Control of Just-in-time*, Harvard Business Review, 67:5, 122–131.
10 Webb, G.R., 1963, *The Closure of Uneconomic Rail Lines in Australia*, Economic Record, 39:88, 468–476.
11 Larson, Donald W., Gallagher, Paul W. and Reynold P. Dahl, Eds., 1998, *Structural Change and Performance of the U.S. Grain Marketing System*, Department of AED Economics Paper, The Ohio State University, Columbus, OH. Reviewed by Ginder, Rogers, 2000, *Agribusiness*, 16:4, 503–505.
12 Canadian International Grains Institute (CIGI), 1993, *Grains and Oilseeds: Handling, Marketing, Processing*, Winnipeg, Vol. 1, 4th Ed., Ch. B-1.
13 Zekulich, Michael, 1997, *The Grain Journey – The History of The Grain Pool of WA*, The Grain Pool of Western Australia, Perth, WA, pp. 5–6.
14 Zekulich, Michael, 1997, *The Grain Journey – The History of The Grain Pool of WA*, The Grain Pool of Western Australia, Ch. 1.
15 Canadian International Grains Institute (CIGI), 1993, *Grains and Oilseeds: Handling, Marketing, Processing*, Winnipeg, Vol. 1, 4th Ed., pp. 106–107.
16 Zekulich, Michael, 1997, *The Grain Journey – The History of The Grain Pool of WA*, The Grain Pool of Western Australia, Perth, WA, p. 12.
17 Canadian International Grains Institute (CIGI), 1993, *Grains and Oilseeds: Handling, Marketing, Processing*, Winnipeg, Vol. 1, 4th Ed., p. 106.
18 Hughes, Alex and Suzanne Reimer, Eds., 2004, *Geographies of Commodity Chains*, Routledge, London.
19 Zekulich, Michael, 1997, *The Grain Journey – The History of The Grain Pool of WA*, The Grain Pool of Western Australia, Perth, WA, pp. 24–26.
20 Zekulich, Michael, 1997, *The Grain Journey – The History of The Grain Pool of WA*, The Grain Pool of Western Australia, Perth, WA, p. 14.
21 Myers, Robert J., Richard J. Sexton and William G. Tomek, 2010, *A Century of Research on Agricultural Markets*, American Journal of Agricultural Economics, 91:2, 376–402.
22 Brennan, D., 1995, *Deregulation and Efficiency in Grain Transport*, Australian Journal of Agricultural Economics, 39:2, 129–141.
23 Rau, A., 1957, *Agricultural Policy and Trade Liberalization in the United States, 1934–1956: A Study of Conflicting Policies*, Librairie E. Droz, Geneva, pp. 93–121.

24 Modelled by John Strachey who was the post-war British Minister of Food and then the Secretary of State for War in the early 1950s.
25 Friedmann, Harriet, 1994, *The International Relations of Food*, in Barbara Harriss-White and Sir Raymond Hoffenberg, Eds., *Food*, Blackwell, Oxford.
26 Friedmann, Harriet, 2002, *The International Political Economy of Food: A Global Crisis*, in Carol Counihan, Ed., *Food in the USA: A Reader*, Routledge, New York.
27 Whitwell, Greg and Diane Sydenham, 1991, *A Shared Harvest – The Australian Wheat Industry 1939–89*, Macmillan Education Australia, Melbourne, Ch. 12.
28 Canadian International Grains Institute (CIGI), 1993, *Grains and Oilseeds: Handling, Marketing, Processing*, Winnipeg, Vol. 1, 4th Ed., pp. 116–117.
29 Whitwell, Greg and Diane Sydenham, 1991, *A Shared Harvest – The Australian Wheat Industry 1939–89*, Macmillan Education Australia, Melbourne, pp. 36–39.
30 These domestic merchant functions will be discussed in Chapter 10.
31 Morgan, Dan, 1979, *The Merchants of Grain*, The Viking Press, New York.
32 FitzSimmons, M., 1986, *The New Industrial Agriculture: The Regional Integration of Specialty Crop Production*, Economic Geography, 62, 334–353.
33 Kenny, Martin, Lobao, Linda M., Curry, James and W. Richard Goe, 1989, *Midwestern Agriculture in US Fordism: From the New Deal to Economic Restructuring*, Sociologia Ruralis, 29:2, 131–148.
34 Bureau of Transport and Communications Economics, 1992, *Relative Efficiencies in the Transportation Of Commodities*, Australian Government Publishing Service, Canberra.
35 Whitwell, Greg and Diane Sydenham, 1991, *A Shared Harvest – The Australian Wheat Industry 1939–89*, Macmillan Education Australia, Melbourne, Ch. 13.
36 Whitwell, Greg and Diane Sydenham, 1991, *A Shared Harvest – The Australian Wheat Industry 1939–89*, Macmillan Education Australia, Melbourne, pp. 264–274.
37 Harvey, Katherine, 2004, *Supply Chain Analysis of Traceability Systems within the Grain Industry*, Department of Agricultural Sciences, La Trobe University, Melbourne.
38 Canadian International Grains Institute (CIGI), 1993, *Grains and Oilseeds: Handling, Marketing, Processing*, Winnipeg, Vol. 1, 4th Ed., Ch. B-10.
39 Whitwell, Greg and Diane Sydenham, 1991, *A Shared Harvest – The Australian Wheat Industry 1939–89*, Macmillan Education Australia, Melbourne, p. 114.
40 Whitwell, Greg and Diane Sydenham, 1991, *A Shared Harvest – The Australian Wheat Industry 1939–89*, Macmillan Education Australia, Melbourne, p. 110.
41 Whitwell, Greg and Diane Sydenham, 1991, *A Shared Harvest – The Australian Wheat Industry 1939–89*, Macmillan Education Australia, Melbourne, p. 111.
42 McColl, Jim, 1988, *Royal Commission into Grain Storage, Handling, and Transport*, Australian Government Publishing Service, Canberra, p. xxv.
43 Eyre Peninsula Development Board library sources.
44 Whitwell, Greg and Diane Sydenham, 1991, *A Shared Harvest – The Australian Wheat Industry 1939–89*, Macmillan Education Australia, Melbourne, Ch. 11.
45 Marriott, Richard S. and Gus McLean, 1991, *Going with the Grain: A History of Queensland's State Wheat Board*, The State Wheat Board, Toowoomba, p. 50.
46 Zekulich, Michael, 1997, *The Grain Journey – The History of The Grain Pool of WA*, The Grain Pool of Western Australia, p. 24.
47 Whitwell, Greg and Diane Sydenham, 1991, *A Shared Harvest – The Australian Wheat Industry 1939–89*, Macmillan Education Australia, Melbourne, Ch. 11.
48 McColl, Jim, 1988, *Royal Commission into Grain Storage, Handling, and Transport*, Aust. Gov. Publ. Service, Canberra, p. xxv.
49 Whitwell, Greg and Diane Sydenham, 1991, *A Shared Harvest – The Australian Wheat Industry 1939–89*, Macmillan Education Australia, Melbourne, p. 149.
50 McColl, Jim, 1988, *Royal Commission into Grain Storage, Handling, and Transport*, Aust. Gov. Publ. Service, Canberra, p. xxv.
51 McColl, Jim, 1988, *Royal Commission into Grain Storage, Handling, and Transport*, Aust. Gov. Publ. Service, Canberra, p. xxv.

52 Whitwell, Greg and Diane Sydenham, 1991, *A Shared Harvest – The Australian Wheat Industry 1939–89*, Macmillan Education Australia, Melbourne, p. 97.
53 Based on the skewness in Figure 1.3 in Chapter 1.
54 Whitwell, Greg and Diane Sydenham, 1991, *A Shared Harvest – The Australian Wheat Industry 1939–89*, Macmillan Education Australia, Melbourne, Ch. 9.
55 McColl, Jim, 1988, *Royal Commission into Grain Storage, Handling, and Transport*, Aust. Gov. Publ. Service, Canberra, p. xxv.
56 McColl, Jim, 1988, *Royal Commission into Grain Storage, Handling, and Transport*, Aust. Gov. Publ. Service, Canberra, p. xxv.
57 Zekulich, Michael, 1997, *The Grain Journey – The History of The Grain Pool of WA*, The Grain Pool of Western Australia, p. 39.
58 Navarro, Shlomo and Ezra Donahaye, Eds., 1993, *Proceedings of an International Conference on Controlled Atmosphere and Fumigation in Grain Storages – Winnipeg*, Caspit Press, Jerusalem.
59 Hall, D.W., Haswell, G.A. and T.A. Oxley, 1956, *Underground Storage of Grain*, Colonial Research Studies, No. 21, HMSO, London.
60 Gilman, G.A. and R.A. Boxall, 1974, *The Storage of Grain in Underground Pits*, Tropical Stored Products Information, 28, pp. 19–38.
61 Boxall, R.A., 1974, *Underground Storage of Grains in Harar Province, Ethiopia*, Tropical Stored Products Information, 28, pp. 39–48.
62 Navarro, Shlomo and Ezra Donahaye, Eds., 1993, *Proceedings of an International Conference on Controlled Atmosphere and Fumigation in Grain Storages – Winnipeg*, Caspit Press, Jerusalem.
63 Navarro, Shlomo and Ezra Donahaye, Eds., 1993, *Proceedings of an International Conference on Controlled Atmosphere and Fumigation in Grain Storages – Winnipeg*, Caspit Press, Jerusalem.
64 Whitwell, Greg and Diane Sydenham, 1991, *A Shared Harvest – The Australian Wheat Industry 1939–89*, Macmillan Education Australia, Melbourne, p. 115.
65 The provision for carry will be discussed in Chapter 10.

Further readings on bulk handling

Atkin, Michael, 1995, *The International Grain Trade*, 2nd Ed., Woodhead Publishing, Cambridge, Chapters 2 and 8.

Brennan, D. and R. Linder, 1991, *Investing in Grain Storage Facilities under Fluctuating Production*, Australian Journal of Agricultural Economics, 35:2, 159–178.

Broehl, Wayne G., 1992, *Cargill: Trading the World's Grain*, University Press of New England, Hanover, NH.

Canadian International Grains Institute (CIGI), 1993, *Grains and Oilseeds: Handling, Marketing, Processing*, Winnipeg, Vol. 1, 4th Ed.

Chicago Board of Trade, 1992, *Grains: Production, Processing, Marketing*, CBOT, Chicago.

Department of Agriculture, Fisheries and Forestry, 2009, *National Residue Survey 2008–2009 Export Grains Program Chemical Residue Monitoring Results*, Canberra.

Ferris, William G., 1988, *The Grain Traders: The Story of the Chicago Board of Trade*, Michigan State University Press, East Lansing.

Francis, John, 2006, *An Analysis of Grain Storage Bags, Sealed Grain Silos and Warehousing for Storing Grain*, Holmes Sackett and Associates, Wagga Wagga, NSW.

Fussell, Betty, 1992, *The Story of Corn*, Alfred A. Knopf, New York.

Hughes, Alex and Suzanne Reimer, 2004, *Geographies of Commodity Chains*, Routledge, New York.

Kaplan, Steven Laurence, 1984, *Provisioning Paris: Merchants and Millers in the Grain and Flour Trade During the Eighteenth Century*, Cornell University Press, Ithaca, NY.

Navarro, Shlomo and Ezra Donahaye, Eds., 1993, *Proceedings of an International Conference on Controlled Atmosphere and Fumigation in Grain Storages – Winnipeg*, Caspit Press, Jerusalem.

Porter, Michael E., 1985, *Competitive Advantage – Creating and Sustaining Superior Performance*, The Free Press – Macmillan, New York.
Sarris, Alexander and David Hallam, 2006, *Agricultural Commodity Markets and Trade: New Approaches to Analyzing Market Structure and Instability*, FAO and Edward Elgar, Cheltenham.
Sewell, Tom, 2005, *The Global Grain Market – From Present Day to the Year 2008/09*, Protea Publishing, Colchester.
Tye, William B., 1990, *Theory of Contestable Markets*, Greenwood Press, New York.
Viller, P., Navarro, S. and T. De Bruin, 2008, *Development of Hermetic Storage Technology in Sealed Flexible Storage Structures*, CAF Conference Paper.

6

GOVERNMENT INTERVENTION FAILURES IN EXPORTING COUNTRIES

A perception of under or over performance in agricultural exports often led to government intervention, with an inordinate optimism in the ability of public authorities to solve 'problems', despite the numerous paradoxical outcomes. Regardless of whether agricultural export intervention arose from monetary and fiscal problems or foreign objectives, it generally distorted food prices, supply chains, and decision-making behaviour. Foreign policies often led to trade isolationism, protectionism and frequently resulted in commoditized commingling of agricultural production, with subsequent dependency on bilateral government-to-government disposal transactions. The final managing director of the government-mandated export single-desk seller in Australia made a lasting observation – 'AWB (Australian Wheat Board) was afflicted by confused governance rules and confused ownership which then led to confused performance and confused culture and confused attitudes towards risk'.[1]

Government ownership of inefficient railway, port, and bulk-handling infrastructure frequently resulted in statutory intervention in food supply chains for the purpose of debt repayment through levies and taxes, but this inadvertently destroyed 'trade' markets and cluster industries. Whenever free riders used constitutionally legal methods, such as interstate road transport in Australia during the mid-to-late 1930s, to avoid such physical high-cost imposts and debt-levy repayments, governments responded with draconian national-selling authority legislation, despite the increasing failure of this response in Nazi Germany during 1933–35.[2] Farmers were 'startled at the elaborate, costly, and despotic scheme of bureaucratic control proposed'.[3]

Just as governments were coping with earlier railway intervention disasters,[4] the same governments were making new interventions directly into food supply chains. Any lessons learnt from the railway disaster and the German Reich Food Estate catastrophes were conveniently swept aside under a new wave of Keynesian interventionist economics, which was consolidated in the Cold War (1947–72)[5] through tariff protection, import–export embargoes, price controls, price 'stabilization' schemes, subsidies, food taxes, and production quotas.[6]

Protectionism and welfare

The opening up of the Mississippi Basin in the early nineteenth century as the fourth global food bowl directly threatened European economies. Instead of undertaking both economic and trade reform, most European countries had retreated into protectionism and increasing isolationism by the middle of the nineteenth century.[7] The advent of the Atlantic clipper ships was merely symptomatic of US trade expansion and increasing loss of European international competitiveness. As a result, much European unemployed and under-utilized labour migrated to the USA.

It has been argued that Bismarck's increasing state welfare and dependence on centralized government was a response to stem this migration exodus to the USA.[8] However, this belies the evidence that there had been increasing trade protectionism, isolationism, and a lack of economic reform during much of the nineteenth century in Continental Europe, which had led to high unemployment and the need for migration. Bismarck's increasing government debt and demand-stimulus social welfare payments were designed to promote economic growth and prevent stagnation that had become increasingly evident during the 1870s' European economic recession.

Decreasing international competitiveness due to rising labour costs and inflation in Continental Europe resulted in increased import dependency and loss of currency reserves.[9] As a desperate alternative to economic reform and retaliatory currency devaluations, European countries fixed the price of gold and then pegged their currency against this fixed gold price rather than a relatively strong currency such as the British pound. The gold mining industry collapsed. However, the gold standard was no substitute for trade protection if economic reform was not undertaken. Britain was not part of the European food bowl region and it chose economic reform, while most food bowl Continental countries actually increased trade barriers in lieu of reform, which led to resentment and jealousy over liberal British freer trade successes.[10]

Linking currency to a fixed gold price was merely a temporary abeyance for Continental countries, with those countries tied to the gold standard staggering on until war began based on an axis of protectionism and isolationism. Much of Europe's gold stocks were then transferred to the US either as a safe haven or war-debt repayment, whilst inflation grew rampant.[11]

Despite the global war, which only increased the economic power of the USA, very little economic and trade reform occurred in Europe, while Britain lost its free trade leadership because imports had to be controlled to enable the repayment of a massive war debt to the USA. The lack of enforcement of competition resulted in private monopolies, and when they failed due to inefficiency, there was increased demand for public authorities to replace them because of the 'failure of capitalism'. If the lesson of trade-competition reform was not learnt, then economic depressions and global wars were to become inevitable, particularly when the USA began its own retaliatory trade protection policies with the introduction of the McKinley tariff in 1890 and the Smoot-Hawley tariff legislation in early 1929.[12]

The real problem is the centricity of each of the four global food bowl regions to protect themselves from illusory fears. Historically, there have been periodic economic power shifts between China, Subcontinent India, and Continental Europe, with the United States being a recent addition.[13] There is a high probability that the current economic power shift from the United States to China will ultimately shift back to India. Paradoxically, in the pursuit of food security protectionism and isolationism, China, Subcontinent India, and Continental Europe have directly caused the deaths of hundreds of millions of its citizens from severe famine and a similar number in associated wars.

This situation might be tolerable if it did not affect the rest of the world. However, because of the population dominance in the four food bowl regions, their interventionist, protectionist, and isolationist policies frequently become the norm for non-food bowl countries. The twentieth century was characterized by most countries in both the Northern and Southern Hemispheres following the four food bowl countries in their pursuit of trade protection and intervention, which then inexorably led to economic problems, the welfare state, and rising net imports including food.

Production-push policies

To ease currency pressures when currency exchange rates were fixed required government policies to promote more exports and discourage less imports. The alternative was for the Central Bank to sell domestic currency when exports were weak and imports were too strong, and to buy domestic currency when exports were too strong and imports were weak. But this required enormous Central Bank financial and gold reserves, something that had been eroding over time for trade deficit countries.

In lieu of Central Bank direct intervention to support the currency fix, there was an increasing tendency for governments during the 1930s and 1950s to push for new production and exports to overcome serious trade deficits and imbalances in their balance of payments. Correcting subsequent difficult microeconomic inefficient cost consequences in the economy generally lagged in implementation priority.

The implication was that commodities such as grain could be quickly expanded within a 12-month timeframe and the increased export revenues would narrow the trade deficit to more manageable levels to support the currency fix. A classic example occurred in the USSR where Trofim Lysenko was attempting to push-start agricultural production[14] after Molotov's disastrous farm collectivization of the early 1930s.

Under production-push policies, governments can either subdivide previous government-owned land into smaller freehold or perpetual leases, or force private freehold land to be subdivided. In extreme cases such as Zimbabwe, private freehold farmland was allowed to be confiscated by non-farmers.[15] Problems will always arise when the new leasees or owners are inexperienced farmers, particularly when the land has only marginal productivity. Urban unemployment and poverty is merely transferred into rural unemployment and poverty.

Production-push policies such as the 1918–50 soldier-resettlement schemes in some countries such as Australia dramatically affected farmers when government and private pastoral land was subdivided.[16] The problem was that the new land was very marginal in cropping productivity, which explained why it had previously been used as pastoral land. This led to an increasing array of expensive government rural welfare support and exit programmes, price support schemes, and income subsidization plans, which only prolonged the poverty of many farmers.[17]

Paradoxically, governments undertake many rural land subdivisions after good seasonal and price conditions on the delusion that such favourable conditions are now permanent. Historically, there may have been very good reason why land was 'locked up' under government control or under huge private pastoral leases. Such marginal subdivisions usually result in regrettable circumstances when good seasons or high price periods finish. Government subsidy-support dependency and unsustainability problems often then continue for decades or centuries into the future.[18]

The existence of large pastoral properties is usually an indication that such land is not profitable for subdivision. If large private pastoral properties were suitable for such subdivision, then family succession pressures would have forced such subdivision in the past. This is generally overlooked when governments decide on private land subdivision. The outcome is frequently unprofitable small farms that are not financially viable and environmentally sustainable, which then become the legacy for future generations.

Direct subsidies to farmers are usually another attempt to increase production or at least maintain existing production.[19] They still remain as the cornerstone of the US Farm Bill, funded by debt, taxpayers, and higher interest rate pressures. Usually these subsidies are paid according to farm size or average production, which results in larger farmers getting bigger relative to the smaller farmer. This can act as a countermeasure to family succession subdivisions.

It can be argued that farm subsidies are beneficial for the government through supporting surplus production and lower inflation, providing food 'security' for the home country, and having some foreign policy power over food importing countries through programmes such as Public Law 480. Farmers benefit from the external cash inflow, which then discourages spot-cash sales and encourages on-farm hoarding. This private hoarding alleviates the need for any government reserves or 'buffer' stocks in food security schemes.

The downside of farm subsidies often is prolonged surplus production, low prices, and vulnerability to poverty.[20] Farmers can then use this as leverage for lobbying for increasing farm subsidies, which has occurred under the Common Agricultural Policy (CAP) of the EU. Governments that respond to such continual cycles of subsidy pressure incur increasing fiscal debt, while farmers become increasingly dependent on the government for income. Subsidy dependency lowers individual initiative and can be a disincentive for improvements to farm efficiency and productivity, which has consequences throughout the food supply chain and on export competiveness.

If income support is made dependent on a threshold, many growers will deliberately ensure that they are below this threshold either through accountancy procedures or action to reduce agricultural production. Income support can act as a de-motivator for farm efficiency, productivity, and even less output.[21] Marginal farmers are discouraged from leaving marginal production regions, which can create long-term environmental and dependency issues.[22] Having commenced income-support payments, politically it is very difficult to wean farmers off such payments, particularly when bank loans are made dependent on such aid, and when property values are artificially supported and distorted.

Governments can attempt to encourage production by ensuring low priced farm inputs, such as payments to fertilizer and farm equipment manufacturers. In lieu of lowering import tariffs, bounties theoretically are supposed to encourage manufacturing production and lower farm production costs. The problem is that inefficient industries become even more reliant on government for income and demand even higher levels of import protection. Bounty distortions may also occur with agricultural production becoming more dependent on artificial fertilizers and less on organic soil improvement.

Statutory selling imposition

Rarely would a government impose a statutory selling authority onto a supply chain without some previous intervention failure. Successive Canadian Wheat Boards in 1919 and 1935 were introduced after earlier cooperative failures[23] due to many farmers continuing to sell differentiated product into US food supply chains. The second Australian Wheat Board in 1939 was a wartime measure to minimize collateral damage from the 1938 price 'equalization' scheme, soldier resettlement policies, marginal production-push fiascos and railway intervention failures, and the need to apply levies, recoup infrastructure debt, and ensure commodity food for troop deployment.[24] What commenced as an Australian wartime measure then continued for another 70 years, with many farmers convinced that the regulated system existed to 'protect their interests'.

The first Australian Wheat Board that operated between 1915 and 1921 arose from similar wartime food supply needs, railway-port debt recoupment, farmer cooperative failures, and supposedly 'free-rider' problems. After the Board was established, farmers then complained of inefficient accounting methods, long delays in payment, inadequate storage facilities, and government prices lower than international prices, which caused the government to disband the Board.[25]

Canada first introduced a federal Board of Grain Supervisors in 1917, which controlled the purchasing and pricing of wheat, and export destinations that were decided upon through government-to-government negotiations.[26] This Board ended abruptly in 1918 amidst uncertainty over operations inadequacies and questionable functional performance. The first Canadian Wheat Board commenced in 1919 but was authorized merely to sell wheat in export markets at 'world prices'. This differed substantially from the price controls for both domestic sales and exports of the previous Board of Grain Supervisors.[27] By 1919, the Canadian government had at least acknowledged that it had no power over export prices and that the domestic price had to reflect the global commodity market reality. It was no coincidence that the dramatic rise of US merchants such as Cargill into international status originated from this period. The government again abruptly disbanded the Board in 1920 because of a lack of government intervention justification.

In the wake of both the Australian and Canadian first board experiences came a proliferation of new cooperative selling movements that generally ignored all the failings experienced by their predecessors and post-war statutory authorities. The inherent characteristics of equality and uniformity within the cooperative membership structure were the antithesis of what was required in product differentiated supply chains. There is a conflict of goals that cannot be easily remedied under a cooperative organizational structure. The usual end result of cooperative membership policy is commingled undifferentiated price-taking commodity pooling that is usually non-aligned to any end-user demand in the food supply chain, which then entrenches poverty for the cooperative members. The inherent weaknesses with voluntary cooperative pooling, marginal production regions, and the dependence on inefficient debt-ridden railways were three commonalities for Canada and Australia.

Cooperative commodity pooling was increasingly plagued by a lack of supply chain power, trade barriers, and pool intake delivery uncertainty that was caused by both production risk and the voluntariness of deliveries.[28] Other problems included financial difficulties when advanced payments to farmers were higher than the final price, the inherent weaknesses of the traditional commingled bulk handling method bundled with increasingly inefficient railways, inadequate supply chain networks, and a general lack of global selling and business experience, skills, and resources. Administrative costs associated with pooling costs increased rapidly through inflation especially when there were no competitive benchmarks.[29]

Government intervention was increasingly perceived to be the answer for cooperative problems such as the elimination of 'free-riders' who also avoided government railway usage, the provision of convenient asset security through commodity vested ownership that guaranteed advanced payments to farmers,[30] as well as being able to centralize tax-levy collections from farmers. From a government perspective, mandated commodity pools were perceived to add government sovereign power in foreign policy negotiations, particularly at a time of trade protection retaliation and bilateral government trade-offs.

One huge problem associated with commodity pooling was commingling of different grades and quality impurities into a meaningless 'fair average quality' category.[31] Even by 1927 there were instances when higher quality was sold on the export market and poor quality was forced onto domestic end users, old crop stocks mixed with the new crop pool, blending mixed qualities, and quality selling deception, all of which caused much buyer angst and food manufacturing failures.[32]

Commodity pool commingling disadvantaged higher grade farmers, which acted as a disincentive to achieve higher grades. Many farmers added fertilizer in soils to produce higher quality, only to discover that commingling did not reward them with cost recovery. Then there were market discriminations between farmers in export-orientated production regions compared to domestic-orientated regions,[33] with financial losses having to be mutually shared.[34]

As a direct result of little demand from end users for commodity pooled product, pool administrators resorted to speculating for a higher price through hoarding, often in a surplus market, which rarely achieved anything except even lower prices, especially when commodity quality was deteriorating.[35] Farmer levies were often required to buy out the carryover from the previous old crop pool. The alternative was a lower first advance payment on the new crop so that the previous old crop pool could be paid out.[36]

In the case of Australia, there was finally an admission in 1929 that a state monopoly statutory selling authority could not set prices independent of the market,[37] which was ten years after a similar Canadian admission. By 1930, there was even an admission that statutory authorities had no control of compulsion over flour milling in the wheat market.[38] The schism between farmers wanting multi-purpose varieties and millers wanting specific milling varieties was never going to be solved by statutory selling authorities.

Climate, soils, weather, and varietal limitations often precluded suitable production that could be aligned to end-user product specifications. No amount of market 'power' through statutory selling or mandatory pooling could overcome this biological problem. Quite simply, the farmer was producing commodity that end users did not want, and low prices were the consequence. Statutory selling merely deferred the inevitable change to alternate farm enterprises, and undermined research efforts to find more suitable plant varieties for end users.

The Canadian government decided in 1935 on an amalgam of the early 1930s' Soviet and German models in the formation of the second Canadian Wheat Board. Wheat production would be vested (statutory ownership of pre-harvested wheat was similar to the nationalization in the Soviet model) to prevent cross-border sales and to enforce delivery compulsion (similar to both the German Reich and the Soviet models), and all wheat in the supply chain was to be under centralized statutory directives (similar to the Berlin-based German Reich model, but based in Winnipeg).

What differentiated the second Canadian board operations from its earlier counterpart was its commodity ownership through its vesting powers that made it a pseudo-merchant, yet it was to operate as a selling agent by receiving commission from farmers for selling into international and domestic supply chains.[39] This division of functions was an acknowledgement that the statutory board did not have the actual selling skills, networks, and experience of a merchant, even though it had the ownership functions of a merchant. As a financial backstop, any losses incurred by the Board were to be guaranteed by general taxpayer revenue.

The Canadian Wheat Board's dependency on international merchants to undertake international commodity sales was paramount to its operations. As a domestic and international agent of commodity boards, Cargill's business growth was now to be exponential.[40] Cargill grew to become a key selling agent for the Canadian Wheat Board and the Ontario Wheat Producers' Marketing Board that was established in 1958, as well as becoming a major exporter of Canadian oilseeds.[41]

In contrast, the second Australian Wheat Board appointed two global regional selling committees. The European committee used Berry Barclay in London as its selling agency, which then arranged sales with merchants[42] such as Toepfer, Continental, and Tradax, which was the European division of Cargill. For Eastern Asia, it used another committee consisting of three international merchants, namely Bunge, Louis Dreyfus, and Knipschildt-Eskelund.[43] Even by 1989 after 50 years of selling experience by the Australian Wheat Board, 40 per cent of the export surplus in the statutory wheat export pool was still sold through international merchants such as Louis Dreyfus, Cargill, and Bunge.[44] The bulk of the remainder was sold through import government tendering.

The paradox was that international merchants could sit in air-conditioned city offices with full public knowledge of domestic tonnages received, stored, and available for export, and negotiate from a position of strength direct with the statutory selling authority, with little local capital investment and no contact with the farmers. An even greater advantage was that the farmer now bore all product accumulation charges, all up-country and port storage costs, and all associated handling and transport costs to a free-on-board origin-port basis, as well as the associated risks.

Reasons why statutory authorities used international merchants included their specialized skills in the global food chains, widespread international trade networks, and usually better resources and contacts in the importing government-tendering processes. The downside for farmers was that international merchants knew export tonnages in advance of international orders and could trade in front or behind of these export orders from the statutory authority. As well, the transparency of export tonnages through statutory selling meant that competing exporting countries as well as importing countries knew exactly what was on offer. Tactical advantage in pricing was lost from a statutory selling perspective.

Public knowledge of the sales and storage data of the Canadian and Australian Wheat Boards was often used to adjust Chicago wheat futures prices as well as both on-farm hoarding and merchant warehouse storage strategies in the USA. Early knowledge of price adjustments enabled the USA to gain a competitive advantage in international sales.

This created the unusual situation whereby US farmers and international merchants wanted the continuation of statutory wheat selling authorities in both Canada and Australia because of the quantity–quality transparencies it provided, as well as the international business it transacted for international merchants. In effect, both Canada and Australia were playing a commodity game with everything on open display to its international competitors. It was a bonanza to the USA because wheat could be privately hoarded on-farm in the USA, whereas in Canada and Australia, there was full disclosure in the national pools.

Regulated selling created the illusion of monopoly selling with one seller and many buyers. The reality was a commission agency using many different sellers, usually international merchants, at a time when the number of buyers was decreasing through government import interventions. What was often perceived by farmers as 'their monopoly with selling power', was often a monopsony with one buyer with market power, and many farmer-sellers all pooled together in a surplus supply chain with no power. To worsen this selling problem associated with low market power, some buyers only purchased commodity commingled product as the 'last resort', and actually preferred such statutory accumulative arrangements as a risk management stopgap.

Ironically, there was a real danger of local 'dictatorship in management' of statutory selling authorities,[45] despite the lack of market power. Some farmers sought political power aggrandizement through statutory positions, whilst some political parties became dependent on this illusion of power. The reality in commodity supply chains was that sellers only have market power when acute shortages infrequently occur, and that this market power was likely to occur regardless of the size of the seller.

Delays in selling resulted in delayed payments to farmers because of poor demand due to product not being aligned with end usage. This had caused the demise of many early farmer selling cooperatives. These payment delays devastated a farmer's cash flow and individual debt repayment. Banks were extremely reluctant to provide pre-sale loans because of credit risk associated with the uncertainty of commodity sales and price, and pooled product ownership. Even when statutory legislation guaranteed minimum prices for Australian wheat growers in 1930, banks would not provide grower payments that were based on double current market prices.[46] Pooled commingled unsold commodity was perceived by banks as extremely poor financial security value and high credit risk even when guaranteed by government.[47]

Vested ownership powers provided some credit security for government financial guarantees and to banks when they lent payments to farmers in advance of the pooled commodity being sold. Under such commodity ownership, governments and taxpayers were then able to offer repayment guarantees to the banks to allow such pre-sale payments to farmers. However, this necessitated government and/or bank representation on the selling board to safeguard government guarantees and bank interests.

Mandatory pooling particularly when statutory authorities have vested ownership powers generates the perception amongst farmers that pricing and hedging is the responsibility of statutory authorities. If farmers perceived responsibility for both production and selling themselves, then they would be more likely to be motivated to take individual responsibility for pricing and hedging, particularly before delivery. By enforcing statutory marketing, government intervention disrupted the good pricing and hedging practices undertaken by some individual growers, thus favouring price speculative activity.[48]

Merchants in a deregulated food supply chain usually conduct integrated back-to-back transactions in 'trade' markets. In the unlikely event that back-to-back transactions do not occur, then their net exposure is usually hedged on a futures market. Statutory intervention rarely is so integrated in the commodity chain. The Australian Wheat Board normally only had 40 per cent hedge ratios including physical forward contracts, because of basis risk, currency risk, transaction-margin cost risk, uncertain production quality, illiquidity in distant months when hedging, and large market position limits imposed by futures exchanges.[49] This meant that the Australian statutory wheat pool often incurred up to 60 per cent speculation during the life of the pool.

Despite all the analyses during the period of supply chain intervention, there was still little evidence that farmers received anything more than an averaged global parity price in a falling market, while the farmer's averaged pooled price in a rising market was always suspect when compared with other non-interventionist pricing and hedging strategies. Pool prices were highest compared to other selling and pricing methods only in the year immediately following price peaks.[50]

Commercial premiums due to grade, varietal specifications, quality, shipping freight rate advantages, credit provisions, and currency relativities accrued to both statutory and non-statutory organizations.[51] So-called induced 'premiums' from price-taking commodity were either an illusion, or shipping-freight advantages that were available to all trade participants, or as a result from corruption such as the oil-for-food scandal.

Because statutory authorities merely acted as selling agents, they often became very indifferent to farmers, end-users, and consumers. Organizational administrators operating under statutory selling received sale commission on tonnages sold, which usually made them disinterested in peripherals such as the sale price achieved, the performance rendered, farmer receipt maximization, and service provision to farmers, either because there was no incentive or they lacked competitive pressure to drive service contestability. Farmer levies often generated secondary income sources.

Statutory accumulation and selling cannot occur without government regulatory directives and controlling supervision, otherwise free riders will undermine such control. However, it was constitutionally illegal to control free riders in some countries such as Australia (Section 92) because of the provision of 'free trade between the States',[52] except in wartime. This inter-state trading made enforceable by constitutional law eventually destroyed statutory wheat selling in the Australian domestic market. The forces of free trade and constitutional law ultimately won, but it took 50 years after supposedly temporary wartime legislation.[53] It was to be another 23 years before Canada disbanded its statutory wheat board.

Price control policies

Attempting to achieve monetary and fiscal 'stability' in an era of trade protection and fixed currency invariably meant some attempt to control commodity prices because of the inherent inflation problem. Statutory authorities usually have been the price administrative vehicle in an attempt to control the many outlying variables such as 'free riders'. Despite many 'draconian' methods, such statutory authorities have invariably failed because of the lack of control over biological supply, the hoarding problem, and fickle consumer demand that frequently expressed itself in civil unrest.

Food supply chain price control schemes by governments in producing and exporting countries include direct price controls, price 'stabilization' schemes, and price equalization policies. Supply diversion, production quotas and import controls usually provide secondary support for price 'stabilization' schemes, while all methods usually have some price discrimination.

There is no simple mechanism to calculate the 'correct' price, as evidenced by Taylor's theoretical meanderings.[54] The parameters to establish a theoretical price 'equilibrium' include rigidity of both demand and supply, both of which are unrealistic assumptions in a biological food supply chain.

Agricultural yield and production cannot be controlled despite known acreages. The kulaks may have been responsible for hoarding under Soviet price controls, but they were not largely responsible for crop yields that were weather determined.[55] As well, forcing consumers into standardized uniform eating behaviours with fewer product lines and less consumer choice can result in dire health outcomes, as currently evidenced in North Korea.

Price controls effectively remove price signals for farmers, merchants, end users, and consumers.[56] Farmers will be discouraged from producing and investing, and debts will accumulate without the occasional high price windfall. Consumers will be reluctant to clear surpluses from supply chains without the occasional low-price windfall.

Controlled low prices generally have impoverished farmers and their communities in the Soviet Union, China, North Korea, and India by demoralizing entrepreneurship and destroying risk-taking investment, which usually made these countries dependent on food imports.[57] There is a real danger that price controls can cause real or artificial food shortages, particularly when alternate higher priced 'black' markets are available, or when long-term hoarding can occur.

If price control administrators raise price during shortages, this encourages hoarding in the expectation of further price rises. Alternatively, price controls on the high price side could encourage over-production and place downside pressure on the price control administrators, unless there was government purchasing with subsequent food-stock quality issues. There is a no-win situation for the price control administrators in terms of determining a 'correct' price when there is little control over biological production or food consumption.[58] It frequently encourages insider knowledge, manipulation of government decision making, game planning and corruption.[59]

Governments of exporting countries with price controls could only 'guarantee' domestic prices to farmers. They had no control over international commodity prices, despite many inter-government wheat and sugar cartel attempts and failures during the late 1920s to the 1950s.[60] The most that statutory authorities could provide was a taxpayer-funded guarantee that supported a minimum price for farmers.[61]

On the demand side, price 'stabilization' schemes usually assumed relatively constant consumer demand or relatively high consumer behaviour predictability. However, on the supply side, there were usually regulations to control supply either through 'buffer' hoarded stocks, mandated commodity pooling, supply-diversion schemes, or quotas on production, usually combined with import controls. It was to be a delicate balance of opposing domestic forces, with an unknown inherent private hoarding problem. Ultimately, the pressure of international price movement combined with price distortions in the domestic food supply chain caused price 'stabilization' schemes to fail under arbitrage pressure.[62]

Usually there was a two commodity pool mechanism for price 'stabilization' schemes with one pool for domestic supply and one pool for export destination, each of which averaged prices over the period of the pool. Averaged prices were periodically paid to farmers from each pool separately. However, there was nothing stabilized merely through price averaging.

By withholding supply through a 'buffer' hoarded stock in a mandated commodity pooling system, domestic prices were forced to remain higher than if the commodity was all sold immediately. Such 'stabilization' schemes therefore discriminated against end users and consumers, which often necessitated patronage through multiple domestic pricing, which could favour some end users such as sugar for processed foods, but discriminate against retail sugar destined for consumers.

'Stabilization' schemes with their associated price discrimination outcomes were possible only when the commodity could be sold in supply chains with different elasticities of demand through different buyer responses to price, and where domestic supply chains could be isolated from exports.

Governments had a vested interest to minimize any guaranteed price so that taxpayer revenue would not be called upon for payment if prices fell. It was also usual for 'stabilized' prices to be established at the low end to control inflation, protect consumers and to prevent riots over high food prices. Commodity prices could never be 'stabilized' at the high end, much to the chagrin of farmers.[63]

In lieu of mandated commodity pooling, there could be provision for supply diversion in an attempt to control supply and support prices for farmers, but discriminate against consumers and end users. This has been used by the US government when commodity surpluses and low prices have occured. Public Law 480 provides the US legislative directive for such commodity food supply diversions, usually as international food 'aid'.

Dumping is one form of supply diversion, and usually results in force selling in an overseas supply chain at a price that is significantly lower than either domestic or international prices. It could refer to heavy discounting of poor quality commodity in international markets, such as the mouldy old-crop US corn in May–August 2010. However, it could also be used deliberately to remove competitors, or to undermine domestic producers in an overseas country and make them more reliant on food imports.[64]

Production quotas were usually implemented by governments in a desperate final attempt to control supply so that higher commodity prices could be realized for farmers, many of whom wanted high price 'stabilization'. Production could be aligned to scarce natural resources such as available water supply, limited arable land, or restricted to delivery-processing capability such as cane allocated for a sugar mill. Quotas diminished the role of price signalling in the supply chain, whilst 'stabilizing' structures led to inefficiency, commoditization and food supply chain isolation.

Paradoxically, government expansionist monetary policies usually undermined any attempts to 'stabilize' commodity prices. Land policies usually encouraged expansion, particularly greater subdivision of much larger government-leased pastoral holdings, which generally caused increased surpluses and prolonged periods of low prices. Production quotas could follow from previous government mistakes in production-push policies.

Government policy coordination in price 'stabilization' schemes was difficult, particularly from an international perspective. In 1930, the US Federal Farm Board recommended a 10 per cent reduction in wheat sowing in the US, yet the Australian government recommended an increased planting of wheat at the same time.[65] The probability of a successful outcome from price 'stabilization' schemes was therefore extremely low.[66]

The general outcome of any higher 'stabilized' prices was an incentive for farmers to increase production, which then increased commodity surpluses and imposed downside pressure on the 'stabilized' price. This tendency towards increased production surpluses undermined the very pricing mechanism designed to protect farmers. The farmer might be encouraged to increase production when international prices could be signalling to decrease production due to global surpluses.

Production quotas generally failed because of highly complex legal administration, 'free riders' and constitutional constraints, particularly in countries such as Australia under the free-trade provisions of the Federal Constitution, whereby there could be no legal restraints on individual state acreage or production that was destined for interstate trade.

Normally, 'stabilization' or home consumption price schemes required added government intervention by controlling imports through tariffs, duties or import embargoes. This enabled a higher artificial domestic price to be implemented compared with the international price. Domestic consumers could be discriminated against when compared to overseas consumers who had access to cheaper food products.

Whereas price 'stabilization' schemes generally attempted to control supply, price 'equalization' schemes attempted to 'equalize' prices through tax discrimination. Price 'equalization' schemes assume that there are no close substitutes and that consumers have inelastic constant demand that is not dependent on price, which are quite unrealistic assumptions.

In 1938, the Australian government introduced both a domestic consumption price for wheat and flour, together with a tax on flour manufacturing.[67] There was also to be a tax on wheat exports when prices were high, to be used to support farmer returns when export prices were low. The timing just before a major war could not have been more disastrous for consumers and end users, with the latter deferring investment decisions. In the disastrous aftermath, the Australian government quickly nationalized the whole wheat supply chain in September 1939 under wartime legislation.

The objective of price 'equalization' is cross-subsidization, whereby there is a cash transfer usually from consumers to farmers while decreasing the domestic–international competitive advantage for end users and restricting growth in the domestic food supply chain. The outcome for farmers is increased dependency on government and disincentives for productivity improvement. Such schemes generally fail because of the added incentives for increased production surpluses with prolonged periods of low price, thus making the 'equalized' price unsustainable without added taxation.

Export controls assistance

Commodity exports can be controlled by government through embargoes, taxes, licences, or subsidies.[68] If governments can be characterized by conflicting policy reversals in the food supply chain, historically export controls have been a multifarious and contradictory nightmare in countries such as India, Argentina, and Russia.

An export ban occurs when exports are not permitted by government. Embargoes are more severe export bans, usually involving not only complete suspension of all trade of the commodity to all countries, but also government power to seize ships suspected of undertaking illegal trade. A government could ban exports to a particular importing country for reasons of foreign policy, trade retaliation, UN sanctions, war, manipulation, corruption, or fear of domestic shortages.

Rice exports from India, Egypt, Bangladesh, Myanmar, Vietnam, and Thailand were banned in 2008, supposedly to protect domestic food supplies at a time of expected global shortage. However, from an importing country perspective, it was perceived to be hoarding by export countries in an attempt to force international prices even higher.

The effect of export bans on international bilateral and multilateral trade relationships can be quite devastating. They can destroy international food supply chain alliances, cause contract default and induce subsequent legal proceedings, distort domestic supply chains, and cause mistrust with domestic producers. Many export governments over-reacted in 2008 to what was only a short-term supply aberration in the international rice market, mostly induced through hoarding. In its wake was more incentive to hoard combined with a destruction of trust in the food supply chain and distress by consumers in importing countries.

Domestic farmers suffer from these export bans because they rely on the occasional windfall profit opportunity to repay debt with perhaps some surplus cash flow for future investment improvement, and to recoup financial losses caused by average prices not matching annual inflation. Government export bans prevent this occasional windfall profit, thereby entrapping their own farmers into commodity poverty and discouraging future production.

Alternately, the importing country can incur complete shutdowns and possible destruction of their food supply chains when exporting countries impose bans. Immediate food shortages can occur in these importing countries until longer-term alternate food supply chains can be organized or substitutes found.

Export taxes might not be as severe on the food supply chain alliances and relationships as export bans, but they can still cause major distortions to domestic supply chains and decrease export competitiveness. Domestic producers incur the brunt of commodity export taxes through lower income, because export commodity prices must reflect the prevailing international supply chain conditions. If there are no offsetting advantages through currency exchange movement, export taxes can often result in a major loss in trade and relationship trust.

If the objective of the export tax is to ensure 'adequate' domestic supply and dampen inflationary price pressures instead of using interest rates rises, then the tax is a disincentive for the farmer to produce and will more likely result in longer-term decreased production and a major loss of investment confidence. Alternatively, if the objective is merely fiscal revenue, damaging export revenue in such a way could lead to worsening balance of payments problems, depreciation of currency, possibly more dependency on imports, and could lead to economic recession.

Government vetoes over agricultural exports through export licensing via vague and subjective criteria such as 'public interest' or 'domestic protection' are likely to destroy good relationships and jeopardize international food supply chains contracts. It presumes that government authorities can determine the needs specifications of all global importers. Export licensing decision making may be reduced to subjectively picking export winners and losers, which can lead to patronage and corruption. The one exception may be food safety concerns over chemical residues.

If the commodity exports have low quality, then normal international contract law suggests that what is delivered by the seller must meet the buyer's contract minimum specifications. Otherwise the contract is made null and void, and the delivery rejected with no payment. To prevent such international trade on the grounds that a government in an exporting country wants to preserve some 'quality' image damages both the exporter who needs to sell surpluses and the importer who may want to buy cheaply. Government export licences that prevent low-quality exports can damage the formation of niche market supply chains whilst aggravating domestic surpluses and low prices. Export licensing presumes that if governments undertook exports, then the product would always be of the highest-grade quality, which is rarely the case.[69]

The objective of most export subsidies is to increase exports, and they are often disguised as export assistance-enhancement policies designed to promote trade, remove domestic surpluses, or meet monetary goals such as reducing trade deficits and improving the balance of payments. Export subsidies have been an integral component of the EU Common Agricultural Policy (CAP),[70] which has been characterized by the encouragement of more domestic production surpluses that then require more export subsidies to remove them.

Governments may be forced to decide which food industries should have export assistance-enhancement policies and which do not, as if the government was the ultimate consumer and the purveyor of good taste. It presumes objectivity when selection is actually reduced to subjectivity.

Export subsidies can often cause the transfer of resources from more efficient domestic industries to more inefficient export industries, which can lead to lower economic growth, persistent high unemployment, and pressure for higher import protection. Subsidy dependency can be a disincentive to seek longer-term international food chain alliances.

Import protection

Trade protection, fixed currency regimes, import substitution, price control policies and stabilization schemes all necessitate import protection, which usually involve an array of tariffs, import embargoes, customs duties, and deliberate bureaucratic delays and administrative complexity. Some governments choose import barriers rather than increase supply chain competition and efficiency because of its fiscal tax advantage, ease of implementation, and its immediacy.

Apart from import controls decreasing payment outflow pressure on fixed currency exchange, it has been argued that import controls protect farmers from more competitive imports.[71] However, the higher cost of imported farm inputs, the erosion of domestic competition, inflation, and the increased dependency on government support can cause an overall loss of international competitiveness for domestic farmers and lower economic growth and exports.[72]

Import protection does not solve domestic industry uncompetitiveness that can arise from a country's comparative disadvantage in land, labour, capital, resources, skills, or knowledge. Existing industries could be unnecessarily prolonged with import protection, while new industries could be protected under infant industry or diseconomies of scale arguments despite a lack of basic comparative advantage. Import policy failures can result in even more intervention to rectify previous intervention-induced problems.

Adjustment to change is generally prevented with import protection, combined with reduced operational flexibility and a general lack of cluster industry support. Industry efficiency is often compromised for the sake of non-competitive monopoly or oligopoly vested interests having persuasion over government policy.[73] Government action is reduced to picking 'winning' industries and rejecting 'losing' industries, whilst decision-making subjectiveness is open to public corruption.

Import protection is characterized by being production based, which explains why this policy generally has been the foundation-stone of socialist governments. It is the antithesis of consumerism and actively discriminates against consumers. As more consumer orientation in food supply chains occurs, it would be expected that the need for import protection would decline.[74] Alternatively, as more consumer alienation occurs, it would be expected to lead to greater food chain inefficiency, insecurity, and shortages.

There is little justification for import protection under flexible currency exchange rates in a competitive deregulated economy. If exchange rates are high, this will generally cause a decrease in exports because of price uncompetitiveness in international markets and the exchange rate will fall. If exchange rates are low, this will usually cause an increase in exports because of the price competitiveness in international markets and the exchange rate will rise. The opposite occurs for imports. Any structural impediments or inflexibility within the economy can prevent this occurring.[75]

Import protection can be divided into physical barriers such as import bans or quotas, enforcing higher import prices through tariffs, the imposition of added costs via customs duties and taxes, or non-physical barriers such as bureaucratic delays or questionable quarantine restrictions. Voluntary trade restrictions can occur under certain circumstances.

Import bans are a more severe method of preventing imports for the purpose of 'infant-industry' development, import substitution, or domestic industry protection. Embargoes usually involve some urgency because of the power of ship seizures. Instead of undertaking economic, trade, and competition reform, the government initiates a total ban on certain imports for short-term relief. Such bans cause supply chain distortions, favour selected industries, discriminate against domestic consumers, and are open to corruption and foreign policy directives. Dependency of the country on domestic producers increases, which adds to food insecurity when domestic disasters occur.

Import quotas are tonnage restrictions on imports that may be part of several related policies that may include 'infant industry' development, import substitution, or domestic industry protection. They may be combined with customs duties to allay administration costs. Some import quotas are not triggered until certain prescribed quantity limits are reached. Again, they are a substitute for economic, trade, and competition reform, and can cause supply chain distortions, favour selected industries, and are open to corruption and foreign policy directives.

Tariffs, customs taxes and duties have the effect of raising the price and cost of imports for the purpose of protecting home country producers. Such dependency on local production again increases food insecurity when domestic disasters occur. Tariffs reduce the amount of imports into the home country, raise consumer prices, decrease domestic product choice, and reduce export competitiveness inefficiency through higher input costs and inflexibility in the economy.[76] Misallocation of resources and supply chain distortions can act as a disincentive for domestic investment and hinder economic growth. When the US legislated in 1929 to increase its tariff on 20,000 imported goods through the Smoot-Hawley Tariff Act of 1930, global trade declined 33 per cent, which precipitated the Great Depression.[77]

Bureaucratic delays circumvent other more obvious import restrictions, but the impact is largely the same. The importer bears most of the cost imposition and is therefore discouraged to trade with the host country. As such, it has the capacity to ruin international supply chain relationships, prevent alliances, disrupt inventory flows, and cause much conflict and legal dispute.

Quarantine and phyto-sanitary restrictions may be valid in some instances to prevent the transfer of diseases and insects, but sometimes they can be misused by governments either as an import embargo or an unnecessary bureaucratic delay and cost. Such import restrictions might be the easy solution rather than investing into research programmes to develop disease resistance, genetic improvement or immunity. Prolonged genetic isolation can increase food supply vulnerability to new diseases, which is synonymous with food insecurity. Inbreeding of plants and animals without occasional external hybrid vigour can decrease food productivity.

Voluntary import restraints occur when exports are only allowed if a certain local content usage is included in the domestic manufacturing process. It purports to protect local producers regardless of their quality, efficiency, cost, and competitiveness. However, it can cause serious economic distortions and resource misallocation, whilst undermining international competitiveness for export industries.

Overseas ownership

Statutory exporters sometimes resort to investment in the importing country through mergers, takeovers, or joint ventures in their desire to stimulate international sales of commingled undifferentiated and usually unwanted commodity, in lieu of long-term food supply agreements. It can be a last-resort selling method.

Overseas ownership may be used to circumvent the import-government tender system or it may complement the tender process using investment as a negotiating tool to influence tender decisions. Alternatively, it can be used in countries that lack effective food supply chains after decades of government intervention or warfare.

The Australian Wheat Board owned bread shops in Moscow and Tianjin during the late 1980s,[78] both of whom had less discerning consumers after decades of government controls. It also had a 30 per cent investment in the Five Star Flour Mills in Egypt during the 1990s, with the expressed Australian government intention to 'achieve long term supply agreements under the Egyptian Government tendering system'. It failed to achieve this objective because the Egyptian government statutory authorities preferred spot purchases via import tendering, especially from exporters who had a freight advantage over Australia, such as the EU, Ukraine, Russia, and Kazakhstan.

Exporter cartels

There were many attempts to establish international government selling cartels for wheat and sugar during the period 1930–1965.[79] It was an era when producer cartels were permitted in global trade, but buyer–consumer cartels were discouraged. Gold, diamond, and crude oil had successful international seller cartels during this period, and these acted as imitation models for agricultural commodities. However, all attempts to form government-authority agricultural commodity selling cartels failed, including tea cartel.[80]

Many conflicting interests existed amongst agricultural commodity export governments that prevented an agreement on an international selling cartel, particularly when trade protectionist policies dominated global trade and international currencies remained fixed to the US dollar and gold conversion. There were the usual divisions between the highly protected food bowl exporters and the less protected marginal-production exporters, which was erroneously described as the 'North–South' hemisphere divide. The main cause of conflict was the uncertainty over agricultural production with a general unwillingness to implement the required production quotas and hoarding via government 'buffer' stocks. In contrast, the success of the cartel movement for gold, diamonds, and crude oil lay in its ability to control supply that was not weather-quality dependent.

Countertrade

Countertrade occurs when international trade is not conducted according to the payment–currency desires of the exporter. Unconfirmed indications are that up to 60 per cent of global trade is not conducted in the desired payment method of the seller. This can result in a forced exchange of a product that has little perceived value to the exporter. Barter trade is one form of countertrade.

General causes of countertrade include shortages of international currency reserves by the host country central bank or cash-liquidity problems of the importer. Statutory import monopolies may have more market power to demand 'countertrade' arrangements whereby the seller is paid in kind, but not in the currency of their preference.

Private food companies may adapt better to countertrade than statutory commodity exporters. A food company that was fully integrated into the food value chain might exchange wheat for biscuits, and assist in the development of both the wheat and biscuit supply chains. Alternatively, countertrade may be better managed by international merchants than single-item food exporters because of their flexibility, backroom settlement operations and networks.

The alternative for statutory sellers and private exporters is to arrange for a third party such as the World Bank to negotiate the exchange. If crude oil was exchanged for rice, then the crude oil could be then sold by the World Bank, and through currency conversion, the rice exporter is paid by the World Bank in the preferred currency. The problem is that countertrade operates outside the cash market, often outside fiscal tax revenue flows, and frequently causes corruption scandals, such as the Iraq 'Oil-for-Food' debacle with US$65 billion in Iraqi crude oil exchanged but for very limited food.[81]

Case study – USSR farm collectivization

The Bolshevik takeover in October 1917 introduced a government that was production based, protectionist and partly centralized with little regard for the consumer or even for the labour workforce. Without constitutional constraints, it quickly led to an 'Oil-Tsar' dictatorship.[82] Private farm land that had been allocated under Stolypin's 1906 reforms was abolished, thereby reverting farmers back into communal peasantry with no decision-making independence and no production incentive.[83] Many previous farm workers took over farms without the organization skills or capacity to operate them. Farm investment initiative stagnated because of government debt and a lack of decision-making direction. Lenin even ordered the redirection of genetic seed stocks for human consumption after he thought that the peasants were hoarding grain.

Although the Politburo initially allowed these peasant farmers to sell their surpluses in open markets, it controlled all prices. New low commodity prices in 1928 encouraged on-farm hoarding in the hope of forcing the Politburo to raise prices. However, the Politburo responded by enforced requisition, which then caused the farmers to produce less in 1929, which resulted in more requisition enforcement.[84] Underground burials of grain occurred and the black market flourished.

Stalin's answer in November–December 1929 to the food supply crisis was enforced farm collectivization, which was totally completed by 1940.[85] Molotov sent 25,000 urban workers and 'shock brigades' into the paddocks to replace the Kulaks and form large landholdings (kolkhozy) and 'agrotowns'. Between 1929 and 1932, huge falls in USSR agricultural production occurred at a time when other global grain exporting countries were experiencing large surpluses. Soviet famine became widespread, despite the Ukraine traditionally being one of Europe's major grain exporters.

Much of the 1933 Ukraine wheat crop was requisitioned and sold internationally to generate finance for the new Soviet Five Year Plan and to remilitarize the USSR in response to Hitler taking over Germany. Seven million Ukrainians died in the subsequent famine, including three million children, which represented 25 per cent of the population.[86] Five million farming families were deported by Molotov to Gulag forced-labour camps, and 24 million disappeared from rural areas.[87]

Stalin admitted to Churchill in 1945 that 10 million people had died as a direct result of the farm collectivization scheme, which only begged the question as to how many died from indirect causes such as in the Gulags and from food-related diseases. Many more people died from hunger in Stalin's Five Year Plan and Mao's Great Leap Forward than all the deaths in the Second World War, yet both events remain relatively hidden from history.

Notes

1 Davis, Gordon, 2010, AWB Managing Director, quoted in *Stock & Land*, Melbourne, 28 October.
2 See case study in Chapter 7.
3 Copland, D.B. and C.V. Janes, Eds., 1938, *Australian Marketing Problems: A Book of Documents 1932–37*, Angus and Robertson, Sydney, pp. 262–263. It referred to headlines in the *Melbourne Age* newspaper in 1931 regarding the proposed formation of a compulsory national wheat pool and a selling/exporting authority in Australia.
4 Webb, G.R., 1963, *The Closure of Uneconomic Rail Lines in Australia*, Economic Record, 39:88, 468–476.
5 Matusow, Allen, 1970, *Farm Policies and Politics in the Truman Years*, Atheneum, New York.
6 Kenny, Martin, Lobao, Linda M., Curry, James and W. Richard Goe, 1989, *Midwestern Agriculture in US Fordism: From the New Deal to Economic Restructuring*, Sociologia Ruralis, 29:2, 131–148.
7 Bairoch, Paul, 1989, *European Trade Policy*, in Peter Mathias and Sidney Pollard, Eds., *The Cambridge Economic History of Europe*, Vol. 8, Cambridge University Press, Cambridge.
8 Lee, W. Robert and Eve Rosenhaft, Eds., 1990, *State, Society and Social Change in Germany, 1880–1914*, Oxford University Press, Oxford.
9 Kunz, Andreas, 1990, *The State as Employer in Germany, 1880–1918*, in W. Robert Lee and Eve Rosenhaft, Eds., 1990, *State, Society and Social Change in Germany, 1880–1914*, Oxford University Press, Oxford, pp. 37–63.
10 Feldman, Gerald, 1997, *The Great Disorder: Politics, Economics, and Society in the German Inflation 1914–1924*, Oxford University Press, Oxford.
11 Trentmann, Frank, 2008, *Free Trade Nation: Commerce, Consumption, and Civil Society in Modern Britain*, Oxford University Press, Oxford.
12 Taussig, Frank W., 1964, *The Tariff History of the United States*, Capricorn, New York.
13 Berstein, William, 2009, *How Trade Has Shaped the World*, Grove Atlantic, London.
14 Soyfer, Valery N., 1994, *Lysenko and the Tragedy of Soviet Science*, Rutgers University Press, New Brunswick, NJ.
15 Chitsike, Francis, 2003, *A Critical Analysis of the Land Reform Programme in Zimbabwe*, paper presented to the 2nd FIG Regional Conference, Marrakech, Morocco, December 2–5.
16 Whitwell, Greg and Diane Sydenham, 1991, *A Shared Harvest – The Australian Wheat Industry 1939–89*, Macmillan Education Australia, Melbourne, Ch. 3.

17 Whitwell, Greg and Diane Sydenham, 1991, *A Shared Harvest – The Australian Wheat Industry 1939–89*, Macmillan Education Australia, Melbourne, Ch. 9.
18 FAO, 1984, *Studies on Agrarian Reform and Rural Poverty*, Economic and Social Development Series, No. 27, Rome. The 1918 soldier resettlement subdivisions are still incurring poverty and land degradation problems in Australia.
19 Anderson, Kym, Martin, Will and Ernesto Valenzuela, 2006, *The relative importance of global agricultural subsidies and market access*, World Trade Review, 5:3, 357–376. Sometimes farmers are paid not to grow crops on former agricultural land, such as depleted cotton soils in the USA.
20 Rasmussen, W.D. and G.L. Baker, 1966, *A Short History of Price Support and Adjustment Legislation and Programs for Agriculture, 1933–65*, Agricultural Economic Research, 18, 69–78.
21 Thompson, Wyatt, Mishra, Ashok K. and Joe Dewbre, 2009, *Farm Household Income and Transfer Efficiency: An Evaluation of United States Farm Program Payments*, American Journal of Agricultural Economics, 91:5, 1296–1301.
22 Dewbre, Joe, Anton, Jesus and Wyatt Thompson, 2001, *The Transfer Efficiency and Trade Effects of Direct Payments*, American Journal of Agricultural Economics, 83:5, 1204–1214.
23 Canadian International Grains Institute (CIGI), 1993, *Grains and Oilseeds: Handling, Marketing, Processing*, Winnipeg, Vol. 1, 4th Ed., p. 106.
24 Whitwell, Greg and Diane Sydenham, 1991, *A Shared Harvest – The Australian Wheat Industry 1939–89*, Macmillan Education Australia, Melbourne, Ch. 3.
25 Whitwell, Greg and Diane Sydenham, 1991, *A Shared Harvest – The Australian Wheat Industry 1939–89*, Macmillan Education Australia, Melbourne, p. 42.
26 Canadian International Grains Institute (CIGI), 1993, *Grains and Oilseeds: Handling, Marketing, Processing*, Winnipeg, Vol. 1, 4th Ed., p. 106.
27 Canadian International Grains Institute (CIGI), 1993, *Grains and Oilseeds: Handling, Marketing, Processing*, Winnipeg, Vol. 1, 4th Ed., p. 106.
28 Zekulich, Michael, 1997, *The Grain Journey – The History of The Grain Pool of WA*, The Grain Pool of Western Australia, Chs. 3, 5, and 8.
29 Whitwell, Greg and Diane Sydenham, 1991, *A Shared Harvest – The Australian Wheat Industry 1939–89*, Macmillan Education Australia, Melbourne, pp. 50–58.
30 Marriott, Richard S. and Gus McLean, 1991, *Going with the Grain: A History of Queensland's State Wheat Board*, The State Wheat Board, Toowoomba, p. 44.
31 Whitwell, Greg and Diane Sydenham, 1991, *A Shared Harvest – The Australian Wheat Industry 1939–89*, Macmillan Education Australia, Melbourne, pp. 264–273.
32 Marriott, R.S. and Gus McLean, 1991, *Going with the Grain: A History of Queensland's State Wheat Board*, The State Wheat Board, Toowoomba, pp. 43–45.
33 Marriott, R.S. and Gus McLean, 1991, *Going with the Grain: A History of Queensland's State Wheat Board*, The State Wheat Board, Toowoomba, p. 63.
34 McColl, Jim, 1988, *Royal Commission into Grain Storage, Handling, and Transport*, Australian Government Publishing Service, Canberra, p. xxv.
35 Marriott, R.S. and Gus McLean, 1991, *Going with the Grain: A History of Queensland's State Wheat Board*, The State Wheat Board, Toowoomba, p. 43.
36 Marriott, R.S. and Gus McLean, 1991, *Going with the Grain: A History of Queensland's State Wheat Board*, The State Wheat Board, Toowoomba, pp. 47–50.
37 Marriott, R.S. and Gus McLean, 1991, *Going with the Grain: A History of Queensland's State Wheat Board*, The State Wheat Board, Toowoomba, p. 54.
38 Marriott, R.S. and Gus McLean, 1991, *Going with the Grain: A History of Queensland's State Wheat Board*, The State Wheat Board, Toowoomba, p. 56.
39 Canadian International Grains Institute (CIGI), 1993, *Grains and Oilseeds: Handling, Marketing, Processing*, Winnipeg, Vol. 1, 4th Ed., pp. 105–112.
40 Brochl, Wayne G., *Cargill: Trading the World's Grain*, University Press of New England, Hanover, NH, 1992.
41 Canadian International Grains Institute (CIGI), 1993, *Grains and Oilseeds: Handling, Marketing, Processing*, Winnipeg, Vol. 1, 4th Ed., p. 33.

42 Zekulich, Michael, 1997, *The Grain Journey – The History of The Grain Pool of WA*, The Grain Pool of Western Australia, Perth, WA, p. 69.
43 Whitwell, Greg and Diane Sydenham, 1991, *A Shared Harvest – The Australian Wheat Industry 1939–89*, Macmillan Education Australia, Melbourne, pp. 59–60.
44 Whitwell, Greg and Diane Sydenham, 1991, *A Shared Harvest – The Australian Wheat Industry 1939–89*, Macmillan Education Australia, Melbourne, p. 2.
45 Harold S. Hunter, 1964, paper presented to annual conference of Co-operative Union of Queensland, 8 April, Brisbane.
46 Marriott, R.S. and Gus McLean, 1991, *Going with the Grain: A History of Queensland's State Wheat Board*, The State Wheat Board, Toowoomba, p. 59.
47 Zekulich, Michael, 1997, *The Grain Journey – The History of The Grain Pool of WA*, The Grain Pool of Western Australia, p. 12.
48 Williams, John, 2007, *Single Desk Keeps Wheat Growers in Commodity Poverty Trap*, Australian Farm Journal, August (Part 1), pp. 48–51, and September (Part 2), pp. 6–8.
49 Bond, Gary E., Thompson, Stanley R. and Jane M. Geldard, 1985, *Basis risk and hedging strategies for Australian wheat exports*, Australian Journal of Agricultural Economics, 29:3, 199–209.
50 Gellert, Simon, 2006, *Analysis of Producer Pricing Strategies*, Department of Agricultural Sciences, La Trobe University, Melbourne.
51 Williams, John, 2007, *Single Desk Keeps Wheat Growers in Commodity Poverty Trap*, Australian Farm Journal, August (Part 1), pp. 48–51, and September (Part 2), pp. 6–8.
52 High Court of Australia, 1936, *James v Commonwealth*, 17 July.
53 Whitwell, Greg and Diane Sydenham, 1991, *A Shared Harvest – The Australian Wheat Industry 1939–89*, Macmillan Education Australia, Melbourne, Ch. 11.
54 Taylor, Fred M., 1929, *The Guidance of Production in a Socialist State*, The American Economic Review, 19:1, 1–8.
55 Conquest, Robert, 1986, *The Harvest of Sorrow: Soviet Collectivization and the Terror-Famine*, Oxford University Press, Oxford.
56 See the functions of price in Chapter 3.
57 Butler, Nick, 1986, *The International Grain Trade*, Croom Helm, London, p. 58.
58 Bale, Malcolm D., 1979, *Distributional Aspects of Price Intervention*, American Journal of Agricultural Economics, 61:2, 348–350.
59 Rose-Ackerman, Susan, 1999, *Corruption and Government: Causes, Consequences and Reform*, Cambridge University Press, Cambridge.
60 Whitwell, Greg and Diane Sydenham, 1991, *A Shared Harvest – The Australian Wheat Industry 1939–89*, Macmillan Education Australia, Melbourne, p. 56.
61 Townsend, Robert M., 1977, *The eventual failure of price fixing schemes*, Journal of Economic Theory, 14:1, 190–199.
62 Salant, Stephen W., 1983, *The Vulnerability of Price Stabilization Programs to Speculative Attack*, Journal of Political Economy, 91:1, 1–38.
63 Miranda, Mario J. and Peter G. Helmberger, 1988, *The Effects of Commodity Price Stabilization Programs*, The American Economic Review, 78:1, 46–58.
64 Peterson, Willis L., 1979, *International Farm Prices and the Social Cost of Cheap Food Policies*, American Journal of Agricultural Economics, 61:1, 12–21.
65 Marriott, R.S. and Gus McLean, 1991, *Going with the Grain: A History of Queensland's State Wheat Board*, The State Wheat Board, Toowoomba, p. 57.
66 Whitwell, Greg and Diane Sydenham, 1991, *A Shared Harvest – The Australian Wheat Industry 1939–89*, Macmillan Education Australia, Melbourne, Ch. 9.
67 Whitwell, Greg and Diane Sydenham, 1991, *A Shared Harvest – The Australian Wheat Industry 1939–89*, Macmillan Education Australia, Melbourne, pp. 53–58.
68 Mitra, Siddgartha and Tim Josling, 2009, *Agricultural Export Restrictions: Welfare Implications and Trade Disciplines*, International Food and Agricultural Trade Policy Council, IPC Position Paper, Agricultural and Rural Development Policy Series, January.

69 Whitwell, Greg and Diane Sydenham, 1991, *A Shared Harvest – The Australian Wheat Industry 1939–89*, Macmillan Education Australia, Melbourne, p. 111.

70 Hill, Berkeley, 2011, *Understanding the Common Agricultural Policy*, Earthscan, Abingdon, Chs. 10–12.

71 Cochrane, Willard W. and Mary E. Ryan, 1976, *American Farm Policy, 1948–1973*, University of Minneapolis Press, Minneapolis.

72 Agarwala, Ramgopal, 1984, *Price distortions and growth*, Finance and Development, 21:1, 33–37.

73 Hilmer, Frederick G., 1993, *National Competition Policy: Hilmer Report*, Australian Government Publishing Service, Canberra.

74 Rozelle, Scott, Park, Albert, Huang, Jikun and Hehui Jin, 1997, *Liberalization and Rural Market Integration in China*, American Journal of Agricultural Economics, 79:2, 635–642.

75 Australian Productivity Commission, 1999, *Impact of Competition Policy Reforms on Rural and Regional Australia*, Canberra.

76 Baumol, William J. and Andrew E. Burke, 2001, *Competition Policy in Dynamic Markets*, International Journal of Industrial Organization, 19:5, 613–634.

77 Madsen, Jakob B., 2001, *Trade Barriers, and the Collapse of World Trade During the Great Depression*, Southern Economic Journal, 67:4, 848–868.

78 Whitwell, Greg and Diane Sydenham, 1991, *A Shared Harvest – The Australian Wheat Industry 1939–89*, Macmillan Education Australia, Melbourne, p. 2.

79 Whitwell, Greg and Diane Sydenham, 1991, *A Shared Harvest – The Australian Wheat Industry 1939–89*, Macmillan Education Australia, Melbourne, Chs. 3, 4, and 9.

80 Gupta, Bishnupriya, 2001, *The International Tea Cartel During the Great Depression, 1929–1933*, Journal of Economic History, 61:1, 144–159.

81 United Nations, 2005, *Volcker Report – Independent Inquiry into the United Nations Oil-for-Food Programme, Chapter III – Humanitarian Goods Transactions and Illicit Payments*, New York.

82 Montefiore, Simon Sebag, 2004, *Stalin: The Court of the Red Tsar*, Phoenix-Orion, London.

83 Hubbard, Leonard E., 1939, *The Economics of Soviet Agriculture*, Macmillan, London.

84 Conquest, Robert, 1986, *The Harvest of Sorrow: Soviet Collectivization and the Terror-Famine*, Oxford University Press, Oxford.

85 Montefiore, Simon Sebag, 2004, *Stalin: The Court of the Red Tsar*, Phoenix-Orion, London, Ch. 3.

86 Conquest, Robert, 1986, *The Harvest of Sorrow: Soviet Collectivization and the Terror-Famine*, Oxford University Press, Oxford.

87 Montefiore, Simon Sebag, 2004, *Stalin: The Court of the Red Tsar*, Phoenix-Orion, London, Ch. 3.

Further readings on export intervention

Boehlje, Michael and Steven Griffin, 1979, *Financial Impacts of Government Support Price Programs*, American Journal of Agricultural Economics, 61:2, 285–296.

Bohman, Mary, Carter, Colin A. and Jeffrey Dorfman, 1991, *The Welfare Effects of Targeted Export Subsidies: A General Equilibrium Approach*, American Agricultural Economics Association, 73:3, 693–702.

Broehl, Wayne G., 1992, *Cargill: Trading the World's Grain*, University Press of New England, Hanover, NH.

Carter, C. and V. Smith, 2001, *The Potential Impacts of State Trading Enterprises on World Markets: The Exporting Country Case*, Canadian Journal of Agricultural Economics, 49, 429–439.

Carter, C., Loyns, R. and D. Berwald, 1998, *Domestic Costs of Statutory Marketing Authorities: The Case of the Canadian Wheat Board*, American Journal of Agricultural Economics, 80, 313–324.

Cole, Harold L. and Lee E. Ohanian, 2003, *New Deal Policies and the Persistence of the Great Depression: A General Equilibrium Analysis*, Research Memo, February, Dept. of Econ., University of California.

Dormois, Jean-Peirre, 2005, *Classical Trade Protection 1815–1914*, Routledge, New York.
Dunsdorfs, Edgars, 1956, *The Australian Wheat Growing Industry 1788–1948*, Melbourne University Press, Melbourne.
Fowke, Vernon C., 1957, *The National Policy and the Wheat Economy*, University of Toronto Press, Toronto.
Hanson, David, 2010, *Limits to Free Trade – Non-Tariff Barriers in European Union, Japan, and United States*, Edward Elgar Publishing, Cheltenham.
Krugman, Paul R., 1990, *Rethinking International Trade*, MIT Press, Cambridge.
Leung, Suiwah, Ben Bingham and Matt Davies, 2010, *Globalization and Development in the Mekong Economies*, Edward Elgar Publishing, Cheltenham.
Magnusson, Lars, 2004, *The Tradition of Free Trade*, Routledge, New York.
McCalla, Alex F. and Andrew Schmitz, 1979, *Grain Marketing Systems: The Case of the United States versus Canada*, American Journal of Agricultural Economics, 61:2, 199–212.
Morriss, William E., 1987, *A History of the Canadian Wheat Board*, Reid More Books, Edmonton, AB.
Pomeranz, Kenneth and Steven Topik, 2005, *The World that Trade Created – Society, Culture, and the World Economy, 1400 to the Present*, M.E. Sharpe, New York.
Porter, M.E., 1990, *The Comparative Advantage of Nations*, The Free Press, New York.
Rothgeb, John, 2001, *U.S. Trade Policy*, CQ Press, Washington, DC, pp. 30–33.
Sandrey, Ron and Russell Reynolds, Eds., 1990, *Farming Without Subsidies: New Zealand's Recent Experience*, New Zealand Ministry of Agriculture and Fisheries, Wellington, NZ.
Solomon, R., 1982, *The International Monetary System, 1945–1981*, Harper and Row, New York.
Vinig, G.T. and R.C. van der Voort, Eds., 2005, *The Knowledge Spillover Theory of Entrepreneurship and Economic Growth*, in *The Emergence of Entrepreneurial Economics*, Elsevier, Oxford, pp. 37–54.
Williams, D.B., Ed., 1982, *Agriculture in the Australian Economy*, University Press, Sydney, Ch. 15.
Wilson, James, Ed., 1980, *The Politics of Regulation*, Basis Books, New York.
Wolken, John D. and Frank J. Navratil, 1984, *The valuation of subsidies in kind and their effect on costs*, Journal of Economics and Business, 36:2, 217–232.

7

HOARDING

Hoarding is any deliberate withholding of agricultural supply by farmers, merchants, commodity pool managers, or government authorities, with the objective of achieving higher prices at a later time. The act of hoarding is therefore characterized as being opportunistic and speculative, with the hope of windfall gains, but it frequently results in financial losses. There is more market power for the hoarder when shortages exist, but little power when there are surpluses.

Hoarding success is impossible for perishable food products because price falls with quality deterioration. This restricts hoarding to those products that can be readily stored, and there has to be the perception that prices can be allowed to rise. However, hoarding can occur whenever price controls exist in an attempt to force authorities to raise prices.

The choice of hoarding exists with various selling and buying methods. Merchants can buy commodity for hoarding instead of accumulating product for 'trade' market usage. Farmers might hoard after harvest instead of immediate sales. However, all farm storage is not hoarding, particularly when farmers store product on behalf of either a merchant or end user. Governments can store inventory for the institutional catering trade, or hoard commodity either in unofficial storages or in formalized legislated 'buffer' stocks. The objective of most commodity pool managers and authorities is to hoard and eventually sell into spot markets, but there can be some forward selling.

The success of hoarding storable commodities depends on the elasticity of demand when shortages occur. This indicates the response of buyers to either price movement or the supply made available, but much depends on the freedom of price movement and the response time of supply.[1] If demand is inelastic with no close product substitutes and new supply is slow to respond when shortages occur, the ability to withhold enough supply of a storable commodity may cause prices to rise, which results in a higher probability of hoarding success. Alternatively, if demand is elastic with close product substitutes and supply can respond to change quickly, then there is a high probability of hoarding failure regardless of the amount withheld. Wheat has less probability of hoarding success because global harvesting supply extends for 11 months of the year, resulting in quick hoarding supply release response times.

'Cornering' a market may be a particular type of hoarding whereby control over enough global supply is sufficient to force prices higher. The success of 'cornering' has been rare for agricultural commodities because of the difficulty in controlling biological production and supply, in contrast to the cartels for gold, diamonds, and crude oil where output can better be controlled.

Hoarding characteristics have changed over the past century. Previous end users might have been more willing to accumulate early and then store, because of the supply and logistics risks. However, just-in-time policies over the past 40 years have pushed responsibility for storage back in deregulated food supply chains, due to improved logistics and operational management efficiencies. This pushing back of product within supply chains has somewhat encouraged private hoarders who are peripheral to the supply chain, but who today have the added advantage of improved bunkering and storage technology.

For storable commodities, supply can come from new-crop production with no storage quality problems, or it can come from old-crop hoarded storage with high storage quality risk. Supply decisions can therefore be made quite separate to production decisions for storable commodities. However, it is usual for hoarded commodity to be excluded from 'trade' markets because it does not meet the minimum quality specification criteria in the food supply chain. Hoarded commodity is generally destined for the animal feed or the bulk-filler food market, which are supply chains of last resort.

Premium prices may be attached to new crop and discounts to any hoarded old crop, but this has nothing to do with backwardation or contango in 'trade' markets.[2] These are simply a bareme of price schedules relating to product differentials based on quality-product risk within the supply chain.

There is nothing 'virtual' about food storage despite some vague contrary viewpoints.[3] Either the storage of the physical product exists or it does not. Futures markets are not 'virtual' storages or granaries, because they are based on physical forward contracts that deliver specified differentiated product into 'trade' markets.[4]

Private pursuits

The objective of private hoarding by farmers and merchants can be defined as the willingness to withhold supply with the deliberate intention of gaining from price rises. Processors such as sugar mills and cocoa manufacturers can both produce and hoard product. Examples include the hoarding of raw sugar by Indian sugar mills in 2009 and Brazilian sugar mills during 2010. One notable cocoa hoarder was Anthony Ward, alias Choc Finger, who cornered and privately hoarded 15 per cent of the global cocoa market in 2010.

Farm hoarding usage has been boosted over the past ten years by increased fund manager activity,[5] JIT end-user policies, farm subsidies, government price controls, and technology improvements. Fund managers have increased the market power of hoarders, farm subsidies have provided the cash flow to sustain hoarding activities, and technology has advanced hoarding methods,[6] while government price controls provide the incentive. The amount of private hoarding in India might be proportionate to the number of real and expected changes to government prices, but because the Indian hoarding police have great difficulty in finding the hoards, this relationship can only be surmised.

Farmers who do not receive government subsidies usually have a disincentive to hoard because of the need for cash flow through spot-cash sales. Farmers with poor cash flow generally struggle to pay harvest contractors and merchandisers when they hoard most of their crop, causing relationships with such input providers to deteriorate, thus jeopardizing future production. However, this is no justification for farm subsidies.

The association between farm subsidies and on-farm hoarding is particularly strong. Much of the on-farm global hoarding occurs in the USA where farmers receive subsidies and cash flow through the US Farm Bill. The government subsidy in the US peanut (and cotton) industry since 2002 has been actively directed to encouraging hoarding.

However, in non-subsidized countries such as Australia, there are many reasons why 75 per cent of farmers in some geographic regions would store on-farm, including the provision for seed or animal feed (13 per cent), avoid harvest delivery congestion (9 per cent), avoid low price periods during harvest surpluses (30 per cent), storing forward sold product, and silo renting to merchants or end users.[7] Storage by a farmer does not always equate to hoarding, which makes farm surveys on hoarding difficult, even if it was even feasible to undertake accurate response research. However, there may be a core group of commodity farmers who use storage for price speculation, hoping for a better price, but this can vary according to production size, farm income, the particular region, enterprise specialization, farm debt, isolation from merchants or end users, and prevailing prices.[8]

Public hoarding

Governments have historically been active in stockpiling storable food. There can be hoarding by governments of producing and exporting countries such as those in the European Union, by importing countries such as those in the Middle East, and by countries that have large seasonal variations, large populations, and high propensities for natural disasters such as China and India. Sometimes the responsibility for public hoarding is given to semi-statutory authorities or licensed to private operators.

The reasons offered for this public hoarding include fear of future food shortages, to quell civil unrest, and a buffer against natural disasters and during war. Public hoarding is also frequently used to control supply for the purpose of implementing price controls and associated price 'stabilization' schemes. However, there is a high propensity for these public price controls to result in private hoarding in the expectation of price rises. Therefore, private hoarding could occur simultaneously with public hoarding if price controls exist, such as in India.

Public hoarding theory

Public hoarding assumed the indifference of consumers. It took an Irish lawyer meandering in mathematical physics in 1881 to invent consumer indifference in economics and convert it into a 'moral science'.[9] Edgeworth reduced consumer choice to indifference based on the unreal assumptions that consumers should be restricted to only two choices that would be prior ranked in preference, with full recognition of the product range similarities and differences, logical thought processes, strong monotonicity, and that consumers always make purchasing decisions based on marginal rates of substitution calculations. Just as there was no spontaneity in judicial deliberations, consumers were supposed to make mathematical calculations before making a final purchase decision. Indifference curves were increasingly used in central authority decision making to calculate an 'optimum tariff' in trade protection and monopoly-based production and supply decisions, which had important implications for government hoarding.

The significance of Edgeworth's indifference curves was that socialists, trade protectionists, monopolists, and mathematical economists could now create hypothetical 'equilibrium' equations based on the assumed indifference of consumers. These hypothetical assumptions underpinned production and protection-driven economics for the next 90 years, whilst ignoring the realities of people who actually purchased and ate food.

There was a very deliberate obfuscation to combine all consumers together in a homogeneous mass, irrespective of whether they existed in the increasing government institutional catering market or the private market. This was an important foundation for public hoarding because it had to have the image of protecting all consumers and not just those subjected to government institutional catering. Merely inviting private consumers to the public communal table, as did occur in Mao's China in the late 1950s, was just not substantive enough.

However, having the theoretical foundations was no substitute for actual decision-making processes. Lenin provided no clue for such decision making in the USSR because of civil war distractions, and even Stalin provided no decision-making leadership. Whilst the cue might have been the new trade-protection putsch that resulted from the Smoot-Hawley 25 per cent US tariff legislation in 1929 triggering the Great Depression,[10] it was Fred Taylor's production management 'trial and error' central directives later in 1929[11] that finally gave some decision-making direction. It was the importance of controlling supply that projected government hoarding into public decision-making eminence.

After ten million citizens had already died from hunger by 1934 under USSR socialist 'trial and error' directives, and with Richard Darré experiencing failures in the German Reich Food Estate, Lange refined consumer indifference theory by proclaiming the 'determinateness' of consumer utility,[12] with support from the Lausanne School of Economics.[13]

Lange supported Taylor's 1929 'trial and error' method of decision making in 1935 by claiming that there was a Marxian 'economic evolution' that took time to implement.[14] Lange consolidated his advocacy for central planning board authority under socialist directives in 1936,[15] proclaiming the use of a standardized homogeneous product whose supply could be controlled (publicly hoarded) and which assumed consumer indifference.[16]

The phoenix of the Canadian Wheat Board arose in 1935 as a Lange prototype of a 'Central Planning Board', but without the associated Ministry of Socialization, and without any acknowledgement of the consumer. It assumed absolute consumer indifference. Despite having total requisition powers over farmers' production, this central planning board had no control over intake tonnage or quality, no control over price (unlike its predecessor the Board of Grain Supervisors in 1917), and no control over end users.

Operations management of such central planning boards was limited to hoarding of traditional commingled bulk handling storage of old-crop commodity-pooled product that had little relevance to end users. Canadian end users generally departed south across the border so that they could access US segregated specified product. 'Trade' markets were destroyed and potential cluster industries went elsewhere, leaving solitary silos silhouetted on the horizon of the prairies.

The major characteristic of the Canadian Wheat Board operations was hoarding of a largely unwanted product that was very dependent on export sales through international merchants, and increasingly dependent on winning import-government tender contracts. In the meantime, there was the hope that global wheat prices might rise and give the illusion that the central selling board had price-making powers despite being a commodity price taker in a global market. There was no ability or capability or authorization to market differentiated product. Farmer representation added to the delusion that this central planning board acted in the best interest of farmers. The real beneficiaries of this type of central selling board were the increasing number of government institutional catering buyers who now had supply chain power because of the accumulation of surplus product, particularly during wartime.

An alternate to the Canadian central board prototype was the combination of both production and price controls. If price controls were designed to control consumer behaviour, then this policy combination was more akin to Lange's definition of communism,[17] whereas the Canadian Wheat Board was largely based on the socialist–collectivist model, albeit with no intake tonnage quality control.

It was the 'communist' combination of central planning with price controls that had encouraged the Ukraine Kulak hoarding, which then led to the subsequent massacre, famine, and gulag deaths in USSR.[18] Hoarding had also become rife in Germany after Darré's failures in his blood and soil 'Blut und Boden' policy after 1934.[19] Apart from private hoarding wherever it was possible, the only 'economic evolutionary' inevitability about these policies was that it led directly to war.[20]

It was an extraordinary statement by Lange when in 1936 he stated: 'If the bureaucrats want successfully to impose a preference scale of their own for the guidance of production, they have to camouflage the inconsistency of their preference scale with that of the citizens by resorting to rationing in the sphere of producers' goods and of resources'.[21] Lange was claiming by 1936 that (food) rationing was justified in the

USSR to 'conceal the real cost of the programme of industrialisation', and that this was necessary to achieve the longer-term goals involved with 'economic evolution'.

Under socialist regimes, government 'buffer' or reserve stocks must be considered hoarding because they withhold supply until shortages occur and prices are higher. Lange assumed that public authorities could manage supply and demand through a simplistic price-manipulation concept.[22] If shortages occur, then public prices will be raised through demand and this will increase production and produce a surplus. But if surpluses occur, then public prices will be lowered through demand and this will decrease production and remove the surplus. Alternatively, governments could release stocks when prices are high thereby decreasing price, or governments could increase stocks when prices are low thereby increasing price. This was Lange's 'economic calculation' that was to evolutionarily lead to a socialist 'equilibrium'.

The problem was that Lange either did not understand the psychology of the private hoarder, or merely presumed that hoarders could be imprisoned as was the case under Hitler, murdered as was the case under Stalin, or executed as still is the case in China. Private hoarders were always to be solitary decision makers,[23] apart from being the anathema under communism or socialism. If government prices were raised, then longer-term production might be encouraged, but increased private hoarding in the shorter term in the expectation of further price rises was more likely to cause acute food shortages, rationing of government stocks, queues, black marketeering, and famine.

As well, any government purchases or requisitions to increase public 'buffer' stocks also causes the private hoarder to withhold product, again with the expectation of either driving prices higher or selling into the black market. The combination of government 'buffer' stocks and private hoardings then acts as a de-motivator for domestic farmers to produce agricultural crops because of their expectation of prolonged suppression of local prices (caused by the inventory multiplier or bull-whip effect).[24] There is a further de-motivator when these farmers do not own land and there is inadequate public investment into farms, as occurs in China and North Korea.

Similarly, the problem of 'buffer' stocks is in offloading or dumping the hoarded old crop product onto a domestic market at a time when increasing new crop local production is being sold at what is perceived by domestic farmers to be realistic prices. This occurred in Middle Eastern and North African countries during February–April 2009 in the aftermath of the 2008 food 'crisis', and again during March–April 2011 during the citizen-consumer uprisings. The subsequent price fall undermines the confidence of farmers who only undertake high-risk agriculture for the expected occasional financial reward.

The combination of new crop sales and public dumping of 'buffer' old crop stocks can seriously prolong low prices (again through the inventory multiplier or bull-whip effect), which can devastate the income and investment of local farmers, and act as an incentive to stop farming. In the aftermath, there can be an increasing dependency on imports, which drives government insecurity, and leads to the perception of needing more public stockpiling and hoarding. It represents a poverty cycle that has occurred in many countries, particularly in Africa and some Asian countries. This is the real 'economic evolution' resulting from the 'trials and errors' of both Taylor's and Lange's theories.

Problems with hoarding

Both public and private hoarding requires funding. Farmers frequently fund hoarding by delaying payments to creditors or extending bank loans if there are flexible lending and credit policies, despite jeopardizing production-operational efficiency and increasing costs. Public hoarding generally is funded by taxpayers who are also forced to be consumers of the hoarded commodity. However, there can be serious problems with both public and private hoarding, irrespective of the capability to store and the capital required to fund the operations.

Consumers suffer the most from a fumigation-contaminated uniform old-crop product of very low quality, and a lack of choice. Wheat has been hoarded in China's government stocks for more than five years. Such frequently chemically treated commodity stocks are then forced upon end users, which eventually become food products for human consumption. Consumers become so accustomed to poor-quality food that they mostly do not know any better, do not demand contamination checks, and frequently imitate the government in the lack of care with chemical-contaminated food. The stakeholder problem is compounded when the consumers are also taxpayers who have to pay for public hoarding and operational expenses.

If hoarding is occurring through public or private warehousing, then adequate fumigation and occupational health and safety controls might be expected to occur. However, on-farm storage usually is conducted very haphazardly. Even with adequate training, farmers may not have the required storage sealing facilities and operational capabilities to maintain high storage standards. This poses health risks to the farmer through fumigants, quality controls may be inadequate for resale purposes, and insect resistance can rapidly develop due to poor application and maintenance procedures. Insect resistance to chemical treatment can then jeopardize food supply chains through contamination, spoilage, wastage, and unsuitability for end usage.

Price controls by the Indian government led to the hoarding of five million tonnes of sugar during 2009, which was the difference between 14 million and 19 million tonnes of sugar. What was initially desired by government as food security actually resulted in food shortages.

Price controls may have the objective of raising prices for farmers to produce more, but the outcome usually is an encouragement to hoard and less supply of storable commodities. When combined with possible government embargoes over imports or exports, game theory plays an important role in prior-guessing what the government intends to do, and positioning in front of those decisions. It then often leads to insider trading, government patronage, corruption of government officials, and bribery of hoarding 'police' inspectors. This can all occur while food shortages increase.

If the government objective is to lower prices for consumers, there is subsequent discouragement to hoard and a motivation to supply more just in case prices fall even further. But lowering government prices rarely occurs when food shortages actually occur. This was never a part of Lange's 'economic calculation', but von Mises had warned about such central directives and outcomes as early as 1920.[25]

Serious hoarding disappears once price controls are lifted because the incentive to game plan in order to achieve higher prices is removed. The normal skewness of commodity prices precludes frequent hoarding profit opportunities. Therefore, if governments want food supply, then they should refrain from price controls that actively encourage opportunistic and speculative behaviour.

In the US, farm subsidies lessen the farmer's dependence on the supply chain for income. If farmers are not motivated by cash-flow concerns to sell, the result can be a propensity to store speculatively on-farm. Farm subsidies can encourage private hoarding.

Farm subsidies actively discourage farmers to integrate into the food supply chain by encouraging the withholding of supply and undertaking retaliatory opportunistic selling. This further commoditizes and alienates the farmer. It is the antithesis of just-in-time food supply chain integration because hoarding fails to meet any of the specific criteria for delivering into a 'trade' market.[26]

Without farm subsidies, there is a motivation for farmers to actively seek cash flows through alignment with end users. They are forced to align with what the end user needs, or else they are forced to find alternate farm enterprises that are more appropriate to the food supply chain. Farm subsidies exacerbate the problem of a surplus of unwanted product because they encourage non-aligned production, which then often leads to a shortage of needed product.

Hoarding must be considered quite incongruous to a 'trade' market. It is normal practice for hoarded product to be rejected by the 'trade' as unwanted old crop unsegregated commodity because it does not meet minimum specifications demanded for end usage. This then is the dilemma for the hoarder, because when hoarding is undertaken in an attempt to achieve a higher price, the subsequent commodity sale needs a price-discounting 'last-resort' buyer because the product is mostly rejected by the local 'trade' market.

The disequilibrium of 'equilibrium'

As much public hoarding is based on vague objectives of 'equilibrium' and 'stability', the theory requires investigation. Antoine Cournot can be perceived as the French instigator of modern socialist theory during the early nineteenth century through his application of mathematics to economics through his price equilibrium, monopolies and duopolies, and gaming.[27] Such theories of free trade restriction drew heavy criticism from the French Liberal School that had been dominant since the French Revolution, and whose perception of price was more of a linkage between individual consumer–buyer behaviour and a seller's willingness and capability to deliver.

Cournot's use of aggregate supply and demand data for homogeneous commodity products that was made possible only under restricted oligopoly, duopoly, and monopoly conditions laid the foundations for much socialist and Marxian economics. It inexorably led to government control over supply through public hoarded stocks as well as price controls in a manipulative attempt to equate demand to available stock supplies.

The approach by Cournot to perfect competition and marginality was nonsensical. There were theoretical assumptions such as a perfect global homogeneity product that was totally unrealistic for biologically produced agricultural products, no barriers to freedom for entry and exit dismissed the need for capital investment and assumed no global trade barriers, full price knowledge assumed that all private transaction prices are publicly disclosed, which they are not, and low transaction costs, which may have been small in the horse, carriage and sailing ship days but are now dependent on very volatile freight and fuel costs as well as adverse currency movement.

The last factor that a farmer knows in the agricultural production cycle is marginal cost and revenue, by which time the significance to decision making is completely lost. Consumers and buyers do not base their purchase price on the cost marginality of a 'typical' farmer. The constant debate by farm management economists over what is a standard farm-cost calculation method and who exactly is the prototype 'typical' farmer is testament that farmers have difficulty in measuring their own costs of production in any one year, let alone passing this individual farm-cost information to every buyer and consumer every year for the purpose of price determination.

Despite the global commodity market having a price-taking characteristic and Cournot's supposed 'communication of markets', Cournot's law of one price was fallacious.[28] Even the perceptions of duopolies that often underlie the law of one price are often myths. McCalla claimed in 1966 that US wheat prices followed those established by the Canadian Wheat Board,[29] but had to rescind these claims within four years.[30] If true, there would have been large speculative gains from insider trading on the knowledge that Canadian Wheat Board decisions would have an immediate impact on Chicago prices. In 1982, it was found that there was a correlation between Canadian Wheat Board prices and US prices in only eight out of 16 years,[31] and that was for two adjacent countries whose weather conditions were shared across the border and who exported to similar countries.

Cournot reduced price determination to a crossing-over effect of aggregate demand and supply, but with no control over biological production and inelastic demand. There was no method to influence consumer demand even under socialism, except under price control manipulation. Under such circumstances, controlling supply influenced the control of price, which then manipulated demand, but that assumed the necessity for publicly hoarded stocks and exclusion of all private hoarding.

Whilst there was some evidence that Cournot realized his own flaws apart from external criticism, Leon Walrus went further and propounded a general theory of supply and demand equilibrium[32] in conjunction with similar marginal theory, as well as founding the Lausanne School of Economics, which was directed by Pareto. The Walrusian principle that a price would be discovered that cleared surplus supplies was admirable; however, prior knowledge of such a price could be only a theoretical estimation, possibly supported by a law of probability.

The price equilibrium inherent in the Walrusian 'law' supposedly occurred where the consumer-surplus price equals the producer-surplus price, or where there was no excess market demand or supply, but this presupposed that someone could actually calculate aggregate outcomes from all consumers and producers, which in this case would be millions of consumers and tens of thousands of farmers in each country. As well, there was a hindsight problem that prevented prior decision making on price, which was to wreak havoc in socialist government commodity stock-hoarding decision making, and caused more errors for the Taylor–Lange theory than trials.

Only a combined monopoly–monopsony organization could ever hope to achieve this 'equilibrium' state. This limitation paved the way for Marxists and socialists to criticize this weakness of capitalism before the end of the nineteenth century, albeit in its monopoly form, and advocate the replacement of capitalism with planning boards and central directives. Rather than admit that supply chain competition might actually clear supplies during surplus periods and induce production when shortages occur, both Marxist-communists and socialists wanted central controls over production, price, and consumer directives so that this imaginary 'equilibrium' state could be achieved in some egotistical power game play.

Kaldor's price 'equilibrium' assumptions[33] in 1934 included a 'closed' economy through a totally protected economy with no imports or exports, 'perfect' competition that was not possible in 'closed' economy and therefore was contradictory, 'perfect' knowledge as if this was feasible or measurable, 'direct exchange' that had all the hallmarks of a barter economy, 'independent' variables to remain constant over time as if this was valid assumption, and no price changes anticipated. Even value changes would be expected in a barter economy. Presumably everything would be in disequilibrium without these unrealistic assumptions of Kaldor.

Out of price 'equilibrium' theories of the Lausanne School in 1930 arose the cobweb theories[34] in Germany. There were assumptions of seasonal production, competition amongst farmers, no hoarding, static prices within the one year but prices could change from year to year, and relatively inelastic supply from farmers. The last three assumptions were largely unrealistic for farmers in commodity supply chains. Any hoarding, price volatility, or supply elasticity would destroy the continuous or convergent cobweb and result in divergent disequilibrium. Paradoxically the divergent cobweb theory therefore proved that commodity markets were in constant disequilibrium or chaotic,[35] but 'equilibrium' theorists were slow to acknowledge this.

One serious limitation with the 'equilibrium' concept in the cobweb theorem was that it only applied to perishable (non-storable) commodities, although this was rarely acknowledged by equilibrium advocates except through the assumption of no hoarding. Application to corn and agriculturally produced rubber implied that Schultz and Ricci[36] were attempting to apply the cobweb theory to storable commodities, which was a large error of judgement.

In its continuous form, the cobweb theory assumed similar but opposite supply and demand elasticities as represented by the percentage change in quantity against the percentage change in price, which is unrealistic for farmers and end users. Also, it assumed similar supply and demand decision–response time lags,[37] but the decision response time for farmers cannot be the same as an end user.

Disequilibrium could be explained by dissimilar supply and demand elasticities, different time lag responses, and the occurrence of hoarding. However, the instigators of the cobweb theorem perceived 'inherent instability' of commodity prices to be a 'special case', compared to the more 'stable' norm.[38] Any supply and demand curves that did not lead to a stable 'equilibrium' were referred to as 'queer shaped', implying these supply and demand curves were abnormal rather than the norm.

Anti-cyclical behaviour by farmers and end users was conveniently ignored by the equilibrium advocates under the assumption of 'rational' decision making: 'The existence of foresight, however incomplete, will always change the situation in favour of stability so long as the expectations of price-changes are in the right direction (in the direction in which prices actually move), though it will change in favour of instability if expectations are in the wrong direction'.[39]

Somehow, farmers and end users were supposed to predict the 'right' direction of prices so that 'equilibrium' could occur. To do this, there was the assumption of 'constancy of fundamental data', which assumes a world of no change. When this assumption of 'constancy of fundamental data' is removed, 'there is no longer any reason to assume that the expectation of future price-changes will generally be in the right direction. Instability in the real world then appears as the result of wrong anticipations.'[40] This is an admittance that price instability is the norm in the 'real world', and that decision makers are more likely to make irrational decisions because of the inability to forecast future price trends.

If there is 'equilibrium' in storable agricultural commodity markets, it occurs at the low price end, as depicted by price skewness.[41] It favours the end user, occurs when surpluses exist, and is the foundation for 'trade' markets.[42] A normal symmetrical distribution of prices over a period of time simply does not occur in commodities.[43] Farmer hoarding usually is an attempt to avoid this predominance of low price.

Keynes suggested that farmers be granted the freedom to produce whatever they like, but that governments were then required to intervene in supply chains to 'stabilize' prices through control over aggregate supply via public storages as buffer stocks.[44] This was still an advocacy for price equilibrium through public hoarding, but at least there was an admittance that there was little control over biological production.

The problem with the equilibrium and converging cobweb theories is that they would have been quite innocent if they had not affected the lives of people. However, it took the deaths of more than ten million in the USSR, 30 million in China, and more than five million in North Korea to prove that the equilibrium theory was fatally flawed. The equilibrium concept led to closed economies through trade protection and other intervention policies that directly pressured the First World War and aided the Second World War. There was no coincidence that German economists were at the forefront of equilibrium theory including Walrus, Kaldor who was educated in Berlin, and many at the Lausanne School, while others such as Lackmann,[45] Morgenstern with game-utility theories, Frisch,[46] and Hanau contributed to the debate. It is easy to blame the Prussian militarists who were merely adjuncts for the equilibrium seekers. But without the theory, there would have been no world wars.

Case study – the German Reich Food Estate

Following the failed Italian Corporate State experiment, and despite the rising famine due to food supply chain interference in the USSR during the previous three years, the new German government introduced the Reich Food Estate on 13 September 1933. It was supposedly to 'unite farmers, wholesalers, retailers, and consumers in a single chain that would eliminate the exploitation of one by another and ensure a fair deal for all'.[47] There was no mention of merchants or any preamble to acknowledge the function of merchants.

Organization of the Food Estate was centralized in Berlin, with new external trade barriers preventing much of the previously imported agricultural commodities. It thus continued the trade war protectionism that was one of the precursors to the 1930s Great Depression,[48] and relied heavily on import replacement policies.

This food supply chain experiment did not eventuate in commodity pools, possibly because of the independent barge operators on Germany's numerous decentralized waterways and canals. But it did lead to centralization of control over agricultural commodity production and the supply chains, price controls, and strict farm output delivery directives. Production quotas were introduced to control supply, and prices were fixed to control inflation, with lofty ideals that farmers could strengthen their financial position and 'claim their rightful place in a new Germany'.[49]

Production quotas and price controls were akin to Lange's definition of communism. The only real difference between Nazi Germany and USSR was private land ownership, but even that was to be lost in East Germany in 1947. When the Reich Food Estate kept failing from its inception, Hitler increasingly looked for agricultural land 'in the East' to counter the rising food shortages, especially focusing on the wheat fields just across the border in Poland.

New armament manufacturing in Germany squeezed the labour market, forcing up farm wages and created farm machinery shortages and high farm input prices. Potential fertilizers were diverted to chemical explosives industries. With fixed output prices and higher input prices and inflation, production quotas were not met and farmers were squeezed into poverty.

Within 12 months in September 1934, the president of the Reichsbank (Schacht) had launched a food production push that was aimed at making Germany self-sufficient in food, so as to reverse negative trade balances and protect crucial foreign currency reserves. Bulk handling facilities for commingled agricultural commodities were built and integrated with a road–rail transport network, with less dependence on river barge transport. Despite this, grain production failed to reach the levels of 1913, shortfalls of 10–30 per cent occurred in pork and fruit, 30 per cent in poultry and eggs, 50 per cent in butter and margarine, 60 per cent in legumes, and over 90 per cent in vegetable oils.[50]

Food shortages and rampant inflation in the German black market were the outcomes, despite supposed government price controls and the appointment of a Price Commissioner, which was reminiscent of the years immediately prior to the French Revolution when bread and salt prices were rigorously regulated. The Reich Food Estate failed because of its inherent tendency to inflate food prices and create food shortages at a time when controls over inflation and civil unrest were of paramount importance to achieve monetary targets. Paradoxically this occurred when international wheat stocks were at record highs and international grain prices were extremely low (an international wheat conference had convened in London during 1933 to discuss the problem of global wheat surpluses).[51]

Hoarding in Germany increased, both by farmers who wanted higher prices and by government to ensure future supply. Laws to compulsorily merge German farms into larger sizes were introduced on 26 June 1936.[52] On 26 October, a compulsory price freeze was instigated, and rationing of butter, margarine, and fats commenced on 1 January 1937, with fruit and coffee rationing occurring during the spring of 1939.

There were many efforts by the German government after the early Reich Food Estate failures in order to boost agricultural production, including the provision of cheap loans, price reductions for fertilizers, price increases for farm output, and the attempt to align production such as flax, vegetable oils, and fat with industrial demand.[53] Young people and prisoners were conscripted into farm work to meet the shortfall when foreign workers were removed.

However, in addition to poor grain harvests in the mid-1930s, foot and mouth disease decimated cattle numbers during 1938. Government controls over bakeries attempted to minimize the effects of the shortage of wheat and rye. White bread could only be purchased on the presentation of a medical certificate.[54] Thirty per cent of animal fodder still had to be imported in 1938, whilst 15 per cent of Germany's food supplies still had to be imported in 1939, mostly through new trade agreements with countries in Southeast Europe.

By March 1939, there were no eggs in Munich, no lard to bake cakes in Elberfeld, no fresh or frozen meat in Berlin, fruit supplies were cleared from markets by 7 am each morning, Ruhr workers could only eat meat once a week, regular long queues and rationing existed, and warehouse police with price regulators abounded in an attempt to control hoarding and illegal selling activities. In April, just as winter wheat was emerging from dormancy in Poland, Hitler made the decision to invade Poland five months later. By May 1939, German people were 'suffering a great deal from the shortage of all kinds of foodstuffs and respectable solid clothing'.[55]

Just as the large Polish wheat fields were ready to be reaped in autumn, German tanks and the army crossed the border on 1 September 1939. Infantry soldiers that very morning were marching through ripened wheat fields and a food bonanza. Germany's military commanders may have been under-prepared, but government survival depended on the stomachs of the people!

The increasing food crisis in Germany was well known to Stalin from 1934–39.[56] As the Polish wheat crop was tillering in June 1939, Stalin became aware of

Hitler's intention to take control over the matured wheat crop by early September, just as it was ready for harvest. Stalin hinted at a Soviet–German alliance, as Polish wheat would also placate the Russian masses who were suffering from food shortages and famine in the Soviet Union. German Foreign Minister Ribbentrop arrived in Moscow on 23 August 1939, approximately ten days before the Polish wheat was ready for harvest. The two socialist countries with the common bond of experiencing food shortages after dismally failed government supply chain interventionist schemes then neatly divided the current new wheat crop between themselves, with Stalin winning eastern Poland, Latvia, Lithuania, and Estonia, while Hitler received western Poland. It was the great wheat paddock carve-up, known officially as the Non-Aggression Pact.[57]

Speed was required as Hitler invaded eastwards and Stalin westwards in a pincer movement to prevent grain disappearing into hoarded stocks. As fast as the armies supervised the Polish sickles and scythes reaping the wheat into stooks, dried, threshed, and bagged, it was onto the horse carts and into the train wagons on route either west or east to meet the urgent demand of teeming millions of people who were dying of starvation as a result of misguided socialist 'equilibrium' pricing policies. The irony was that Stalin was to offer Lange a post-war position in the Polish cabinet.

The military authorities in both Germany and the Soviet Union were horrified by the planned invasion of Poland, as both armies were under-prepared for a war of this magnitude. There were military assassination attempts on Hitler after the Polish invasion plan became known. It explained why Hitler was forced to invade Poland when he did, to take control of the Polish wheat harvest as it was about to mature, and why there was a large time interval before the invasion of Denmark and Norway in April 1940, after war in the west had been declared and the German military was forced into serious war preparation.

Production, acreage, and supply

Production is not synonymous with either acreage or supply for storable commodities. Uncertain yield separates production from acreage, while hoarding separates much of production from supply. It is only for perishable commodities that supply might be aligned to production because of time limitations to storage. What is produced must quickly be supplied to the food chain.

If a farmer harvests a perishable commodity such as lettuce, there are only a limited number of hours to deliver it to market. Assuming that there are no prior pricing arrangements, the farmer has to accept whatever buyer demand exists at the market at the time of delivery. If the current bid price is $1 each for lettuce and the offer price is $1.50, then a farmer with an immediate truckload of lettuce might get $0.80 because of the shock to the market of receiving a whole truckload of lettuce.

Once the farmer sold the truckload of lettuce at $0.80, the bid price might return to $1 each for the lettuce and the offer price might return to $1.50, but that depends entirely on continually changing supply and demand factors. There is nothing 'equilibrium' about price for a perishable food commodity in a deregulated market, given substitute products, changing consumer demand, weather influences on supply and demand, and completely different supply and demand psychologies because of time, management decision-making differences, and agronomic considerations.

However, when the storage time is extended for non-perishable commodities, the supply curve depicting the willingness to sell as price changes may display very little relationship to production. There are two separate decision-making processes for storable commodities based on acreage and supply. The difference between what is finally produced and what is actually supplied can be a measure of hoarding. As the technical ability to successfully hoard has improved over the last 70 years, so the gap between the production and supply of storable commodity foods has widened.

Acreage curves can differ between price and income decisions. A farmer could decide on acreage based on price, as indicated in Figure 7.1. The farmer could increase acreage because there is a current high or increasing price. This assumes the capability exists to increase the number of acres planted, given no limitations to land, labour, capital, agronomic, or machinery factors. However, farmers could decrease their acreage by adopting the anti-cyclical approach or undertake crop rotation, thus making the decision psychology more difficult.

Figure 7.1 indicates that low prices will tend to result in low acreages planted (A), whereas high prices will tend to result in high acreages planted (B). Resource and agronomic limitations and minimum price-acreage responses may prevent this curve from having a linear relationship. Rather the curve has a higher probability of being concave.

In contrast, a farmer could be motivated by income maintenance to increase acreage (Figure 7.2). At low prices, a farmer may be forced to increase the acreage planted to maintain average income, assuming that the farmer has the capability to

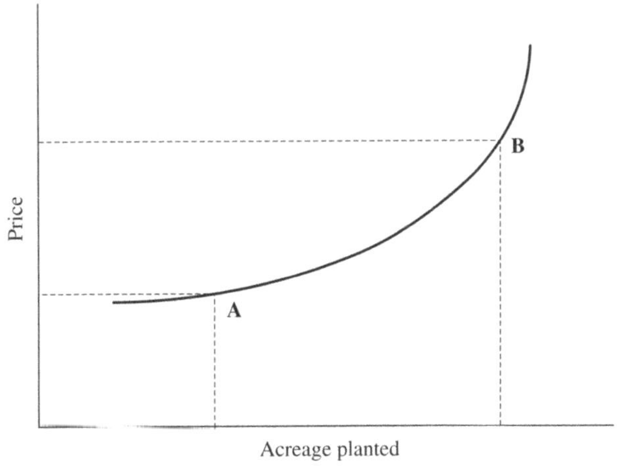

FIGURE 7.1 Theoretical acreage curve response to price
Source: John Williams

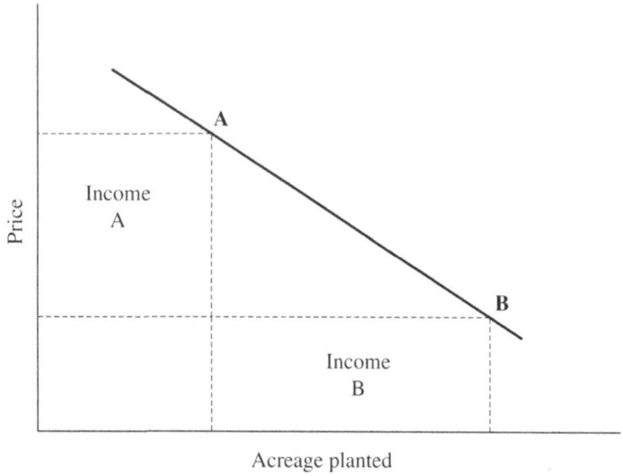

FIGURE 7.2 Possible acreage curve response to income
Source: John Williams

increase acreage. At high prices, farmers might decide to plant less acres and undertake crop rotation, with the objective that average income would be highly likely to be achieved.

Figure 7.2 indicates that at low prices, a farmer may decide to increase acreages planted for the purpose of income maintenance (B), while at high prices, a farmer might decide to limit acreages to achieve some agronomic objective (A). Both incomes (A and B) could remain relatively the same.

If final production is outside the control of farmers or governments because of yield uncertainty, then there can be no certainty of production response to price as was theorized in the cobweb theorem. Response implies human decision making. To claim that production is a response to price is to assume prior knowledge of final production, which is affected by time lags. If production was known in advance, then price might be known within a price range. However, there is little to suggest that this would be an equilibrium situation, particularly when both supply and price can be influenced by hoarding.

Production does not equate to supply in storable commodity markets because of hoarding, regardless of whether it be short or long term. It is erroneous to claim in non-perishable commodity markets that there is a direct relationship between price and production, or that production equals supply. Price is more dependent on actual supply within 'trade' markets, rather than on occurrences in non 'trade' storages.[58]

There is an inverse relationship between price and supply for storable commodities (Figure 7.3). As price rises, less commodity is supplied (A) in the expectation of further price rises. Alternatively, if prices fall, more commodity is supplied (B) in the expectation of further price falls.

Above A on the hoarder's supply curve, the less likely will any supply be released as price increases because of the supplier's expectations of further price rises. The lower the price, the more likely the supplier will sell because of expectations of further price falls, as well as falling prices decreasing the profitability of hoarding.

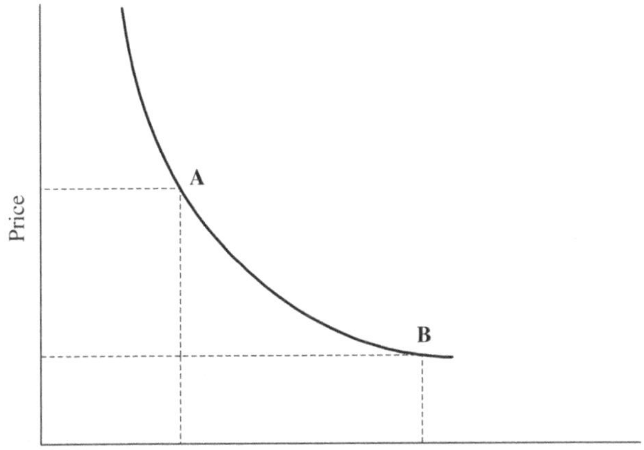

FIGURE 7.3 Storable commodity supply curve - hoarder's response curve
Source: John Williams

This has important ramifications for governments that attempt each year to manipulate commodity supply through price controls. When commodities become short in supply, governments usually raise commodity prices. This causes more hoarding in the expectation of further price rises in the near future. What may have been a temporary shortage, which should have been corrected by allowing imports into the food supply chain, the government transforms into a major food crisis.

Demand curves

Figure 7.4 depicts the hoarder's supply response curve against the demand curve of an end user to indicate where shortages and surpluses are likely to occur.

Above E, shortages will occur because demand exceeds supply, whereas surpluses will occur below E because supply exceeds demand. At price P1 in Figure 7.4, the hoarder will supply quantity Q4, while the end user will only be demanding quantity Q3. The effectiveness of hoarding is thus removed in surplus markets.

However, at price P2, the hoarder will restrict the response to supply at quantity Q1 in the hope of forcing prices up further, while the end user will restrict demand from quantity Q3 to Q2 by using substitutes or cutting back food manufacturing.

At some price above P2, the end user will usually be forced to rapidly offer higher prices to attract supplies to prevent closures at manufacturing plants and staff retrenchments. The problem is that such price rises are exactly what the hoarder wants and this encourages even greater withholding of supply.

A peak price is reached when the end user does lay off staff and closes manufacturing plants, or shifts to alternate manufacturing, and these actions result in a complete reversal of supply chain power. Alternatively, prices could fall in the anticipation of huge new crop production that has been a response by farmers to the high prices. The hoarder is left with unwanted undifferentiated commodity that has to be dumped or discounted

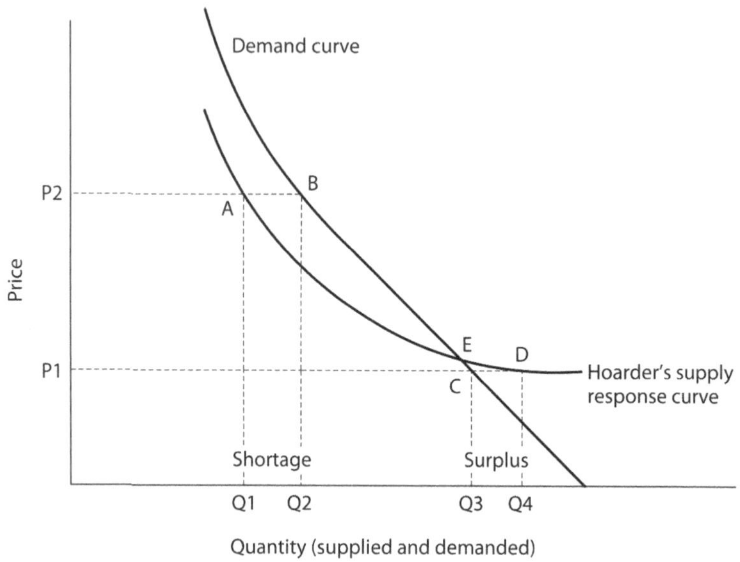

FIGURE 7.4 Buyer's demand curve with hoarder's supply response curve
Source: John Williams

into a non 'trade' supply chain with collapsing prices in front of new crop production, which results in supply being greater than demand. This occurs at prices below E.

Supply chain power for the hoarder only occurs when end users raise price. If end users reduce their bid prices or lose buying interest, then the hoarder loses supply chain power. There is a delicate balance between supply chain power and panic for the hoarder, because hoarding usually involves unwanted undifferentiated commodity that is yet to be sold. Figure 7.4 assumes that the end user actually wants to buy the hoarder's commodity, which may not be the case because of quality.

Perceptions of shortages are often more important than the reality of surpluses. Hoarders generally benefit from a rumoured perception in the supply chain, and lose when the reality becomes known. The more effective the rumour, the greater will be consumer panic.

The problem of shortages when hoarding occurs can usually be resolved quickly by governments allowing imports, but this is something that governments generally fail to do either because of farmer protection policies, hoarding vesting interests, or protection of government-hoarded commodity stocks that could gain from a windfall. It could be argued that the hoarding problem might never have started if the government had allowed freer trade to occur. Hoarding is the dilemma of governments that restrict trade, as occurred in the German Food Reich.

This more realistic approach to the impact of hoarding on price determination undermines Haldane's mathematical approach[59] to trade cycles and price equilibrium, and Ezekiel's theoretical cobweb theorem[60] that simply ignored hoarding. Hoarders exist regardless of whether they fit with econometric modelling, and regardless of government attempts to exterminate them.

The rise of speculative bunkering

The rise of hermetically sealed non-fumigated bunkers has opened up new, bigger, and more convenient methods for hoarding, while large maintenance costs such as fumigation and quality losses can be reduced. It has increased the ability of individuals to accumulate and act as speculative merchants.

Bunkering meant that expensive traditional warehousing infrastructure could be by-passed. It also meant that commodity could be hidden better from competitors, supply chain intermediaries, and anti-hoarding 'police' inspectors. There is a lack of both public and private transparency with private bunkering, which is ideal for hoarding conditions. This explains why global commodity food stocks-to-usage ratios have seemingly decreased as successive governments have deregulated their supply chains.

Improved bunkering provides farmers with greater price speculative opportunities when increasing shortages occur, but the downside is further alienation from 'trade' markets and increased commoditization. However, if farmers can master the art of hermetically sealed bunkers or silos, the possibility of better alignment with organic 'trade' markets may exist, which would decrease the incentive to hoard.

Hoarding psychology

Agricultural commodity hoarders have been found to be high risk takers, acknowledge that they speculate on price by hoping for higher prices, are more likely to make decisions alone and spontaneously, and are more likely to be very reluctant and sceptical in the adoption of new selling and pricing ideas.[61] They are more likely to have risk-attitude flexibility, with changes to their risk profile depending on their role in the family, farm, and business. There appears to be strong character traits of risk taking and individual decision making, but perhaps with a contradiction between flexibility and a reluctance to change.

Farmers who use storage as part of their selling method are more likely to achieve perceived 'realistic' target prices when compared with forward contracting, cash sales at harvest, or pooling.[62] This supports general perceptions of hoarding success, despite the evidence that hoarding ranks low when compared to other selling and pricing strategies, even in shortage years such as 2007–08.[63] There are no ready explanations for this difference between perception and reality, other than storage users having the flexibility of when they sell, and this is likely to occur whenever prices are perceived to be 'realistic'. The corollary suggests that farmers who do not hoard were less likely to achieve target prices that were perceived to be realistic, but this is not borne out in strategy comparison and testing reality.

Larger and more diversified farmers who are less dependent on any one farm enterprise are more likely to hoard than smaller farmers.[64] This is consistent with larger cash flows decreasing the urgency to sell. Smaller farmers would be expected to sell a higher percentage of their production quickly in order to generate sufficient cash flow for farm operations and to repay debt. In contrast, there have been instances in Australia when very large farmers have delayed paying contractors and merchandisers for 11 months because of a lack of cash flow resulting from hoarding. Farm size may also influence the timeliness to repay debts.

Theoretically, hoarders would be expected to have low debt and high cash reserves. However, hoarders' debt levels were found to be relatively high.[65] There is consistency in that hoarders perceive themselves as risk takers, with actual high risk taking for price, debt, and cash flow. There would be psychological inconsistency if hoarders somehow had low debt and high cash reserves. Alternatively, it could be argued that the reason why farmers have low debt and high cash reserves is because they did not undertake hoarding.

This risk-taking evidence by hoarders was also supported by farmers with high drought frequency also using much more storage and much less cash sales at harvest. High farming risk resulted in a greater propensity to take on even more risk.

Geographic location and transport distance may be important considerations for hoarding. Farmers who were near major merchant or end-user activity were much more likely to hoard.[66]

Notes

1 Supply response times are far more important than longer-term supply elasticity.
2 Backwardation and contango characteristics in 'trade' and futures markets will be discussed in Chapter 10.
3 von Braun, Joachim and Maximo Torero, 2009, *Implementing Physical and Virtual Food Reserves to Protect the Poor and Prevent Market Failure*, Policy Brief 10, International Food Policy Research Institute, Washington, DC.
4 Carter, Colin, 1999, *Commodity Futures Markets: a Survey*, The Australian Journal of Agricultural and Resource Economics, 43:2, 209–247.
5 Pace, Noemi, Andrew Seal, and Anthony Costello, 2008, *Has Financial Speculation in Food Commodity Markets Increased Food Prices?* Lancet, No. 371, 1648–1650.
6 See Chapter 5.
7 Williams, John, 2009, *Decision-Making in Australian Wheat Marketing and Price Risk Management*, School of Land and Environment, University of Melbourne, p. 111.
8 Williams, John, 2009, *Decision-Making in Australian Wheat Marketing and Price Risk Management*, School of Land and Environment, University of Melbourne.
9 Edgeworth, Francis Ysidro, 1881, *Mathematical Psychics: An Essay on the Application of Mathematics to the Moral Sciences*, N.P., London.
10 Madsen, Jakob B., 2001, *Trade Barriers, and the Collapse of World Trade During the Great Depression*, Southern Economic Journal, 67:4, 848–868.
11 Taylor, Fred M., 1929, *The Guidance of Production in a Socialist State*, The American Economic Review, 19:1, 1–8.
12 Lange, Oskar R., 1934, *The Determinateness of the Utility Function*, The Review of Economic Studies, 1:3, 218–225.
13 Shibata, Kei, 1933, *Marx's Analysis of Capitalism and the General Equilibrium Theory of the Lausanne School*, The Kyoto University Economic Review, July.
14 Lange, Oskar R., 1935, *Marxian Economics and Modern Economic Theory*, The Review of Economic Studies, 2:3, 159–192.
15 Lange, Oskar R., 1936, *On the Economic Theory of Socialism, Part One*, The Review of Economic Studies, 4:1, 53–71.
16 Allen, Roy G.D., 1934, *The Nature of Indifference Curves*, The Review of Economic Studies, 1:2, 110–121.
17 Lange, Oskar R., 1936, *On the Economic Theory of Socialism, Part One*, The Review of Economic Studies, 4:1, 60.

18 See case study in Chapter 6.
19 See case study later in this chapter.
20 von Mises, Ludwig, 1944, *Omnipotent Government: The Rise of the Total State and Total War*, Yale University Press, New Haven, CT.
21 Lange, Oskar R., 1936, *On the Economic Theory of Socialism, Part One*, The Review of Economic Studies, 4:1, 71.
22 Lange, Oskar R., 1936, *On the Economic Theory of Socialism, Part One*, The Review of Economic Studies, 4:1, 57–68.
23 Williams, John, and Bill Malcolm, 2012, *Farmer Decisions About Selling Wheat and Managing Price Risk in Australia*, Australasian Agribusiness Review, 20, Paper 1, 1–10.
24 Lee, Hau Leung, Padmanabhan, V., and S. Whang, 1997, *The bull-whip effect in supply chains*, Sloan Management Review, 38:3, 93–102.
25 von Mises, Ludwig, 1920, *Economic Calculation in the Socialist Commonwealth*, Die Wirtschaftsrechnung Im Sozialistischen Gemeinwesen, Archiv Für Sozialwissenschaften, 47, 86–121.
26 This aspect will be discussed further in Chapter 10.
27 Cournot, Antoine A., 1838, *Researches on the Mathematical Principles of the Theory of Wealth*, Macmillan, New York.
28 Ardeni, Pier Giorgio, 1989, *Does the Law of One Price Really Hold for Commodity Prices?* American Journal of Agricultural Economics, 71, 661–669. (This article was written at the pinnacle of food supply chain regulation and government price controls for monopolies and oligopolies.)
29 McCalla, Alex F., 1966, *A Duopoly Model of World Wheat Pricing*, American Journal of Agricultural Economics, 48, 711–727.
30 McCalla, Alex F., 1970, *Wheat and the Price Mechanism: Or Duopoly Revisited and Abandoned*, paper presented at the Seminar on Wheat, University of Manitoba, October.
31 Spriggs, John, Kaylen, Michael, and David Bessler, 1982, *The Lead-Lag Relationship between Canadian and U.S. Wheat Prices*, American Journal of Agricultural Economics, 64:3, 569–572.
32 Walrus, Leon, 1874, *Elements of Pure Economics*, N.P., Lausanne.
33 Kaldor, Nicholas, 1934, *A Classificatory Note on the Determination of Equilibrium*, The Review of Economic Studies, 1:2, 122–136.
34 Schultz, Henry, 1930, *Der Sinn der statistischen Nachfragen*, Heft 10, Veröffentlichungen der Frankfurter Gesellschaft für Konjunkturforschung, Kurt Schroeder Verlag, Bonn p. 34, and Ricci, *Umberto, Die synthetische Ökonomie von Henry Ludwell Moore*, Zeitschrift für Nationalökonomie,Wien, Band 1, Heft 5, pp. 649–656.
35 Hudson, Richard L. and Benoit B. Mandelbrot, 2004, *The (Mis)Behaviour of Markets: A Fractal View of Risk, Ruin, and Reward*, Basic Books, New York.
36 Kaldor, Nicholas, 1934, *A Classificatory Note on the Determination of Equilibrium*, The Review of Economic Studies, 1:2, 133–134.
37 Kaldor, Nicholas, 1934, *A Classificatory Note on the Determination of Equilibrium*, The Review of Economic Studies, 1:2, 134–136.
38 Kaldor, Nicholas, 1934, *A Classificatory Note on the Determination of Equilibrium*, The Review of Economic Studies, 1:2, 136.
39 Kaldor, Nicholas, 1934, *A Classificatory Note on the Determination of Equilibrium*, The Review of Economic Studies, 1:2, 136.
40 Kaldor, Nicholas, 1934, *A Classificatory Note on the Determination of Equilibrium*, The Review of Economic Studies, 1:2, 136.
41 Refer to Figure 1.3 in Chapter 1.
42 This will be discussed in Chapter 10.
43 Hudson, Richard L. and Benoit B. Mandelbrot, 2004, *The (Mis)Behaviour of Markets: A Fractal View of Risk, Ruin, and Reward*, Basic Books, New York.
44 Keynes, John Maynard, 1938, *The Policy of Government Storage of Food-Stuffs and Raw Materials*, The Economic Journal, 48:191, 449–460.

45 Lackmann, L.M., 1936, *Commodity Stocks and Equilibrium*, The Review of Economic Studies, 3:3, 230–234.
46 Frisch, Ragnar, 1936, *On the Notion of Equilibrium and Disequilibrium*, The Review of Economic Studies, 3:2, 100–105.
47 Evans, Richard J., 2006, *The Third Reich in Power*, Penguin, London, Ch. 4.
48 Madsen, Jakob B., 2001, *Trade Barriers, and the Collapse of World Trade during the Great Depression*, Southern Economic Journal, 67:4, 848–868.
49 Evans, Richard J., 2006, *The Third Reich in Power*, Penguin, London, p. 346.
50 Evans, Richard J., 2006, *The Third Reich in Power*, Penguin, London, p. 347.
51 Whitwell, Greg and Diane Sydenham, 1991, *A Shared Harvest – The Australian Wheat Industry 1939–89*, Macmillan Education Australia, Melbourne, pp. 56–57.
52 Evans, Richard J., 2006, *The Third Reich in Power*, Penguin, London, p. 348.
53 Evans, Richard J., 2006, *The Third Reich in Power*, Penguin, London, p. 348.
54 Evans, Richard J., 2006, *The Third Reich in Power*, Penguin, London, p. 350.
55 Evans, Richard J., 2006, *The Third Reich in Power*, Penguin, London, p. 350.
56 Evans, Richard J., 2006, *The Third Reich in Power*, Penguin, London, Ch. 5.
57 Evans, Richard J., 2006, *The Third Reich in Power*, Penguin, London, pp. 692–693.
58 This will be explained more in Chapter 10.
59 Haldane, John B.S., 1934, *A Contribution to the Theory of Price Fluctuations*, The Review of Economic Studies, 1:3, 186–195.
60 Ezekiel, Mordecai, 1938, *The Cobweb Theorem*, The Quarterly Journal of Economics, 52:2, 255–280.
61 Williams, John, and Bill Malcolm, 2012, *Farmer Decisions About Selling Wheat and Managing Price Risk in Australia*, Australasian Agribusiness Review, 20, Paper 1, 1–10.
62 Williams, John, and Bill Malcolm, 2012, *Farmer Decisions About Selling Wheat and Managing Price Risk in Australia*, Australasian Agribusiness Review, 20, Paper 1, 1–10.
63 Williams, John, and Bill Malcolm, 2012, *Farmer Decisions About Selling Wheat and Managing Price Risk in Australia*, Australasian Agribusiness Review, 20, Paper 1, 1–10.
64 Williams, John, and Bill Malcolm, 2012, *Farmer Decisions About Selling Wheat and Managing Price Risk in Australia*, Australasian Agribusiness Review, 20, Paper 1, 1–10.
65 Williams, John, and Bill Malcolm, 2012, *Farmer Decisions About Selling Wheat and Managing Price Risk in Australia*, Australasian Agribusiness Review, 20, Paper 1, 1–10.
66 Williams, John, and Bill Malcolm, 2012, *Farmer Decisions About Selling Wheat and Managing Price Risk in Australia*, Australasian Agribusiness Review, 20, Paper 1, 1–10.

Further readings on hoarding

Ackerman, F. and A. Nadal, 2004, *The Flawed Foundations of General Equilibrium*, Routledge, London.

Bobenrieth, Eugenio, Bobenrieth, Juan, and Brian D. Wright, 2006, *Theoretical Advances in the modelling of storage and speculation*, in Alexander Sarris and David Hallam, Eds., *Agricultural Commodity Markets and Trade – New Approaches to Analysing Market Structure and Stability*, FAO, Rome, and Edward Elgar, Cheltenham.

Bowley, M., 1973, *Studies in the History of Economic Thought before 1870*, Macmillan, London.

Brennan, Michael J., 1958, *The Supply of Storage*, The American Economic Review, 48:1, 50–72.

De Vroey, M., 2002, *Equilibrium and Disequilibrium in Walrusian and Neo-Walrusian Economics*, Journal of the History of Economic Thought, 24:4, 405–426.

Donzelli, F., 1997, *Pareto's Mechanical Dream*, History of Economic Ideas, 5:3, 127–178.

Fackler, P.L. and M.L. Livingston, 2002, *Optimal Storage by Crop Producers*, American Journal of Agricultural Economics, 84, 645–659.

Fornari, Harry D., 1982, *Recent Developments in the American Grain Storage Industry*, Agricultural History, 56:1, 264–271.

Hicks, Sir John, 1974, *The Crisis in Keynesian Economics*, Basic Books, New York.

Machlup, F., 1963, *Essays on Economic Semantics*, Prentice-Hall, Englewood Cliffs, NJ.
Mandler, M., 1999, *Dilemmas in Economic Theory: Persistent Foundation Problems of Microeconomics*, Oxford University Press, Oxford.
Rothbard, Murray N., 1988, *Chaos Theory: Destroying Mathematical Economics from Within?* The Free Market – The Mises Institute Monthly, VI:3, 1.
Schumpeter, J.A., 1954, *History of Economic Analysis*, Oxford University Press, Oxford.
Weymar, H., 1966, *The Supply of Storage Revisited*, The American Economic Review, 56, 1226–1234.
Williams, Jeffrey C. and Brian D. Wright, 1991, *Storage and Commodity Markets*, Cambridge University Press, New York.
Working, Holbrook, 1949, *The Theory of Price of Storage*, The American Economic Review, 39:6, 1254–1263.

8

CORRUPTION IN THE FOOD CHAIN

This chapter is dedicated to Peter Mulamba[1] and Michael Wansley[2].

The definition of corruption is somewhat vague depending on public and private perspectives. Public corruption can refer to the misuse of a public position for some private gain.[3] Alternatively, there can be many different aspects of private corruption, such as collusion, lobby groups, political 'donations', and employment of previous government officials.

Generally, private sector problems increase when private monopolies or oligopolies occur, which suggests that increased competition is an important factor in minimizing corrupt behaviour in the private sector. The corollary is that those governments which do not enforce competition are more likely to be those with a high incidence of corruption.

Both public hierarchal positions and private monopolies create power, whether that is real or perceived. Perceptions of ultimate power in public hierarchal positions and private monopoly situations can lead to a high incidence of corruption. In food supply chains, it can be argued that only consumers have real market power through choice of purchases, therefore any perceptions of supply chain power are illusory concepts.

The highest level of corruption occurs with developing or transitional countries, with most of these governed by socialist policies.[4] Occurrences of socialist governments and associated centralist-intervention policies with lower per capita income levels or a polarized income distribution are frequent.[5] As well, there is a close association between trade protectionist countries and high incidences of corruption.[6]

Whilst societal culture might be a major contributor to private corruption, efforts to change cultural mores are generally very difficult when there is parallel public corruption. Government interventionist and protectionist policies together with related organizational structures and institutions are important precursors to corruption,[7] probably more so than average income levels or education. This chapter will focus on the causes and consequences of corruption, rather than seek ways to change individual cultural behaviour. It is argued that if public corruption can be minimized through greater competition, then it might be easier to change individual behaviour.

Corruption Indices

There are three main corruption monitoring indices, the Corruption Perception Index (CPI), the Bribe Payers Index (BPI), and the Global Corruption Barometer (GCB). Whilst there has been some criticism that the occurrence of corruption is much worse than that measured in the indices,[8] the relative movement in the indices does provide some indication of global trends.

The Corruption Perception Index (CPI) by Transparency International is an index that regularly monitors public sector corruption. It measures and compares different government departments, as well as comparing overall government performance internationally. Table 8.1 indicates the top ten countries with the least public sector corruption, with New Zealand, Denmark, and Singapore in the top three. These might be characterized as small geographical countries with small populations, medium economic growth rates, and relatively high income levels.

Table 8.2 lists the bottom ten countries with the most public sector corruption, with Somalia, Afghanistan, and Myanmar being in the top three. Again, these might be characterized as relatively small geographical countries with slightly larger populations, but with low economic growth rates and low income levels.

In contrast, the Bribe Payers Index by Transparency International monitors the tendency for private companies from major exporting countries to transact international bribes. Table 8.3 indicates the top five countries that have the least tendency for private companies to transact bribes, with Belgium, Canada, and the Netherlands in the top three. Geographical characteristics of these countries are diverse, but all have relatively high per capita incomes.

TABLE 8.1 Top 10 countries with the least public-sector corruption

Rank	*Country*	*Corruption Perception Index (2009 Best Score)*
1	New Zealand	9.4
2	Denmark	9.3
3	Singapore	9.2
4	Sweden	9.2
5	Switzerland	9.0
6	Finland	8.9
7	Netherlands	8.9
8	Australia	8.7
9	Canada	8.7
10	Iceland	8.7

Source: Transparency International

TABLE 8.2 Top 10 countries with the most public-sector corruption

Rank	*Country*	*Corruption Perception Index (2009 Worst Score)*
1	Somalia	1.1
2	Afghanistan	1.3
3	Myanmar	1.4
4	Sudan	1.5
5	Iraq	1.5
6	Chad	1.6
7	Uzbekistan	1.7
8	Turkmenistan	1.8
9	Iran	1.8
10	Haiti	1.8

Source: Transparency International

TABLE 8.3 Top five countries with the least tendency for company international bribes

Rank	*Country*	*Bribe Payers Index (2008 Best Score)*
1	Belgium	8.8
2	Canada	8.8
3	Netherlands	8.7
4	Switzerland	8.7
5	Germany	8.6

Source: Transparency International

TABLE 8.4 Top five countries with the most tendency for company international bribes

Rank	*Country*	*Bribe Payers Index (2008 Worst Score)*
1	Russia	5.9
2	China	6.5
3	Mexico	6.6
4	India	6.8
5	Brazil	7.4

Source: Transparency International

In contrast, Table 8.4 lists the top five countries with the most tendency for a private company to offer bribes internationally. Russia was the worst, followed by China, Mexico, India, and then Brazil. All of these countries are characterized by large populations and economies, relatively high economic growth rates, and increasing per capita incomes.

Gifts, bribes, and extortion

The provision of gifts between two strangers has a long history in many countries.[9] It was usual to offer gifts to gain trust as a basis for trade or peace. Suspicion was thereby reduced, communicative dialogue was introduced, and friendships established. In some countries and cultures, the non-provision of a gift was an insult and provocative.

It is the rarity of gift offering that makes gifts so powerful. If gifts are offered regularly, then they lose their ability to surprise and appeal. They become so mundane that they are automatically expected. Any disruption to product or service provision that began as a gift might be now perceived insulting or demeaning. The original intention of gift giving as a surprise is reversed. It becomes a surprise not to receive the product or service. Dependency has therefore removed any element of surprise.

The dilemma of distinguishing between a gift and a bribe dates back to at least the ancient Greek philosophers. In the old French language, a bribe related to bread probably because of its scarcity under government controls, and may have implied that the act of bread giving unduly sought to influence the recipient.

Acts of sacrifice for others and the achievement of well being from the donor may disguise deeper emotions of insecurity, and may achieve the opposite to what was initially desired.[10] The act of exchanging gifts may be similar to international countertrade because it rarely gives the person what is really desired. However, the act of gift giving or exchanging gifts is embedded deep into the local societal culture in many countries. In some countries, gift giving has been officially confined to only certain prescribed days of the year.

A gift is rarely given without a purpose and expected outcome. Even the selection of a gift may be biased towards an intention or functionality of the gift. Timeliness tends to reinforce the expectations of certain desired outcomes by the donor. The recipient of the gift rarely displays honesty in expressing real emotions. A gift and bribe are inseparable in many situations, despite the ambiance of innocence.

The gift may be defined as a bribe if it creates dependency by either the giver or taker, seeks favouritism from the other person, seeks to influence the other person, or if there is an extortion component through an action that is dependent upon reciprocating exchange or payment. There might be tax avoidance issues whenever a gift is a pecuniary alternative such as a substitute for a financial debt repayment, or whenever there is a barter exchange of 'gifts'.

Dependency

The type of gift is important in assessing the impact of dependency. If the gift was a rarity that was not produced locally, then the longer-term impact might not be considerable despite the occurrence of dependency. However, if the gift replaced something already being accomplished by the individual or a local group, then the ramifications of dependency on the gift taking become more identifiable. Even worse, the ramifications on the gift taker might have been expected by the gift provider with an ulterior motive. Many enemies have been defeated by the provision of 'gifts' before a battle.

Gifts that are regularly provided and create dependency, especially those that replace local products and services, tend to reduce the incentive to work and produce locally, and remove the motivation to create and develop new ideas. This results in more dependency. Regardless of the motive of gift giving, whether it be sinister or an exuberance of 'doing good', the results may be devastating for the recipient.

Rarely would a gift be provided without some encumbrance. Usually there is a subtle expectation that the recipient will be overwhelmed with appreciation for the donor and that the latter will either maintain or increase social status acceptance. Patronization is intricately linked to giving, with issues of self-esteem and craving for social status. The altered psychology of the recipient is rarely considered, except when a change to the recipient was the original intention.

Basic causes for corruption

Government intervention in the economy or markets has been identified as the primary cause for public corruption.[11] Many corrupt practices arise from government organizations and agencies that are associated with such intervention. However, there are a myriad of associated causal factors that can entrench and prolong corruption.

Wherever there is extreme dependency by private individuals on discretionary public decision making, the probability of corruption is extremely high. This type of corruption is highly correlated with bargaining power,[12] with a strong likelihood of corruption continuation whilst this power remains. The size of the graft payment demanded usually depends on the ability of the private sector to pay.

Merely dissipating or decentralizing public decision making often results in inefficient decision making, which can perpetuate corruption, albeit in a different form. The decision-making objective needs to be challenged. Instead of issuing a government import procurement tender or implementing an import quota for the private sector, it may be better to remove structural impediments for the private sector to trade internationally.

Market or supply chain intervention can often be unnecessary, ineffective, or counterproductive. Policies that seek economic 'stability' can often result in greater uncertainty and volatility. These can quickly result in private sector frustration, particularly when there are competitive pressures in the private sector and none in the public sector. This encourages corruption to either obtain the required approval quickly or to circumvent the system.

If the objective of the public regulations was tax revenue generation, then an inefficient public collection system is likely to slow or prevent transactions occurring and be self-defeating through declining tax receipts. Unless the taxation revenue collection from the private sector was efficient, there would be added burdens to the private sector in terms of additional transaction costs, as well as generating tax revenue that was eroded by high administrative and collection costs.

Alternatively, if the inefficient regulations were deliberately aimed at preventing imports, exposing the public sector to corruption might be the wrong method. It might be far better to remove structural impediments to export industry efficiency, thereby enabling domestic industries to compete favourably with imports. Bureaucratic inefficiency is unlikely to overcome a country's lack of comparative advantage in land, labour, capital, resources, skills, or knowledge; rather it will tend to worsen it.

Human behaviour may not be the primary cause of corruption in global food chains but rather a secondary outcome. The organizational structure may create opportunities for corruption of the individual rather than the individual corrupting the organization. Any public or private monopoly–monopsony organizational structure can attract corrupt practices.[13]

Wherever there is discretionary public control, or the perception of real or illusory market power, there will be grounds for corrupt practices.[14] This again suggests that corruption does not have a public versus private divide,[15] rather it is the organizational structure combined with the perception of discretionary power that ultimately causes the potential corruption of the individual.

It can be argued that if market power is sufficiently dispersed or if there was no propensity for unitary discretionary decision making, then a supply chain might function more efficiently and effectively. However, if discretionary decision-making power is centralized, then there is a higher probability that corruption could occur. Both frustration and opportunism may be the key motivators to corruption.

Due diligence is required when either property rights are not clearly distinguishable or if there is shared ownership of common property. There may be little market power associated with commingled commodity pool management, but this has not prevented individuals within such management structures becoming 'dictators'.[16]

Governments rarely have sufficient information to make informed selections and rejections of private companies for particular undertakings, other than in public tenders. Public decision makers generally do not have the flexibility to adjust to change whenever private company organizational structures and operations adapt to competitive market changes. Usually the paradox prevails, which means private companies selected for favourable public treatment can undergo much anguish because of the higher probability of subsequent failure or discredit.

At worst, government patronage favours certain private companies at the expense of potentially more viable alternate companies. Such subjective selectivity is corruption, which can be subsequently entrenched into government dynasties akin to the powerful ancient clan and nobility societal structures.

Protecting infant industries under import protection and replacement policies is another form of government patronage, irrespective of the ability of the new industry to succeed. New industries are sometimes supported despite a lack of comparative advantage, which leads to general economic inefficiency. Industry and supply chain efficiency, particularly for potential export industries, is compromised for the sake of vested interests having persuasion over government policy.[17]

What may have began in Brazil during June–July 2010 as sugar hoarding by millers ended in a mire of congestion and corruption by late July. It had been an attempt to withhold sugar supplies until three weeks prior to Ramadan, which began on 12 August. Instead, rain and grossly inadequate road and port infrastructure resulted in total Brazilian port congestion, which took months to resolve with more than 110 ships continually waiting off the ports to load sugar. Port-loading belts were exposed to the rain and therefore could not operate in adverse weather.

Corruption occurred when ship owners bribed Brazilian port officials to facilitate queue-jumping. It also occurred when truck drivers were bribing each other as well as road-port officials to also jump the congestion queue. There are times when hoarding is futile, but to undertake mass selling when there was grossly inadequate public infrastructure was a recipe for disaster and a banner for corruption, even for the world's sixth largest economy.

Fiscal expenditures such as subsidies can often be reduced to patronage and largesse. They can cause economic distortions because larger organisations usually get the most benefit regardless of their efficiency, which then prevents potentially more efficient smaller organizations from developing and increasing in size.[18] Subsidies can distort trade, either enabling exporters to undercut competitors or facilitating international dumping under the guise of food 'aid'.

Apart from being discriminating and supply chain distorting, price controls can be corrupting through favouritism and creating dependency on government. Protection to one group, such as either high prices favouring producers or low prices favouring consumers, disadvantages the other group. This creates powerful vested interests in maintaining the status quo.[19]

Subjectivity enters the decision-making processes involved with choosing who is allocated government licences, or which industries should benefit from import embargoes, or which companies should get tariff protection, or which individuals should get a quota. Government preferential treatment, political donations, and powerful linkages with the private sector are all forms of corruption in restricting or allocating outcomes from government policies.[20]

Once government restrictions or allocations occur, then this tends to create powerful vested interests in maintaining the status quo, even to the extent of ensuring that governments get re-elected. Political parties are often started to ensure that government largess, protection, and allocations are maintained.[21] Democratic processes are thus undermined through parochial vested interests.

As well, government licences, permissions and allocations can often be on-sold, thus creating a market in public largess under the guise of using the 'market mechanism'. A water allocation created by government authority can be subsequently sold and make huge profits for the recipient of such allocations. In Australia, some farmers make more income from trading government annual water allocations than they do from actual farming. This creates dependency on the government to continue the system, as well as causing resource misallocation. Serious corruption over the distribution of public water contracts has occurred in India.[22]

Wherever public corruption occurs, there is a high probability of lobbyists who act as intermediary links to government officials, with payment through clandestine commissions. In situations where public corruption is widespread, a large lobbying industry becomes ensconced around public administrative centres. The 2006 Cole Inquiry into corrupted Australian wheat sales to Iraq under the Oil-for-Food Programme reported that professional introductory middlemen associated with Iraqi officials do not add any value to the food chain.[23]

Lobbying has sometimes been considered better than bribing because it can seek to remove the institutional or bureaucratic obstacle, it can affect all stakeholders collectively whereas bribery is confined to individual behaviour, and the outcome might be more permanent rather than the short-term bribe solution.[24] Counter-arguments include the shift of corruption to other sectors, and the encouragement by vested interests for governments to continue to intervene with protectionist policies. The longer-term solutions of less intervention and protection policies, and the introduction of competition to minimize corruption, are rarely the result of initiatives from lobbying groups. Such groups have a vested interest in the status quo, otherwise they soon would fail.[25]

Restrictive UN trade sanctions were imposed against Iraq in the aftermath of the Gulf War in 1990–91, which was followed by the introduction of the Oil-for-Food Programme supposedly for humanitarian reasons in 1995. What started as a typical counter-trade situation to alleviate consumer suffering in Iraq ended up in a mire of corruption of governments and private companies, involving at least 49 countries. By manipulating the price that oil was exchanged for food, Iraqi corruption skimmed up to US$4 billion or 10 per cent of the total value of the scheme.[26]

One example of the Oil-for-Food transactions was wheat that was exchanged in Iraq during 1995–2003 for crude oil that then had to be on-sold to the World Bank for payment to the exporter in US dollars. This simple commodity oil-for-wheat bartering system was 'open to corrupt practices', 'compromised Australian Government Foreign Affairs-Trade officials', and 'did not lead to long-term food value-chain integration'; while 'premiums' that were so widely touted as justification for the continuation of Australia's wheat export single-desk were actually 'corruption payments'.[27] Iraqi government officials and ministers were rewarded through payments via a clandestine Jordanian trucking company that had 'connections' with the Iraqi president.

Unless wartime sanctions are involved, attempting to enforce restrictive trade sanctions against particular countries has rarely been successful. Unintended consequences include corruption, new clandestine secondary markets, the development of new international alliances, and possibly war.

The rationale that governments can direct exporters to selected trade partners based on moral and ethical issues can lead to patronage and corruption and may reinforce or entrench the perceived 'bad' policy in the rejected country. Gaining some international 'benefit' via moral, ethical, or trade advantage through a foreign policy initiative should not come at the expense of increasing domestic inefficiency and international uncompetitiveness of domestic companies.

There have been instances where governments have multiple exchange rates for their currency, based on industry sectors such as tourism, importing, exporting, and investing.[28] This creates opportunities for arbitraging, either legal or otherwise. It facilitates windfall profits by banks arbitraging between the various exchange rates. But it also encourages black market private trading in currency to overcome shortages in one sector. Currency dealing proliferates when exchange between the sectors is legal, while corruption is encouraged when inter-sector exchange is made illegal. A thriving currency exchange in the domestic market distorts resource allocation and distracts from more productive pursuits.

Wherever governments have public ownership of major resources, this concentrates power to a few individuals. Patronage and largess proliferate under such market power control, usually under the guise of 'democracy'. In contrast deregulation and privatization dissipate market power and lessen the possibility of such concentrations of power.[29]

The association between low wages and corruption is ambiguous and problematical. Earlier theories[30] that suggested low wages in the public sector led to a higher incidence of corruption have been challenged through a lack of evidence.[31] The extent of the bribe and whether anti-corruption laws are enforced are major variables. Whenever government controls are lax, the probability of bribery to boost any level of income can be high.[32]

Higher wages can actually increase expectations of higher bribes.[33] However, the corollary that higher wages may be due to lower corruption has not been proven. Using aggregate or average wage data has distorted research outcomes.[34]

Government import tender fraud

Corruption in the actual public procurement process is rampant even in developed countries, through insider knowledge, low transparency, and high opportunism.[35] The issuing of public tender contracts should meet strict tendering legal guidelines,[36] but they are usually given to those who can best 'rort the system'.[37] Public tendering creates opportunities to extort bribes and divert funds. As well, corruption can continue in the subsequent public supply chains, particularly when private competition is stifled.[38]

International sellers who are involved or privy to the import government tender fraud can be either public or private.[39] However, it has been difficult to establish criminality that is beyond reasonable doubt in criminal trials against public import tender officials because of the complicity of international organizations.

It has been a very surreal situation for many importing governments since the 1950s. Regardless of whether traditional bulk commodity handling practices were coerced by exporting countries or deliberately chosen by the importing country, governments found themselves responsible for a huge accumulation of stockpiled bulk product that was generally commingled and undifferentiated, with little specificity to end-user and consumer requirements.

From the beginning, there was a general lack of knowledge as to how bulk food supply chains were supposed to behave. Bureaucrats were inexperienced in dealing with bulk problems that arose with completely new food supply chains. There was a vacuum of management direction in most cases. Traditional family dynasties assumed control in many instances in developing countries, simply because there was a void of knowledgeable leadership and a scramble for authoritative positions.

If farmers in more developed countries had serious problems in the adoption of traditional commingled bulk handling methods, most farmers in less developed countries did not have the finance or capability to change to bulk handling. Many farmers simply were excluded from the new bulk commingled supply chains in developing countries, thus causing an increasing dependence on imports of what were previously locally grown traditional food ingredients.

With centralized government-owned bulk commingled commodity facilities, public officials had power over the domestic supply chain, which excluded merchants. Consumers were made dependent on food that only the government could provide. This created an environment for corruption, particularly with the import tender system, but the cause actually lay with the adoption of exclusive traditional commingled bulk handling methods by the exporting country that then made the importing country very dependent on this type of logistics system.

Government corruption with bulk commodity import tendering usually evolved from the existence of import power through its general domestic exclusion of farmers and merchants, dependency of end users and consumers on discretionary government decisions, large bank loan inflows for infrastructural import development and maintenance, huge cash flows between numerous government authorities, non-transparent negotiations, and a lack of adequate auditing and surveillance controls. This occurred at a time when the requirement for government institutional catering was expanding rapidly.

Many African and Asian government purchasing agencies still issue bulk commodity import tenders for contracts based on the lowest purchase price. This buying method results in 'last resort' commodity sellers who traditionally have included many statutory selling authorities queuing for tender orders, discounting price to improve queue positioning, sometimes offering generous credit provisions, and generally offering bribes to tender officials to jump the queue.[40]

Whenever there are claims of statutory 'power over prices', suspicions of corrupt practice arise, simply because of commodity buyers and sellers being global price takers. When overseas end users pay more when plentiful supplies exist, or when farmers receive less payment when shortages occur, then thorough public audits are required to examine why these events occurred under commodity price-taking conditions.[41]

Three categories of fraud relating to government procurement tenders have been identified. Fraud can occur prior to the tendering process, during the tendering process, and then during the execution of the contract for supplies.[42]

Buying more than required, otherwise known as inflating the quantity, can be done to deliberately force a discounted price. The commodity is then on-sold domestically at international market prices.[43] Differences between the international market price and the discounted price are skimmed off by public officials, with the international market price being recorded on the public expenditure ledgers.

Limiting the list of potential import tender suppliers based on discretionary pre-qualifications can result in import prices being quoted artificially higher. Import price manipulation can be a result from either bid fixing, which occurs when procurement specifications are tailored to suit a favoured supplier, or tender flexibility that enables procurement specifications deliberately to be made ambiguous and tailored to suit a favoured supplier. A portion of the higher priced payment is syphoned off by public tender officials.

Deliberately causing delays in registering new suppliers or intentional confusion over the release of details of the public import tender can cause seller frustration and lead to fraud to resolve the issues quickly. Bribes can be pecuniary or in kind.

Tender orders can be split to avoid audit controls. Monitoring and auditing of small regular-occurring tenders can be more difficult to audit than a single large procurement. Order splitting has a higher probability of avoiding centralized scrutiny and control systems.

Whilst some or all of the tender suppliers can collude on higher price maintenance, some of the price benefits can be paid to the public tender officials.[44] Tender contracts can be issued to fictitious suppliers. Subsequent payments supposedly to these fictitious suppliers are syphoned off by the public tender officials.

Import tender officials can issue tender notices to the subsidiaries of favoured suppliers. What appears to be a competitive tender offering may be just one supplier with many subsidiaries. Fraud is necessary to maintain secrecy about the parent supplier.

Government officials who issue the import tender can revise the contract specifications after awarding the tender that unduly favours the awarded supplier.[45] Import specifications can be adjusted after the import order is issued. For this contract flexibility privilege, the exporter pays a bribe to the tender officials. Government officials could also revise the tender payment details after awarding the tender. A higher price can result in payment skimming for the public tender officials.

A larger problem of corruption occurs when there is no delivery of the product in collusion with procurement officials. Payments then occur amongst the parties involved despite no delivery. The receipt of the product or service is falsely certified. Alternatively, there can be recirculation of part of the quantity supplied, which results in the favoured supplier being paid twice, albeit with kickbacks. Because of the larger nature of this fraud, there needs to be adequate reconciliatory audits of many accounts.

Fiscal benefits may flow to the winner of the tender to offset a lower price in the tender acceptance. There may be customs duty or tax exemptions that unduly favour the awarded supplier. Government officials then on-sell in the domestic market at a higher price. Differences between the low import price and the higher domestic selling price are skimmed off by tender officials.

'Emergency' provisions may override proper tender processes and audit procedures. Making quick decisions can attract fraudulent transactions, particularly when food 'aid' is involved.[46]

Consequences from corruption

Corruption indices support the evidence that less-developed countries are usually those with high corruption levels.[47] Many of the arguments that suggest corruption is the result of less economic development can be refuted. It is argued that it is not low wages, education, or lack of food 'aid' that retards economic development and growth, but rather the evidence suggests that one of the major barriers to economic development is high corruption occurrences.[48] Low wages and low education are the effect from corruption, not the cause.

Corruption can lead to some individuals receiving most of the income, but these opportunists are usually not the risk-taking entrepreneurs who stimulate employment and growth of a country. Rather they are more likely to shift savings off-shore into unproductive money-laundering pursuits.[49] The lack of income of the majority of the population prevents motivation into potential productive investment that is required for growth, while economic and social improvement is undermined by corruption-induced disincentives.

Regulatory circumvention can lead to expensive transaction costs.[50] These costs are similar to a tax on transactions, but they are still a business cost. Sometimes it is difficult to prove whether it is the government regulations that cause officials to demand bribes for regulation circumvention, or whether the regulations were designed by corrupt bureaucrats to extort bribes.[51] Causality remains disputed.

If domestic costs rise through protection of inefficient domestic industries, or if corruption directly adds to the cost of inputs, then potential exporters are disadvantaged and may lose international competitiveness. This has serious longer-term ramifications through increasing trade deficits and balance of payments problems, which can then often result in even higher protection levels. It can be an ever-tightening cycle leading to lower domestic investment, frequent economic recessions, and periods of prolonged high structural unemployment.

Successful businesses are more at risk from extortion than less successful ones.[52] If external corruption increases with internal business success, then this acts as a disincentive to improve and compete profitably. Entrepreneurs may organize simply to avoid public service dependency, or to develop only those enterprises that result in quick flexible outcomes to avoid closure threats.[53] More resources might be devoted to achieve government largess and patronage rather than to productive activities.[54] High corruption returns might even attract private entrepreneurs into the public sector.[55]

Corrupt agreements are not enforceable by law. This can result in reneging and defaulting of dubious contractual agreements undertaken under a corrupt environment, with one party to the agreement not fulfilling the contractual obligations. It is difficult to undertake organizational management operations when this law enforcement uncertainty and the potential for high financial losses exists.[56]

There may be direct tax evasion or improper tax exemptions.[57] It is highly unlikely that illegal bribes would be declared for income taxation purposes because of the high probability of detection. Some governments introduce broader based tax collection methods rather than confront the more serious problem of corruption.

If corruption erodes the tax base, then the government is highly unlikely to remain solvent and provide essential services with adequate infrastructure. Government debt accumulates while the general population often do not have enough savings to fund this debt through government-issued bonds, nor do they trust a corrupt government for bond investment. Vital infrastructural programmes become delayed, which then hinders efficiency, increases costs, and destroys the ability of potential exporters to compete internationally.

The higher the level of corruption, the more likely is it that vested interests will create an industry within themselves. Corruption tends to breed more corruption.[58]

This tends to cause investment to be directed only to those areas that will maximize corruption payments.[59]

The role of competition

There is a positive association between corruption and market regulation[60] because of the existence or perception of power, which is then translated into an ability to extract bribes.[61] If much corruption has occurred as a result of government food supply chain intervention through traditional commingled bulk handling, the corollary would suggest that deregulation of the food supply chain plus a return to privatized differentiated product container-size marketing should decrease public corruption.

However, much still depends on creating a competitive environment in the private sector that is enforced through objective over-vision, combined with both public and private determination to minimize corrupt practices.[62] Competition without adequate law enforcement merely makes smaller organizations more vulnerable to collateral damage from corruption.

Policies such as deregulation, privatization, and decentralization of public utilities have had a positive impact on lessening the bureaucratic powerbase in more developed countries, but only when strident anti-corruption measures have been implemented.[63] Striving for international competitiveness can act as a strong motivator in questioning the role and objectives of government regulation, which may lessen the propensity for corruption. The corollary suggests that those countries that pursue nationalism, protectionism, interventionism, and isolationism are more likely to incur the consequences of high corrupt practices.[64]

Corruption is more likely to occur with the larger windfall profits usually associated with rent-seeking monopolies[65] than for smaller competitive-based profits that are extremely sensitive to added costs.[66] Whenever large windfall profits are accrued by public monopolies, or by private monopolies that are protected by government,[67] there will be envy by those who perceive that they are missing out on the profit sharing.

Alternatively, competition policies that remove monopoly structures and market power can decrease much of the foundation and motivation for corruption.[68] Wherever multiple individuals or organizations can contest the right to provide government goods and services, there is likely to be less corrupt practices.[69]

Competition would be expected to reduce the polarization of high income versus low income that occurs in many developing and transitional countries. Income would then be spread over a wider section of the population, rather than it being concentrated in a small sector that has a high probability of overseas money laundering.[70] A more competitive environment would be expected to lead to greater risk taking by entrepreneurs, more investment-led growth, and higher employment.[71]

The level of corruption has decreased in countries such as Singapore and Hong Kong where legal and financial intermediaries have been made more competitive and independent, combined with independent anti-corruption agencies.[72] Litigation may produce positive results under such circumstances.[73] Wherever organizational standards can be established, then monitoring of public service performance can be made much easier.[74] Singapore and Hong Kong anti-corruption measures can act as international models for benchmarking purposes.[75]

Delegating public responsibilities to the private sector can result in increased service performance with the provisos that this sector has sufficient competition and is not corruptible,[76] and that there is public hiring integrity and subsequent performance monitoring.[77] However, the problem of actually defining what is the responsibility of government, and its role regarding the food supply chain, still exists. Any large government institutional catering sector will overhang policies relating to privatization, competition, and anti-corruption, and could undermine or corrupt such positive initiatives in the longer term. The size of government institutions is therefore an important consideration.

Notes

1 Peter Mulamba was Deputy General Manager of Malawi's grain marketing board who disappeared in September 2004 from his car on an open road in Malawi. He was a key state witness in two fraud and corruption cases against former finance minister Friday Jumbe and former president Bakili Muluzi, The court cases were instigated by the International Monetary Fund over missing grain assignments when many people in the country were starving.

2 Michael Wansley was an Australian accountant and auditor who was murdered over a 'discovery' in March 1999 while negotiating a debt-organizational restructure on behalf of bank creditors of the Kaset Thai Sugar Company at Nakhon Sawan in Thailand.

3 Gerasimova, Ksenia, 2008, *Can Corruption and Economic Crime be Controlled in Developing Countries and If So, Is It Cost-effective?* Journal of Financial Crime, London, 15:2, 223.

4 Svensson, Jakob, 2005, *Eight Questions About Corruption*, The Journal of Economic Perspectives, 19:3, 19–42.

5 Ades, Alberto and Rafael Di Tella, 1999, *Rents, Competition, and Corruption*, American Economic Review, 89:4, 982–993.

6 Sachs, Jeffrey D. and Andrew M. Warner, 1995, *Economic Reform and the Process of Global Integration*, Brookings Papers on Economic Activity, 1, 1–118.

7 Svensson, Jakob, 2005, *Eight Questions About Corruption*, The Journal of Economic Perspectives, 19:3, 19–42.

8 Donchev, Dilyan and Gergely Ujhelyi, 2010, *What Do Corruptions Indices Measure?*, University of Houston, Economics Department, June.

9 Klitgaard, Robert, 1991, *Gifts and Bribes*, in Richard Zeckhauser, Ed., *Strategy and choice*, MIT Press, Cambridge, MA, Ch. 9, pp. 212–239.

10 This aspect will be discussed further in Chapter 9.

11 Mauro, Paolo, 1997, *Why Worry About Corruption?* International Monetary Fund, Economic Issues No. 6, February.

12 Svensson, Jakob, 2005, *Eight Questions about Corruption*, The Journal of Economic Perspectives, 19:3, 32.

13 Burke, Ronald J. and Cary L. Cooper, 2009, *Research Companion to Corruption in Organizations*, Edward Elgar, Cheltenham.

14 Mauro, Paolo, 1997, *Why Worry About Corruption?* International Monetary Fund, Economic Issues No. 6, February, 1–13.

15 Fleming Peter and Stelios C. Zyglidopoulos, 2009, *Charting Corporate Corruption*, Edward Elgar, Cheltenham.

16 Hunter, Harold S., 1946, paper presented to annual conference of Co-operative Union of Queensland, 8 April, Brisbane.

17 Varshney, Ashutosh, 1998, *Mass Politics or Elite Politics? India's Economic Reforms in Comparative Perspective*, The Journal of Policy Reform, 2:4, 301–335.

18 Acemoglu, Daron and Thierry Verdier, 2000, *The Choice Between Market Failures and Corruption*, The American Economic Review, 90:1, 194–211.

19 Lipset, Seymour Martin, and Gabriel Salman Lenz, 2000, *Corruption, Culture, and Markets*, in Lawrence E. Harrison and Samuel P. Huntington, Eds., *Culture Matters*, Basic Books, New York.

20 Nye, Joseph S., 1967, *Corruption and Political Development: A Case-Benefit Analysis*, The American Political Science Review, 61:2, 417–427.

21 Whitwell, Greg and Diane Sydenham, 1991, *A Shared Harvest – The Australian Wheat Industry 1939–89*, Macmillan Education Australia, Melbourne, p. 44.

22 Svensson, Jakob, 2005, *Eight Questions about Corruption*, The Journal of Economic Perspectives, 19:3, 38.

23 Cole, T.R.H., 2006, *Inquiry into Certain Australian Companies in Relation to the UN Oil-For-Food Programme*, Attorney-General's Department, Canberra, November.

24 Harstad, Bard and Jakob Svensson, 2004, *Bribes, Lobbying and Development*, Mimeo, IIES, Stockholm University.

25 Svensson, Jakob, 2005, *Eight Questions About Corruption*, The Journal of Economic Perspectives, 19:3, 31.

26 United Nations, 2005, *Volcker Report – Independent Inquiry into the United Nations Oil-for-Food Programme, Chapter III – Humanitarian Goods Transactions and Illicit Payments*, New York.

27 Cole, T.R.H., 2006, *Inquiry into Certain Australian Companies in Relation to the UN Oil-For-Food Programme*, Attorney-General's Department, Canberra, November.

28 Mauro, Paolo, 1997, *Why Worry About Corruption?* International Monetary Fund, Economic Issues No. 6, February, 1–13.

29 This will be discussed further in Chapters 11 and 12.

30 Svensson, Jakob, 2005, *Eight Questions About Corruption*, The Journal of Economic Perspectives, 19:3, 19–42.

31 Becker, Gary and George Stigler, 1974, *Law Enforcement, Malfeasance and the Compensation of Enforcers*, Journal of Legal Studies, 3:1, 1–19.

32 Rauch, James and Peter Evans, 2000, *Bureaucratic Structure and Bureaucratic Performance in less Developed Countries*, Journal of Public Economics, 75:1, 49–71.

33 Mauro, Paolo, 1997, *Why Worry About Corruption?* International Monetary Fund, Economic Issues No. 6, February.

34 Mookherjee, Dilip and Ivan P.L. Png, 1995, *Corruptible Law Enforcers: How Should They be Compensated?* Economic Journal, 105, 145–159.

35 Dorn, Nicholas and Michael Levi, 2008, *Do European Procurement Rules Generate or Prevent Crime?* Journal of Financial Crime, 15:3, 243–260.

36 Svensson, Jakob, 2005, *Eight Questions About Corruption*, The Journal of Economic Perspectives, 19:3, 19–42.

37 Moyo, Dambisa, 2009, *Dead Aid*, Allen Lane (Penguin), London, p. 50.

38 Olken, Benjamin, 2004, *Monitoring Corruption: Evidence from a Field Experiment in Indonesia*, unpublished manuscript, Harvard University, Cambridge, MA.

39 Borgin, Karl and Kathleen Corbett, 1982, *The Destruction of a Continent: Africa and International Aid*, Harcourt Brace Jovanovich, London, p. 95.

40 Singh, Balwinder, 1999, *India: Procurement Frauds*, Journal of Financial Crime, 7:2, 183–185.

41 Cole, T.R.H., 2006, *Inquiry into certain Australian companies in relation to the UN Oil-For-Food Programme*, Attorney-General's Department, Canberra, November.

42 Singh, Balwinder, 1999, *India: Procurement Frauds*, Journal of Financial Crime, 7:2, 183–185.

43 Singh, Balwinder, 1999, *India: Procurement Frauds*, Journal of Financial Crime, 7:2, 183–185.

44 Singh, Balwinder, 1999, *India: Procurement Frauds*, Journal of Financial Crime, 7:2, 183–185.

45 Singh, Balwinder, 1999, *India: Procurement Frauds*, Journal of Financial Crime, 7:2, 183–185.

46 Borgin, Karl and Kathleen Corbett, 1982, *The Destruction of a Continent: Africa and International Aid*, Harcourt Brace Jovanovich, London.

47 Banfield, Edward, 1958, *The Moral Basis of a Backward Society*, Free Press, Chicago.

48 Wraith, Ronald and Elgar Simpkins, 1963, *Corruption in Developing Countries*, Norton, New York.
49 Svensson, Jakob, 2005, *Eight Questions About Corruption*, The Journal of Economic Perspectives, 19:3, 19.
50 Svensson, Jakob, 2005, *Eight Questions About Corruption*, The Journal of Economic Perspectives, 19:3, 20.
51 Mauro, Paolo, 1997, *Why Worry About Corruption?* International Monetary Fund, Economic Issues No. 6, February.
52 Mauro, Paolo, 1997, *Why Worry About Corruption?* International Monetary Fund, Economic Issues No. 6, February, p. 5.
53 Choi, Jay P. and Marcel P. Thum, 1998, *The Economics of Repeated Extortion*, Columbia University Working Paper, No. 98/99-03.
54 Murphy, Kevin, Andrei, Shleifer, and Robert Vishny, 1991, *The Allocation of Talent: Implications for Growth*, Quarterly Journal of Economics, 106:2, 503–530.
55 Svensson, Jakob, 2005, *Eight Questions About Corruption*, The Journal of Economic Perspectives, 19:3, 37.
56 Shleifer, Andrei and Robert Vishny, 1993, *Corruption*, Quarterly Journal of Economics, 108:3, 599–617.
57 Mauro, Paolo, 1997, *Why Worry About Corruption?* International Monetary Fund, Economic Issues No. 6, February, p. 5.
58 Andvig, Jens and Karl O. Moene, 1990, *How Corruption May Corrupt*, Journal of Behaviour and Organization, 3:1, 63–76.
59 Mauro, Paolo, 1997, *Why Worry About Corruption?* International Monetary Fund, Economic Issues No. 6, February, p. 6.
60 Svensson, Jakob, 2005, *Eight Questions About Corruption*, The Journal of Economic Perspectives, 19:3, 34.
61 Shleifer, Andrei and Robert Vishny, 1993, *Corruption*, Quarterly Journal of Economics, 108:3, 599–617.
62 Bliss, Christopher and Rafael Di Tella, 1997, *Does Competition Kill Corruption*, Journal of Political Economy, 105:5, 1001–1023.
63 Rose-Ackerman, Susan, 1999, *Corruption and Government: Causes, Consequences and Reform*, Cambridge University Press, Cambridge.
64 Tanzi, Vito, 1994, *Corruption, Governmental Activities, and Markets*, IMF Working Paper 94/99, International Monetary Fund, Washington, DC.
65 Murphy, Kevin, Shleifer, Andrei, and Robert W. Vishny, 1993, *Why is Rent Seeking so Costly to Growth?* American Economic Review, 83:2, 409–414.
66 Rose-Ackerman, Susan, 1978, *Corruption: A Study of Political Economy*, Academic Press, New York.
67 Ades, Alberto and Rafael Di Tella, 1999, *Rents, Competition, and Corruption*, The American Economic Review, 89:4, 982–993.
68 Tullock, Gordon, 1967, *The Welfare Costs of Tariffs, Monopolies, and Theft*, Western Economic Journal, 5:3, 224–232.
69 Shleifer, Andrei and Robert Vishny, 1993, *Corruption*, Quarterly Journal of Economics, 108:3, 599–617.
70 Kumar, B.V., 2000, *Corruption: An Indian Perspective*, Journal of Money Laundering Control, 3:3, 266–279.
71 De Speville, Bertrand, 1997, *Hong Kong: Initiatives Against Corruption*, OECD Development Centre Studies, Paris.
72 Heilbrunn, John R., 2004, *Anti-Corruption Commissions: Panacea or Real Medicine to Fight Corruption*, World Bank Institute, No. 37234, Washington, DC.
73 Hay, Jonathan and Andrei Shleifer, 1998, *Private Enforcement of Public Laws: A Theory of Legal Reform*, American Economic Review, 88:2, 398–403.
74 Her Majesty's Treasury, 2005, *2004–2005 Fraud Report: An Analysis of Reported Fraud in Government Departments*, London.

75 Quah, Jon S.T., 1997, *Singapore's Experience in Curbing Corruption*, in Arnold J. Heidenheimer, Michael Johnston, and Victor T. Le Vine, Eds., *Political Corruption: A Handbook*, 4th Ed., Transaction Publishers, New Brunswick, NJ.
76 Svensson, Jakob, 2005, *Eight Questions About Corruption*, The Journal of Economic Perspectives, 19:3, 36.
77 Galtung, Fredrik, 1998, *Criteria for Sustainable Corruption Control*, The European Journal of Development Research, 10:1, 115–117.

Further readings on corruption

Alatas, Syed Hussein, 1980, *The Sociology of Corruption*, Times Book, Singapore.

Amin, Galal A., 1974, *The Modernization of Poverty: a Study in the Political Economy of Growth in Nine Arab Countries 1945–1970*, Leiden, Brill.

Bartos, Stephen, 2006, *Against the grain: the AWB scandal and why it happened*, University of NSW, Sydney.

Boy, James K., 1993, *The Philippines: the Political Economy of Growth and Impoverishment in the Marcos Era*, Basingstoke, Macmillan, in association with the OECD Development Centre.

Bull, Martin J. and James Newell, 2003, *Corruption in contemporary politics*, Palgrave Macmillan, New York.

Klitgaard, R., 1988, *Controlling Corruption*, University of California Press, Berkeley.

Lamour, Peter and Peter Grabosky, 2001, *Corruption in Australia: Its Prevention and Control*, in Gerard R. Caiden, O.P. Dwivedi, and Joesph Jabbra, Eds., *Where Corruption Lives*, Kumarian Press, Bloomfield, CT.

Mendoza, Amado M., 2001, *The Industrial Anatomy of Corruption: Government Procurement, Bidding, and Award of Contracts*, Public Policy, 5:1, 43–71.

Miller, Geoffrey P., 2010, *Economics of Ancient Law*, Edward Elgar Publishing, Cheltenham.

Moore, Stephen, 1997, *Power and Corruption: The Rotten Core of Government and Big Business*, Majestic Books, London.

Overington, Caroline, 2007, *Kick back: inside the Australian Wheat Board Scandal*, Allen and Unwin, Sydney.

Schedler, Andreas, Diamond, Larry, and Marc Plattner, Eds., 1999, *The Self-Restraining State: Power and Accountability in New Democracies*, Lynne Rienner Publishers, Boulder, CO.

Shah, Anwar, 2006, *Corruption and Decentralized Public Governance*, Policy Research Working Paper No. 3824, The World Bank, Washington, DC.

Tangri, Roger, 1999, *The Politics of Patronage in Africa: Parastatals, Privatization, and Private Enterprise*, Trenton, New York.

Vittal, N., 2003, *Corruption in India*, Academic Foundation, New Delhi.

9

FOOD AID PROBLEMS

International 'aid' is a recent phenomenon, having generally evolved only during the twentieth century. It had some origin in providing emergency food or services during wartime blockades, revolutions, and post-war reconstruction programmes. In terms of international public 'aid', it reached its peak during the Cold War with strategic trade-offs with foreign policy and political objectives.

Towards the end of the production-push era that coincided with the Cold War, international 'aid' was mostly linked with large-scale foreign capital expenditure and development projects such as dams, roads, railways, bridges, seaports, airports, power, water, telecommunications, and sewerage.[1] This usually aligned international public aid to achieving foreign policy, trade, and overseas investment objectives.

However, during the past 20 years, there has been a partial shift from public aid to privately sponsored aid, with direct government-to-government 'aid' being replaced by private donor industries.[2] The negative effects of this aid may be worse because of the entrenchment and prolonged time periods involved as well as the interdependency of these donor industries. There was the example of an African mosquito net maker who once employed ten workers with 150 dependants. 'Aid' donation of 100,000 nets to the country destroyed his business.[3]

Religious missionary zeal during the nineteenth century has been replaced by misguided 'charitable' zeal in the twentieth century, or by the exuberance of altruistic youth 'on a mission' before they commence their careers.[4] Some under-employment in developed countries has found an escape in less developed countries, often led by celebrities seeking some 'therapeutic outcome'. Visual electronic media prey on the private guilt of more developed countries and apportion blame without explaining why people in other countries are really suffering. Rarely is it mentioned that 'aid' can encourage corruption, cause increased dependency, entrench income polarization, and undermine the very existence of the people through patronage.[5]

Oblique foreign policy and trade objectives may not be readily transparent, particularly when commodity surplus countries merely want to dump their excess in a foreign country. There is a delicate balance between assistance and increasing the misery of people, particularly when opportunism is combined with underlying psychological paradoxes associated with ulterior motives such as morals and ethics.

There may be moral, ethical, and economic justification for short-term humanitarian relief during emergencies such as natural disasters and war. However, there is a danger when such aid is automatically expected, prolonged, or when 'relief' is perceived to entrench the status quo and prevent changes occurring. The worse outcome is when donor-receiving countries pass the sole responsibility for natural disaster relief management onto international 'emergency relief' agencies. Any prior preparedness plans are thwarted, thus increasing the vulnerableness of the local population.

Another justification for aid could be technological and knowledge transfer to developing countries. However, the issue arises as to why existing local research and development programmes are inadequate, and why venture capital is not flowing into such technology and knowledge-based industries. The worse outcome from such international transfers of resources is the dependence created on imported resources and the impact caused by undermining local initiatives.

One large identifiable problem is that earlier research into the benefits of international aid programmes, which were often funded by vested interests, are still being quoted as justifiable evidence for the continuation of international aid programmes,[6] despite much later evidence suggesting that the longer-term outcomes are extremely damaging. Research validity should suggest less emphasis on earlier research and more emphasis being placed on recent research because of subsequent factor changes, but the reality is often the opposite because of vested interests wanting to maintain the status quo.

Increasing abject famine exists in a world of plenty, as evidenced by prolonged periods of low commodity prices. The scapegoat is frequently population growth. However, there is evidence that high population growth arises as a combat against high rates of death, disease, and malnutrition.[7] It is argued that much of this death, disease, and malnutrition is the consequence of either misguided socialist policies that distort normal supply chain behavioural signals and destroy food chain logistic flows, or a deliberate attempt to protect the social elite regardless of the consequences to the rest of the population.

The real dilemma in this chapter is whether to support these misguided political ideologies and corrupt governments through the giving of 'aid', or whether to let them fail quickly so that the consumers can reassert themselves through privatized food supply chains. 'Aid' can prolong the misery of people who are unfortunate enough to be subjected to bad government. Aid organizations are rarely flexible and most resist change,[8] which paradoxically aligns them with bad government.

Albania might be used as an example whereby the socialist government clung so long to its misguided production-push interventionist principles, but in the end it was forced to allow a consumer revolution because it received little international aid and global financial support. In contrast with North Korea, Albania did not have a strategic geographic position with vast important natural resources, and therefore did not attract international attention. Consumers in Albania are now adequately feeding themselves, whereas there is continuing starvation in North Korea,[9] which is being supported by the Chinese government because of foreign policy, defence, and resource objectives in East Asia.

There is a tendency for socialists to claim the world should feed and assist such socialist-stricken countries, which only panders to a political system that caused the initial problems. Moralists and ethicists suggest that there is an obligation to assist distressed people, regardless of the misery being prolonged under such political systems. The pragmatists point to countries such as Albania where the demise of the previous famine-stricken socialist regime has been quickly driven by a consumer revolution. Marketing specialists would argue that previous failed production-push interventionist government policies should be replaced by more bountiful consumer-driven competitive strategies.

Gift giving and 'aid'

Chapter 8 detailed how a gift can be defined as a bribe if it creates dependency by either the giver or taker, seeks favouritism from the other person, or attempts to influence the other person. Even when 'aid' is dissected as humanitarian aid, emergency aid, charity-based aid, bi-lateral government-to-government aid, or multilateral 'aid', there is often difficulty in defining 'aid' and differentiating it from gift-giving. It can be described as patronizing,[10] perhaps associated with a superior power–knowledge psychological complex and egocentricity.[11]

If it is difficult to separate 'aid' from gift giving, then 'aid' can incur the same criticism such as seeking dependency, favouritism, and power. These three factors create the required prerequisites for financial opportunism, which is often the objective within the foreign policies of donor countries.

One major difference between aid and gift giving is the continuance. Usually with aid there is continuation because there is no aid-exit strategy, which can lead to problems such as dependency, expectancy, and apathy, with no motivation to change.[12] In contrast, gift giving is usually attached with a 'surprise' factor, which is absent from aid giving.

Donor countries are frequently motivated by political opportunism, but lack the experience and competence to manage international aid effectively and efficiently, even in waste minimization.[13] As well, receiver countries frequently stage manage and dramatize isolated events to achieve maximum sympathy from donor countries, which is similar to a spoilt child, with dubious motivations and relatively unknown final beneficiaries.[14]

Aid and corruption

There are many causes and methods of corruption as were detailed in Chapter 8. Whether corruption causes market distortions, or whether market distortions cause corruption, it might be argued that there are multiple cause and effect relationships. The important consideration in this chapter is whether 'aid' prolongs both corruption and supply chain distortions in the host country, which can destroy normal food supply chains and cause increasing occurrences of famine and starvation, as well as encouraging population explosions to counter the high death, disease, and malnutrition rates.

International aid can cause corruption as well as increase corruption,[15] particularly when large amounts of money have to be disbursed quickly or in the aftermath of civil unrest.[16] Graft and monetary 'aid' are likely to be highly correlated.[17]

Unlike tax revenues, international aid generally lacks audit accountability in the host country. Donor organizations can undermine the recipient's accountability controls.[18] Foreign 'aid' can be easily stolen, redirected, or extracted.[19] Much of the monetary 'aid' to Africa ends up being laundered overseas.[20]

Large-scale corruption is mostly intertwined with supply chain distortions caused by government intervention policies.[21] Instead of encouraging change to correct market distortions, 'aid' merely entrenches a culture of corruption[22] and perpetuates poor governance.[23] 'Aid' can 'make a good policy environment bad, and a bad policy environment worse'.[24]

In instances where 'aid' merely perpetuates the continuation of corruption and the status quo regarding failed government interventionist schemes, there can be serious moral and ethical issues concerning the provision of such aid.[25] When such 'aid' supports rapid population growth for continued survival under such corrupt interventionist governments, then there are also serious economic concerns.

Acknowledgement of the impact of foreign 'aid' on corruption has continually occurred. The UNDP Human Development Report 1994 indicated that 'development aid' for social and economic sectors was being directed to fund unproductive and corrupt expenditures. Awareness of this association has caused public international aid to decrease, but in its wake has arisen misguided private international aid. Recipients of largess are unlikely to be critical of the type of donor or their intentions.

There has been some advocacy for 'trade not aid', as if mere international trade can solve the problems that have arisen through corruption and public intervention into food supply chains. When there is no attempt to overcome corruption or to withdraw from supply chain interventionist policies, then international trade only offers windfall gains for corrupt officials and justifies the status quo. More often, this trade will be countertrade and therefore highly open to corruption.[26] Countertrade provides even more reasons why trading with such countries should be avoided.

Dependency

One type of dependency can occur when the 'aid' recipient becomes reliant on the donor for existence or survival.[27] Merely increasing 'aid' may result in increased dependency. It is difficult to separate dependency from patronization.[28]

Increasing dependence by the recipient and patronization through the donor's continual largess are major issues, but there is also interdependence of vested-interest stakeholders. Individual stakeholders cannot survive without the continuation of the whole system.[29] Even retired politicians may become unwitting stakeholders in this interdependence web as others seek to exploit their previous connections.

Increasing the individual dependency on any form of 'aid' undermines the willingness and motivation of recipients to help and improve themselves. Donors may obtain both self-satisfaction and ego gratification because of the subtle dependency and the power thus shifted from the recipient to the donor. But this offers very little positive outcomes for the recipient.

Wherever 'aid' merely props up corrupt governments, there is a high probability that any legal private wealth accumulation will be confiscated through public corruption.[30] Any incentive to rise above the poverty trap is thus removed. The objective of the 'aid' might be poverty reduction, but the outcome could be greater entrenchment of poverty.

Diversion risk

Diversion risk can be defined as the risk of international aid not being used by the recipient for the original intended purpose of the donor. This diversion could be intentional, or it may be due to ignorance, bureaucratic apathy, or lack of accountability and auditing. There is a high probability of diversion risk if the original intention was to achieve some foreign policy objectives. A World Bank study found that 85 per cent of international aid was diverted to purposes other than those originally intended.[31]

The risk of diversion usually increases with corruption, but it could also be due to misguided or inappropriate objectives of the donor. Merely transposing objectives from one country that has achieved successful outcomes to another while expecting similar outcomes may be very inappropriate or patronizing.

The sheer logistical scale of international aid frequently causes loss of control and policy direction. Maintaining control is a major problem especially amongst competing organizations as well as private aid agencies competing with host country bureaucracies.[32]

Conditionality

There have been instances where international aid has been made conditional on reducing government intervention, privatizing previous nationalized industries, liberalizing trade, reforming institutions, reducing the size of the civil service, increased democracy, and better governance.[33] Positive outcomes have been achieved in some Asian and Latin American countries with economic growth and private capital inflows ultimately replacing international aid.[34]

However, conditional aid can be abused and lead to corruption. If international aid is dependent on using specific products or services, employment of foreign staff to the exclusion of recipient country staff, or conditional on the adoption of certain economic or political policies,[35] then it might be perceived as extortion or patronizing. Under such circumstances, diversion risk might be expected to be high.

Conditionality of international aid can often fail because of corruption and poor governance, cronyism, anti-competition, and inefficiency.[36] Having too high an expectation on bureaucratic organizations to achieve desired objectives can cause conditionality to fail. Many African countries have regressed through weak public institutions into more corruption.[37]

Economic distortions

International 'aid' can cause potential economic resources to shift away from more productive activities to less productive pursuits in order to maximize the pecuniary benefits gained from 'aid'.[38] Such serious misallocation of resources can prolong economic and food supply chain distortions, ultimately causing food shortages and famine.[39]

As well, international aid has been found to increase the overall size of government,[40] expand the public sector relative to the private sector,[41] as well as being perceived to lessen the need for good governance in collecting tax revenues.[42] If the tax revenue base is loosened in the recipient country, then governments become more dependent on foreign aid and 'loans' for fiscal revenue, and are more likely to be irresponsible with fiscal expenditures. Any increased size of the public sector and dominance of fiscal expenditures in the economy is likely to discourage both domestic and foreign entrepreneurs, regardless of the looseness of the tax revenue base.[43]

A 'vicious cycle of aid' occurs when foreign aid props up corrupt governments, which undermines efficient resource allocation and investment and leads to inadequate economic growth, unemployment, and greater poverty.[44] This often increases international debt, and a quickening cycle of economic destruction develops. The Heavily Indebted Poorest Country (HIPC) debt relief programme may be symbiotic with debt–poverty interdependence.

There is evidence that international aid has resulted in slower economic growth and more poverty.[45] When international government aid flows were at their peak between 1970 and 1998, poverty in Africa rose from 11 per cent to 66 per cent.[46] This explains why there has been general reluctance by governments of more developed countries to provide international aid.

The negative impact of aid is exacerbated by government interventionist policies.[47] Even the International Monetary Fund has warned that international aid leads to lower economic development.[48] Most aid-dependent countries experienced negative economic growth rates in the period 1978–2008 and a 'dramatic descent into poverty'.[49]

Concessional loans are frequently provided as 'aid' provisions to corrupt governments at below market interest rates, but conditional on floating interest rates and not fixed rates. This is an active encouragement to maximize debt. When interest rates rise, there is an inability to repay because of inadequate tax revenues. Either the loan is then converted into a grant because of the inability to repay,[50] or much foreign monetary aid is diverted into emergency repayment of bank debt. By 1990, debt repayment to foreign banks by less developed countries was greater than aid given to these countries.[51] There is a cycle of debt, then new aid, followed by more debt.

Other economic distortions occur, such as the negative relationship between savings and international aid.[52] Aid reduces savings and investment and increases inflation and interest rates, thereby reducing the international competitiveness of the export sector.[53]

Foreign aid was found to finance unproductive public expenditures, rather than investment.[54] Thrift is replaced by waste.[55] There is evidence that aid can cause logistical bottlenecks and congestion.[56]

International aid can divert people's attention from productive economic activity to politics, which can often weaken the social culture of a country.[57] Aid can end up mostly in the hands of the elite. Given the propensity for much international aid to originate today from private sources in the donor country, it is often a case of taxing the poor in the developed countries to pay for the rich in the less developed countries.[58]

Aid can result in economic inefficiencies. Decisions on aid revenues and expenditures are either delayed because of internal or external bureaucratic hierarchies, or rushed because of rising costs, limited budgets, and end of the financial year allocations that require immediate expenditure, otherwise entitlements are lost.[59] There may be management fees received or provided but no public accountability.[60]

Many countries are selected as international aid recipients because of their good governance, only to have this good governance undermined by corruption-induced aid.[61] The objectives of international aid become seriously questionable when such paradoxical outcomes occur.

Undermining domestic food producers

Resource sustainability frequently underpins local cultures and agricultural societies, regardless of the country's economic development status. Short-term management decisions by outsiders frequently have adverse longer-term consequences in the recipient country. But foreigners can easily depart and leave the responsibility for damage to the local people.

Transposing agricultural practices between countries through international aid programmes can be high risk[62] because of a lack of local experience, and complex local agronomic issues might not be immediately known to outsiders. Inter-country agricultural resource transfer either of capital, labour, 'knowledge', or technology may have the appearance of benevolence, but the transfer may often result in unsustainability[63] and increasing dependency.

There may be economic barriers that prevent private investors and agriculturalists in the recipient country from investment and development in the first instance, or there may be known local agronomic barriers that prevent sustainable agriculture. International aid is highly unlikely to change these barriers. When such 'aid' fails, reinstating previous traditional farming practices and correcting the misallocation of resources may be very difficult, which then creates an environment for corruption. There is a strong argument that food 'aid' supports unproductive investment, causes expenditure waste, and diverts funds from more productive pursuits.[64]

There can be agricultural production resource distortions.[65] If 'aid' favours large farmers at the expense of small farmers, it creates dependency problems for the larger farmers, and acts as a disincentive to improve for the smaller farmers. Efficient management and sustainability may often not be dependent on farm size, rather it may be more influenced by individual entrepreneurship and family support mechanisms, which are undermined by 'aid'.[66]

When the incentive of local farmers is destroyed by artificially cheap alternate supplies, international aid becomes nothing more than highly subsidized food imports. There is a replacement of rural output by food imports.[67] Food imports may be important wherever there is supply chain efficiency. However, aid-induced imports destroy the incentive for local food production, distort domestic food supply chains, and undermine local food importers.

Aid organizations often feel obliged to donate food during times of food scarcity. But it is those times of scarcity that induce farmers to produce more to take advantage of the high price. A normally functioning supply chain with adequate real price signals should have short periods of high prices, followed by prolonged periods of surpluses. That is the natural dilemma of the farmer. But to distort the domestic price through artificially low priced imports via 'aid' agencies merely acts as a de-motivator to produce local food, regardless of the previous comparative and competitive advantages of the farmer. Food scarcities thus become prolonged, and when international aid continues, the country becomes permanently dependent on imports. This has been the dilemma of Africa.[68]

In times of food 'scarcity', there is a need for local people to adapt and be flexible. Under a normally functioning food supply chain, importers would offer a range of alternate food products. Such diverse food product offering is increasingly becoming the role of global supermarket chains around the world. If there are barriers to food imports and supermarket chains, then recipient countries are forced to rely on food aid whenever local production fails.

If the food supply chain is inherently 'fragile',[69] then 'aid' or gift giving will have an immediate impact both on the physical food supply chain and psychological behaviour of food supply chain intermediaries. This undermines local food supply chains, depreciates the value that was previously created within these supply chains, and may even destroy them. Dependence on foreign food supply chains then replaces the local supply chain and all of its associated initiatives and motivation.

When global surpluses of low quality commodity food are deliberately dumped by major producing countries into less developed countries under the guise of 'aid', this completely destroys the price of local farmers and undermines their motivation to produce. The humanitarian objective of US Public Law 480 is destroyed when the dumping of commodity food surpluses merely undermines and removes farmers in the recipient country, thus making future exports to that country more assured. Either the commodity is dumped at no cost, or it is subsidized by the donor country, or there is the provision of cheap finance for the recipient country from the 'donor' country to purchase the commodity.

Public Law 480, which is the alias for the Agricultural Trade Development (USA) Assistance Act, can be implemented to remove domestic commodity food surpluses, with the objective of increasing the price to US farmers and decreasing the obligation of the US government to pay domestic farm subsidies. Eisenhower referred to it in 1954 as 'the basis for a permanent expansion of our exports of agricultural products with lasting benefits to ourselves' while Kennedy subtly renamed it 'Food for Peace'.

This 'Food for Peace' removed international competition in recipient countries, regardless of whether this was intentional or unintentional. It can thus be used to create much global food supply chain power. African countries have used this argument for international financial compensation, but conveniently overlook their own role in causing domestic food supply chain distortions in the first instant. Much of the reason for their countries being targeted for such commodity dumping is because of the existent distortions in the domestic food supply chain, thus enabling dumping to easily appear as humanitarian 'aid'.

There have been misguided attempts to 'rectify' the problem of food dumping associated with international 'aid' programmes. Instead of flooding foreign markets with donor food and pushing local farmers out of business, aid money has been used to buy food from farmers within the aid-receiving country and then distribute that food to the local citizens in need.[70] However, this undermines the normal functioning of the food supply chain. It does nothing to remove the barriers that prevent the normal functioning of the food supply chain. Acting as a pseudo-merchant is unlikely to encourage actual local merchants and supply chain intermediaries to undertake their normal risk-taking activities.

There were some positive results from the 'Green' Revolution in encouraging food production in developing countries during the 1960s. However, increased production without functional local food supply chains leads to the commodity poverty trap, with farmers increasingly dependent on a price-taking commodity, often combined with rampant domestic cost inflation. Without adequate farm profitability, farmers lack sufficient longer-term investment that would sustain land practices, a financial incapability to adapt to change, and a lack of capital to integrate with any food supply chain. Farmers soon lose motivation under such circumstances.

Substitute for trade reform

Farmers in member countries of the Organization of Economic Cooperation and Development (OECD) receive US$265 billion in agricultural subsidies annually.[71] US farm subsidies are approximately US$20 billion annually. Farm subsidies under the Common Agricultural Policy represented 48 per cent of the total 2006 EU budget.[72] It can be argued that these subsidies encourage over-production, which generally suppresses prices,[73] which then creates the need to dump these surpluses in overseas countries through artificial 'aid' programmes. While there are recipient countries willing to accept this dumped surplus, it might be expected that these farm subsidies and 'aid' dumping will continue, thus causing major trade distortions.

It has been estimated that African countries lose US$500 billion annually because of global trade restrictions and subsidy schemes in developed countries.[74] However, the size of this loss can be challenged because major distortions to domestic production and local food supply chains in Africa would prevent much alignment with current global food chains.

In the calculation of the US$500 billion, it is assumed that distorted African food supply chains could meet all the supply chain prerequisites demanded by global food manufacturers, which would not be the reality in most instances. On the presumption that these domestic distortions could be rectified, which is not an easy task, there is a strong case for African countries to claim that trade protection policies and related barriers of the four global food bowl regions of the USA, EU, Asian Subcontinent, and Indo-China do hinder international trade.

The problem with African countries is that their domestic food surpluses have decreased so rapidly over the past 20 years due to production and supply chain distortions and their increasing dependence on food aid. Regardless of whether there was trade reform by the four global food bowl regions, it is highly unlikely that African countries could significantly supply the global food supply chains while such local distortions and high levels of corruption remain.

There is a continuous chain of cause and effect. The governments of highly subsidized global food bowl countries have a vested interest in causing supply chain distortions in non food bowl countries. If international aid dumping contributes to such distortions and converts potential competitors into dependent customers, then that becomes one of the costs of international supply chain regulation and restricted global trade. It achieves certain foreign policy and trade objectives for the global food bowl countries.

Paradoxically, it has been bio-fuel that has done more to rectify the problem than action by either government or trade reformers. Higher usage of agricultural surpluses by bio-fuels and associated higher prices between 2007–2010 lessened the need for both farm subsidies and commodity food dumping under PL 480, as well as decreasing the temptation for food aid agencies to buy commodity food for distribution.

Encourages military conflict

Because international aid giving is an enormous money-exchanging industry, it attracts many types of criminal and corrupt officials. Cash can be syphoned off quickly in overseas money laundering schemes, but it can also be used for usurpation and other military objectives. Trust is removed, conflict becomes rife, and this can foster a military culture.[75] The underlying objective of many rebellions is to assume control of the government for financial advantage that is largely dependent on foreign aid donations.[76]

Aid has been found to make conflict more likely.[77] Somalia's civil wars have been blamed on competition for control of the large food aid programme.[78] International aid programmes that have lowered average incomes and slowed economic growth have contributed have contributed towards many civil wars.[79]

Aid as a vested-interest industry

One of the biggest problems with 'aid' is the international industries that arise to support it in both the donor and recipient countries and in between. 'Aid' giving has become a huge international private growth industry employing hundreds of thousands of people. Apart from emergency humanitarian relief during natural disasters and war, its appeal is based on a false guilt sentiment usually reinforced by media dramatization of singular events, rather than on paradoxical reality in the recipient country. Many people in recipient countries resent such false portrayals of their country, particularly when the drama is reinforced by international actors with stage-managed performances.

It is resource distorting because it offers 'quick solutions' to temporary unemployment in the donor-giving country, whilst perpetuating the supply chain distortions and corruption in the recipient country. Both the international 'aid' industries and the vested interests in the recipient country thrive on the status quo and resist any radical reforms toward a market-based economy that might benefit farmers, supply chain intermediaries, and consumers.

Paradoxical vested interests abound with aid agencies. International aid is frequently required to enable or protect debt repayment from the recipient country to the donor country. The donor has a 'greater need for giving aid than the recipient has for taking it'.[80] Vested interests in aid agencies perceive less aid would jeopardize their jobs. When less aid does occur, private aid organizations then appeal for government support for survival.[81]

Despite being private in a seemingly competitive environment, charity-based aid is frequently bureaucratic and inefficient with ineffective organizational structures, high administrative costs, and dependent on government foreign policy.[82] The occasional success stories are either very rare or portray biased opinion. Rarely is the impact on the whole food supply chain portrayed or examined objectively.

Sometimes there is a political or religious motive attached to international aid. Grand schemes to achieve international redistribution of wealth and equality for all can end with central planning, dependency on state welfare, and further polarization of income. The greatest gift for recipient countries would be for them to have efficient and effective food supply chains. Instead, aid agencies generally introduce a pseudo-substitute food supply chain, which quickly destroys the remnants of any existing one.

Notes

1 Reusse, Eberhard, 2002, *The Ills of Aid – An Analysis of Third World Development Policies*, The University of Chicago Press, Chicago.
2 Reusse, Eberhard, 2002, *The Ills of Aid – An Analysis of Third World Development Policies*, The University of Chicago Press, Chicago, p. 112.
3 Moyo, Dambisa, 2009, *Dead Aid*, Allen Lane (Penguin), London, p. 44.
4 Reusse, Eberhard, 2002, *The Ills of Aid – An Analysis of Third World Development Policies*, The University of Chicago Press, Chicago, p. 85.
5 Hancock, Graham, 1989, *Lords of Poverty*, Macmillan, London.
6 Morton, Kathryn, 1975, *Dependence*, Croom Helm, London.
7 Birdsall, N., Kelly, A.C. and S.W. Sinding, Eds., 2001, *Population Matters: Demographic Change, Economic Growth, and Poverty in the Developing World*, Oxford University Press, New York.

8 Reusse, Eberhard, 2002, *The Ills of Aid – An Analysis of Third World Development Policies*, The University of Chicago Press, Chicago.

9 Demick, Barbara, 2010, *Nothing to Envy – Love, Life, and Death in North Korea*, Random House, New York.

10 Moyo, Dambisa, 2009, *Dead Aid*, Allen Lane (Penguin), London.

11 Reusse, Eberhard, 2002, *The Ills of Aid – An Analysis of Third World Development Policies*, The University of Chicago Press, Chicago, pp. 87–88.

12 Moyo, Dambisa, 2009, *Dead Aid*, Allen Lane (Penguin), London, p. 67.

13 Reusse, Eberhard, 2002, *The Ills of Aid – An Analysis of Third World Development Policies*, The University of Chicago Press, Chicago, pp. 89–94.

14 Reusse, Eberhard, 2002, *The Ills of Aid – An Analysis of Third World Development Policies*, The University of Chicago Press, Chicago, p. 89.

15 Alesina, Alberto and Beatrice Weder, 2002, *Do Corrupt Governments Receive Less Foreign Aid*, American Economic Review, 92:4, 1126–1137.

16 Glennie, Jonathan, 2008, *The Trouble With Aid: Why Less Could Mean More for Africa*, Zed Books, London, pp. 71–72.

17 Moyo, Dambisa, 2009, *Dead Aid*, Allen Lane (Penguin), London, p. 46.

18 Glennie, Jonathan, 2008, *The Trouble With Aid: Why Less Could Mean More for Africa*, Zed Books, London, p. 72.

19 Moyo, Dambisa, 2009, *Dead Aid*, Allen Lane (Penguin), London, p. 52.

20 Moyo, Dambisa, 2009, *Dead Aid*, Allen Lane (Penguin), London, p. 56.

21 Tanzi, Vito, 1994, *Corruption, Governmental Activities, and Markets*, IMF Working Paper 94/99, International Monetary Fund, Washington, DC.

22 Moyo, Dambisa, 2009, *Dead Aid*, Allen Lane (Penguin), London, p. 49.

23 Mauro, Paolo, 1997, *Why Worry About Corruption?* International Monetary Fund, Economic Issues No. 6, February, p. 5.

24 Moyo, Dambisa, 2009, *Dead Aid*, Allen Lane (Penguin), London, p. 40.

25 Moyo, Dambisa, 2009, *Dead Aid*, Allen Lane (Penguin), London, p. 40.

26 Borgin, Karl and Kathleen Corbett, 1982, *The Destruction of a Continent: Africa and International Aid*, Harcourt Brace Jovanovich, London, Ch. 5.

27 Moyo, Dambisa, 2009, *Dead Aid*, Allen Lane (Penguin), London, p. 66.

28 Hancock, Graham, 1989, *Lords of Poverty*, Macmillan, London, p. 22.

29 Reusse, Eberhard, 2002, *The Ills of Aid – An Analysis of Third World Development Policies*, The University of Chicago Press, Chicago, p. 106.

30 Moyo, Dambisa, 2009, *Dead Aid*, Allen Lane (Penguin), London, p. 41.

31 Moyo, Dambisa, 2009, *Dead Aid*, Allen Lane (Penguin), London, p. 39.

32 Reusse, Eberhard, 2002, *The Ills of Aid – An Analysis of Third World Development Policies*, The University of Chicago Press, Chicago, p. 22.

33 Moyo, Dambisa, 2009, *Dead Aid*, Allen Lane (Penguin), London, pp. 21–39.

34 Moyo, Dambisa, 2009, *Dead Aid*, Allen Lane (Penguin), London, p. 22.

35 Moyo, Dambisa, 2009, *Dead Aid*, Allen Lane (Penguin), London, pp. 38–39.

36 Moyo, Dambisa, 2009, *Dead Aid*, Allen Lane (Penguin), London, pp. 38–39.

37 Moyo, Dambisa, 2009, *Dead Aid*, Allen Lane (Penguin), London, p. 22.

38 Papanek, Gustav F., 1973, *Aid, Foreign Private Investment, Savings, and Growth in Less Developed Countries*, Journal of Political Economy, 81:1, 120–130.

39 Dewbre, Joe, Thompson, Wyatt and Joshua Dewbre, 2007, *Consistency or Conflict in OECD Agricultural Trade and Aid Policies*, American Journal of Agricultural Economics, 89:5, 1161–1167.

40 Boone, Peter, 1996, *Politics and the Effectiveness of Foreign Aid*, European Economic Review, 40:2, 289–329.

41 Reusse, Eberhard, 2002, *The Ills of Aid – An Analysis of Third World Development Policies*, The University of Chicago Press, Chicago, p. 17.

42 Moyo, Dambisa, 2009, *Dead Aid*, Allen Lane (Penguin), London, p. 66.

43 Moyo, Dambisa, 2009, *Dead Aid*, Allen Lane (Penguin), London, p. 50.

44 Moyo, Dambisa, 2009, *Dead Aid*, Allen Lane (Penguin), London, p. 46.
45 Moyo, Dambisa, 2009, *Dead Aid*, Allen Lane (Penguin), London, p. 28.
46 Moyo, Dambisa, 2009, *Dead Aid*, Allen Lane (Penguin), London, p. 47.
47 Singh, Ram D., 1985, *State Intervention, Foreign Economic Aid, Savings and Growth in LCDs: Some Recent Evidence*, KTKLOS, International Review for Social Science, 38:2, 216–232.
48 Moyo, Dambisa, 2009, *Dead Aid*, Allen Lane (Penguin), London, p. 47.
49 Moyo, Dambisa, 2009, *Dead Aid*, Allen Lane (Penguin), London, p. 46.
50 Reusse, Eberhard, 2002, *The Ills of Aid – An Analysis of Third World Development Policies*, The University of Chicago Press, Chicago, p. 112.
51 Moyo, Dambisa, 2009, *Dead Aid*, Allen Lane (Penguin), London, p. 49.
52 Reichel, R., 1995, *Development Aid, Savings, and Growth in the 1980s: A Cross-section Analysis*, Savings and Development, 19:3, 279–296.
53 Moyo, Dambisa, 2009, *Dead Aid*, Allen Lane (Penguin), London, p. 61.
54 Boone, Peter, 1996, *Politics and the Effectiveness of Foreign Aid*, European Economic Review, 40:2, 289–329.
55 Reusse, Eberhard, 2002, *The Ills of Aid – An Analysis of Third World Development Policies*, The University of Chicago Press, Chicago, p. 94.
56 Moyo, Dambisa, 2009, *Dead Aid*, Allen Lane (Penguin), London, p. 64.
57 Bauer, Peter, 1954, *Economic Analysis and Policy in Under-Developed Countries*, Cambridge University Press, Cambridge.
58 Moyo, Dambisa, 2009, *Dead Aid*, Allen Lane (Penguin), London, p. 58.
59 Reusse, Eberhard, 2002, *The Ills of Aid – An Analysis of Third World Development Policies*, The University of Chicago Press, Chicago, p. 86.
60 Reusse, Eberhard, 2002, *The Ills of Aid – An Analysis of Third World Development Policies*, The University of Chicago Press, Chicago, p. 85.
61 Reusse, Eberhard, 2002, *The Ills of Aid – An Analysis of Third World Development Policies*, The University of Chicago Press, Chicago, p. 21.
62 Reusse, Eberhard, 2002, *The Ills of Aid – An Analysis of Third World Development Policies*, The University of Chicago Press, Chicago, p. 15.
63 Reusse, Eberhard, 2002, *The Ills of Aid – An Analysis of Third World Development Policies*, The University of Chicago Press, Chicago, p. 17.
64 Mauro, Paolo, 1997, *Why Worry About Corruption?* International Monetary Fund, Economic Issues No. 6, February, p. 5.
65 Hancock, Graham, 1989, *Lords of Poverty*, Macmillan, London, pp. 72–75.
66 Moyo, Dambisa, 2009, *Dead Aid*, Allen Lane (Penguin), London, p. 16.
67 Sahn, David E., 1996, *Economic Reform and the Poor in Africa*, Clarendon Press, New York, Ch. 13.
68 Friedmann, Harriet, 1990, *Origins of Third World Food Dependence*, in H. Berstein, Ben Crow, Maureen Mackintosh and Charlotte Martin, Eds., *The Food Question: Profits versus People*, Earthscan, London.
69 This fragility will be discussed in detail in Chapter 10.
70 International Food Aid Conference VII, 2005, *Strengthening the Food Aid Chain*, 3 May, Kansas City, MO, p. 45.
71 OECD, 2009, *Agricultural Policies in OECD Countries 2009: Monitoring and Evaluation*, Paris.
72 EU Budget Papers, 2006, Publications Office, European Commission, Brussels.
73 Food and Agriculture Organization of the United Nations, 2006, *Can Supply Management Halt the Decline in Agricultural Commodity Prices?* FAO Commodity Market Briefs No. 1, October, Rome.
74 Moyo, Dambisa, 2009, *Dead Aid*, Allen Lane (Penguin), London, p. 115.
75 Moyo, Dambisa, 2009, *Dead Aid*, Allen Lane (Penguin), London, pp. 59–60.
76 Grossman, Herschel I., 1992, *Foreign Aid and Insurrection*, Defence Economics, 3, 275–288.
77 Moyo, Dambisa, 2009, *Dead Aid*, Allen Lane (Penguin), London, pp. 59–60.

78 Maren, Michael, 1997, *The Road to Hell: The Ravaging Effects of Foreign Aid and International Charity*, The Free Press, New York.
79 Collier, Paul, 2003, *Natural Resources, Development and Conflict: Channels of Causation and Policy Interventions*, The World Bank, April.
80 Moyo, Dambisa, 2009, *Dead Aid*, Allen Lane (Penguin), London, pp. 54–55.
81 Hancock, Graham, 1989, *Lords of Poverty*, Macmillan, London, pp. 72–75.
82 Moyo, Dambisa, 2009, *Dead Aid*, Allen Lane (Penguin), London, p. 8.

Further readings on food aid problems

Abbott, John C., 1992, *Politics and Poverty: A Critique of the Food and Agriculture Organisation of the United Nations*, Routledge, London.
Barrett, Christopher B. and Dan Maxwell, 2005, *Food Aid After Fifty Years: Recasting its Role*, Routledge, London.
Berstein, H., Crow, Ben, Mackintosh, Maureen and Charlotte Martin, 1990, *The Food Question: Profits versus People*, Earthscan, London.
Borgin, Karl and Kathleen Corbett, 1982, *The Destruction of a Continent: Africa and International Aid*, Harcourt Brace Jovanovich, London.
Carlsson, Jerker, Gunnar Kohlin and Anders Ekbom, 1994, *The Political Economy of Evaluation: International Aid Agencies and the Effectiveness of Aid*, St. Martin's Press, New York.
Guillaumont, Patrick and Lisa Chauvet, 2001, *Aid and Performance: A Reassessment*, The Journal of Development Studies, 37:6, 66–92.
Hanson, Henrik and Tarp Finn, 2000, *Aid Effectiveness Disputed*, Journal of International Development, 12:3, 375–398.
Jackson, Tony and Deborah Eade, 1982, *Against the Grain: The Dilemma of Project Food Aid*, Oxfam, Oxford.
Karapinar, Baris and Christian Haberli, 2010, *Food Crises and the WTO: World Trade Forum*, Cambridge University Press, Cambridge.
Maren, Michael, 1997, *The Road to Hell: The Ravaging Effects of Foreign Aid and International Charity*, The Free Press, New York.
Michaely, M., 1981, *Foreign Aid, Economic Structure, and Dependence*, Journal of Development Economics, 9:3, 313–330.
Mosley, Paul, Hudson, John and Sara Horrell, 1987, *Aid, the Public Sector, and the Market in Less Developed Countries*, Economic Journal, Royal Economics Society, 97:387, 616–641.
Papanek, Gustav F., 1972, *The Effect of Aid and Other Resource Transfer on Savings and Growth in LDCs*, Economic Journal, Royal Economics Society, 82:327, 934–950.
Sahn, David E., 1996, *Economic Reform and the Poor in Africa*, Clarendon Press, New York.
Shah, Anup, 2007, *Food Aid*, Global Issues, 3 December, 1–25.

10

IMPORTANCE OF 'TRADE' MARKETS AND MERCHANTS

'Commercial trade' markets are the most misunderstood of all markets. Their transactions are private yet these are accumulatively made public wherever futures markets exist. Without 'trade' markets, there can be no futures markets. 'Trade' markets can exist without a futures market, but the forward market for 'trade' markets can be weak without futures markets. They exist together mutually if not symbiotically, but depend on such other markets as the warehouse market.

The evolution and examination of 'trade' markets have been lost in a maze of theories on geographic location,[1] clusters,[2] agricultural 'industrialization',[3] interregional trade,[4] and currently embedded into much integrative supply chain management theory. There have been erroneous comparisons with mercantilism, even though oligopolistic mercantile protectionism[5] was the antithesis of 'trade' markets that only exist in highly competitive deregulated supply chains.

Grave misinterpretations about forward markets and prices seemingly began in 1924 when the US Federal Trade Commission on the Grain Trade reported on page 217 that forward futures prices reflected forward estimations of supply and demand. Whilst that is true for perishable commodities and under shortage conditions, it is certainly not the case for storable products in surplus 'trade' markets. That seemingly innocuous error has permeated literature ever since, turning futures markets into perceived speculative havens, which they are not. Speculators and arbitrageurs are quite peripheral to the underlying forward mechanism of 'trade' and futures markets, although there can be temporary market distortions.

Keynes subsequently expanded on this misguided speculation theme in 1930[6] by introducing 'risk premiums', implying that speculators established the forward market mechanism based on risk perceptions. This was despite the usage of contango and backwardation in forward markets being very old agricultural merchant and 'trade' market terms. Kaldor continued with similar confusion in 1939 by introducing extraneous terms such as 'marginal risk premium' and 'convenience yield of stocks', whilst complicating the issue with arbitrage strategies.[7]

There was no clear distinction between the categories of storage by either Keynes or Kaldor. 'Trade' inventory should have been separated from commodity stocks in private hoards and government 'reserves', and even the latter should have been dissected into institutional inventory and non-institutional stocks. However, economic supply theory continued to be aggregated for the purposes of facilitating equilibrium theory,[8] despite all the aggregative data limitations.[9]

All agricultural futures markets are derived from the underlying physical 'trade' market, which is where the word 'derivative' arose. If the 'trade' market is destroyed by regulation, then the futures market collapses. Futures markets are primarily based on the underlying 'trade' inventory, and the storage charge relating to such inventory is known as the carry cost. These 'trade' and futures markets are not based on any dealings with private hoardings or government stocks, which are quite anathema to futures markets. Futures markets are also not derived from speculation, as both Keynes and Kaldor purported.

Keynes factitiously denigrated the 'trade' market for not wanting to store when he stated that 'the competitive system abhors the existence of stocks' and that it is 'an outstanding fault of the competitive system that there is no sufficient incentive to the individual enterprise to store surplus stocks'.[10] There was no mention or acknowledgement of the accumulation and storage functions of merchants in 'trade' markets. Food was considered as a Marxian-based[11] 'commodity fetishism' that was merely supposed to reflect production and labour costs.[12]

Under Keynes' socialist system,[13] if farmers produce the wrong product or if they produce excess to end-user requirements, private end users are to be blamed for not storing this unwanted commodity. Existence of unwanted commodity was justification for Keynes to advocate for government storage monopolies. Keynes seemingly had no knowledge of forward markets having a price signalling role for farmers. In effect, the forward price mechanism as reflected in the futures market was the only supply chain signal from 'trade' markets for many farmers. It was significant that the only country on the Continent to translate many of the writings of Keynes was Nazi Germany.[14] In exchange, Hicks by 1939 had translated much of the German equilibrium theories into English, attempting at the same time to solve the comparative static mysteries required to achieve such equilibrium.[15]

Holbrook Working had already completed a major 48-year 'trade' market study based on Chicago wheat futures for the July–September forward months between 1885 and 1933, and published the findings in 1934,[16] yet both Keynes and Kaldor chose to ignore the findings and publication. There was a very deliberate move by the 'equilibrium' advocates to obfuscate any knowledge about the proper functioning of both 'trade' and futures markets. Meanwhile, the Roosevelt government's 'ever-normal granary' and Federal Farm Board concept was biblically juxtapositioned onto previous food supply chains in the US.[17] Politics and ancient history was dominating rational economic debate.

Under forward market criticism from both Dow[18] and Hawtrey,[19] Kaldor stated in 1940[20] that his previous forward price determination theory in 1939 was 'inadequate'. However, even despite this recantation, Kaldor's amended version remained very deficient in that it did not acknowledge that deliverable futures contracts were merely standardized 'trade' market forward contracts that were recognized as such by the US Grain Futures Act of 1922, did not account for carry and storage in the forward months under surplus conditions, and did not recognize the significance of end-user demand driving up spot prices whenever shortages occurred.

Working replied to these misconceptions in late 1940,[21] followed by a discussion on the spot–forward price relationship in 1942.[22] A major dissertation was published in 1948 reiterating how backwardation occurred in forward months of 'trade' and futures markets during times of shortages.[23] In the following year, Working explained how a contango market functioned in 'trade' markets through the issue of carry, which was factored into forward prices in 'trade' and futures markets whenever surpluses occurred.[24] This was fundamental mechanical knowledge of understanding how deliverable futures markets and forward prices actually function based on 'trade' markets. He then consolidated the material in further publications in 1954[25] and 1961.[26] Despite all this, Kaldorian economics from the pre-war Berlin school still persists in contemporary microeconomic theory.[27]

Defining a 'trade' market

A 'trade' market is characterized by supply agreements between end-users and merchants usually in a food supply chain regional cluster. These agreements may be relatively short term or they may develop into a longer-term relationship. 'Trade' markets are a prerequisite for efficient food supply chains in both producing–exporting and importing countries.

The main characteristics of a 'trade' market involve merchants creating location value for an end user through specific product accumulation, and the time value created through carry and storage, followed by just-in-time delivery to the end user. This generally restricts 'trade' markets to those food products that can be stored.[28]

'Trade' markets only occur for agricultural output because of the seasonality of final production, which is usually within three months of the year. Most other production-based industries operate through a switchboard, which can control supply according to demand, but this does not occur for biological time-lagged agricultural production that is yield and weather dependent.

'Trade' markets usually exclude perishable foods because retailers want freshness, and that usually means immediate delivery direct from the farmer. Supermarkets do not pay for storage of perishable foods because of shelf-life limitations and turn-over targets. There is usually a simple fresh supply contract between the retailer and farmer who has to meet strict production, preparation, and delivery guidelines. This makes it similar to a 'trade' market; however, the distinguishing characteristic from a 'trade' market is that there are no specific accumulation functions with carry and storage components, thus excluding merchants. Whilst futures markets for perishable commodities can be established, their forward pricing mechanism is tenuous, thus making them vulnerable to speculation as was duly noted by Keynes.

The need for 'trade' markets

The seasonality of agricultural production is a serious food supply chain issue. Perishable products offer no solution to seasonality other than for consumers to do without, or source the food internationally from countries with different seasonal production times. In contrast, non-perishable agricultural products may be abundant but they are not required as yet by the food supply chain.[29]

Without orderly logistics procedures through 'trade' markets, farmers would dump their product at the door of the end user. This would force the end user to be responsible for unwanted and unneeded product. For reasons of quality control, hygiene, and logistics, any form of dumping is an anathema to an end user. Therefore, the end user is required to take control of the logistics, and is thereby forced into paying a supply chain intermediary such as a merchant to source, accumulate, store, and deliver. This creation of both place and time value in the supply chain for end users is really the main function of the domestic merchant. The corollary is that 'in the absence of local merchants, there is no one remaining in the supply chain to take responsibility for the required supply chain functions and services'.[30]

Farmers generally pay for all storage logistics whenever they do not sell into 'trade' markets or if 'trade' markets do not exist. In contrast, an end user in a 'trade' market rewards the merchant (or farmer) by paying a carry charge per month to store specific product.[31] This carry charge is negotiated as part of the 'trade' market supply agreement. It can be contentious, especially when the time value of money, product shrinkage, and insurance is expected to be included in the physical storage or warehouse cost. One offsetting factor for the merchant is economies of scale whereby the unit storage cost decreases with quantity throughout.

Forward 'trade' market prices become established based on the monthly increments of this carry charge, and a public forward price barème benchmark is derived wherever many 'trade' markets are clustered. When such public benchmark prices are formalized, futures markets evolve.[32] Therefore, the essence to establish agricultural futures markets is to have strong local 'trade' markets; however, these can only exist in competitive deregulated supply chain clusters where accurate price signalling occurs.

Agricultural futures markets are perceived to 'discover' forward market prices, but the reality is that these forward prices must reflect local forward 'trade' market prices. Any major permanent distortion between the two markets represents inefficiency. A major problem arises whenever a futures market attempts to straddle two separate 'trade' markets, such as the local and export trade.[33]

A liquid public transparent forward market, as characterized by a futures market, is necessary to manage price risk through hedging that offsets market position.[34] Therefore, to be able to manage price risk, there is a need for a futures market that has been derived from a strong local 'trade' market.[35] Any absence of strong local 'trade' markets prevents the formation of a liquid futures market, and therefore there is no forward market mechanism to offset market position risk. Price risk management becomes impossible. Without a forward market, all transactions are reduced to spot market. Price speculation for farmers, merchants, and end users can initially result from having no forward position risk offsetting mechanism.

Therefore, a lack of strong 'trade' markets necessitates commodity spot buying and selling speculation. This creates uncertainty that prevents organizational linkages in the product flow chain and results in periodic unsustainable financial losses, which then forces governments to intervene to protect both farmers and food security. But the primary underlying problem is the lack of strong 'trade' markets and the inability to manage the associated supply chain risk.

'Trade' markets are responsible for maintaining private food reserves in the supply chain for end users, sourced either from local farmers or from imports. Government supply chain intervention destroys these extremely sensitive 'trade' markets, thus removing the ability and incentive of the private sector to store food reserves destined for the food chain. Removal of these 'trade' markets makes the food supply chain totally dependent on government commingled commodity 'buffer' stocks for food reserves, but which have little end-usage specification. End users quickly depart, leaving a vacuum between consumers and government officials, with consequences only too well known in North Korea.[36]

Characteristics of 'trade' markets

'Trade' markets reflect what consumers need and are currently demanding. Because consumer behaviour is dynamic, 'trade' markets need to be flexible, which explains why many 'trade' supply agreements do not extend beyond 12 months.[37] This affects the forward month prices of futures markets, with liquidity seriously declining beyond 12 months. Usually there is an element of speculation beyond 12 months in forward agricultural futures months because they do not reflect 'trade' market conditions.

One of the main characteristics of 'trade' markets is that they exist mostly in surplus markets. Storable agricultural product prices are skewed[38] with long periods of surpluses and low prices,[39] which favours the formation of 'trade' markets. Product shortages have the tendency to destroy 'trade' markets, and these occurrences need to be minimum, otherwise end users collapse financially. The dilemma for farmers is that the surplus attributes which are desirable for 'trade' markets are not always in the best price interests for farmers. This supply–price standoff becomes the real issue for food security.

The most unusual characteristic of storable agricultural products is their non-edible form, which prevents direct farmer–consumer relationships. Perhaps with the exception of pulses and refined sugar, nearly all other storable agricultural products need further value adding for human consumption, such as processing and manufacturing into final food products.[40] This necessitates storage of agricultural products in the food supply chain prior to being received by the end user, who as a food manufacturer may be positioned at least two intermediary stages before the final consumer in the supply chain.

When transport and infrastructure were problems, the location of storage was usually very near to the end user, which added to urban congestion. As transport infrastructure improved, storage was pushed back in the food supply chain towards the farmer.[41] In deregulated markets, product accumulation and storage were then made the responsibility of the specialist domestic merchant, and the supply agreements between the merchant and the end user became known as 'trade' markets. The merchant met the demand of the end user who met the demand from consumers.[42] If farmers failed to produce the correct product demanded by the supply chain, 'trade' markets failed unless imports were permitted.[43]

The introduction of just-in-time deliveries in many deregulated food supply chains in the 1980s pushed storage even further back in the supply chain,[44] even beyond merchants. Under certain strict guidelines relating to specified product, storage, and delivery, farmers can enter 'trade' market supply agreements for storable products.[45] Whilst these are still rare and depend on the skills, organization, and capital investment of the farmer, it might indicate a pathway for the future. It represents the only method to minimize the number of intermediaries in the food supply chain for storable products. However, there are many associated problems that necessitate operations management solutions.[46] As well, if the farmer cannot or will not tailor production to meet local end-user specific requirements, the result is unwanted price-taking commodity, for which the 'trade' cannot be made responsible.

The foremost consideration of 'trade' markets is that suppliers must meet product specification, which relates to product attributes, quality, and timing.[47] It is usual for merchants to undertake strict quality assurance programmes and implement International Organization for Standardization (ISO) quality controls to meet end-user specification. Farmers are generally excluded from food 'trade' markets because of the high cost of meeting minimum quality assurance and occupational health and safety standards.

Even if farmers meet product attribute requirements, they must do so consistently in order to join 'trade' markets. Because weather-agronomic-dependent product attributes are outside the control of any one farmer, continuity in the supply of consistent product usually is beyond any single grower except in irrigated regions. This makes the role of merchants in accumulating specified product so important in 'trade' markets.

Continuity of supply in 'trade' markets means meeting delivery specifications for exact quantity every time.[48] There is an increasing global trend for end users to demand just-in-time logistics that involve small-size continual deliveries. This usually restricts delivery to road transport or shipping containers, which means that the supplier has to arrange for this delivery specification to occur for the duration of the supply agreement. Specialist regular road or sea transport requirements usually preclude farmers and favours merchant logistics operations.[49]

The supplier in a 'trade' market agreement must arrange the warehousing for product storage, maintain the infrastructure and operations according to ISO and quality assurance requirements, and maybe undertake regular maintenance such as fumigation, which then requires meeting strict occupational health and safety regulations.[50] This requires adequate management, administration, and organizational controls. Sometimes the supplier will delegate such responsibilities and controls to a professional operator under contract. Farmers are poorly organized to undertake such functions.

Suppliers aligned with 'trade' markets must be contented with locking in a price for the duration of the supply agreement. There have been many instances where farmers have broken 'trade' market supply agreements when shortages occur and prices rise. Farmers tend to be opportunistic when price speculation favours them, and this can cause abrogation of contractual obligations. Alternatively, when prices fall, farmers want to be involved with fixed price supply agreements. End users generally exclude most farmers because of this opportunistic behaviour. There is a need for farmers to manage and enhance their price quite separate from the supply agreement, but this necessitates using futures or options markets, which is something that many farmers are not prepared to undertake.

There is no provision made for delivery default on a 'trade' market supply agreement, otherwise the end user would not enter such an agreement. A legal cash settlement might be possible under a contract clause, but the outcome rarely reflects the disruption to the food supply chain, the stand-down losses including labour and machinery costs incurred by end users, subsequent congestion and over-time losses, and the loss of good will to retailers and customers. Because of high production and delivery risk associated with single farmers, end users prefer to spread the risk of delivery default by appointing merchants who have the minimum size required to ensure either local or imported product. Farmers generally lack accumulation skills and the ability to import in order to maintain continuity of consistent supply.

The last characteristic is the provision of service to end users, which sometimes entails 24-hour operations, seven days a week. Individual farmers are unlikely to provide such service and be in a position to respond to urgent end-user demands at short notice. Merchants are more likely to have the flexibility in logistics operations to provide the necessary response mechanisms.

Forward markets

Strong local 'trade' markets are precursors to futures markets, but without a futures market, 'trade' markets have no transparent and unambiguous forward price market. 'Trade' markets can exist without a futures market, but negotiations over a forward price are frequently tedious and often cause conflict between the end user and supplier. Under such conditions, there is a tendency for weak supply chain relationships and linkages.

Pulse 'trade' markets without futures markets are no comparison to grain 'trade' markets with strong futures markets. But perhaps this has more to do with product characteristics, with pulses requiring little product processing while grain needs much product (manufacturing oilseeds and meal require much processing). Strong forward markets may therefore be related to the need for manufacturing or processing. The Chicago futures market only arose after tin-can food manufacturing began locally during the 1840s.

'Trade' market transactions without a futures market are frequently reduced to spot price settlements. At best, this is a rudimentary speculative food supply chain existing with high risk with no forward relationships. Commodity cycle shocks can cause this supply chain to easily break down into very opportunistic behaviour. Instead of increasing competition in the food supply chain to correct the price distortions in the supply chain, which are causing weak 'trade' markets to exist, governments frequently intervene further and worsen the price distortions, thus preventing the establishment of high volume (liquid) futures markets.

Without a futures market, suppliers to the 'trade' market may perceive end-user market power in establishing forward prices, regardless of whether this market power is real or not. The existence of a futures market that is characterized by transparency and liquidity lessens the perception of market power of individuals or end users.

A futures market derives its spot price from the spot 'trade' price through spot basis.[51] An efficient forward futures market for storable products should reflect the forward 'trade' market price. From a local perspective, the only difference between spot and forward prices in either the 'trade' or futures markets should be a contango or backwardation occurence.

Forward markets are characterized by orderly marketing, but only when surplases occur. This makes forward markets possibly the most fragile of all markets. They can collapse because of shortages or through distorted price signals. Supply agreements can fail through contract default by merchants or farmers whenever shortages occur.

When surpluses occur in 'trade' markets, end users are forced to pay a higher forward price to merchants because of the storage carry charge reimbursement. The supply agreement binds the end user and merchant together into the future through the provision of carry. It represents a forward premium to cover the added costs involved with creating time value. Historically, this premium was known within the 'trade' market as a contango, which was a charge or percentage paid by the buyer to a seller to delay the delivery of physical inventory.

Local liquid futures markets will reflect the storage carry cost in the forward month prices. This forward market is known as a contango market, which is a market in which a contango is charged per month. It is indicative of a surplus market, and the movement of the size of the contango will indicate spot price direction. An increasing contango will indicate increasing surpluses with spot prices highly likely to fall. Alternatively, a decreasing contango will indicate decreasing surpluses with spot prices likely to rise at some future time.[52]

Whenever the spot price falls or rises in a contango market, the same forward carry structure in the market is usually maintained (Figure 10.1). The forward month prices move in a parallel structure regardless of the spot price movement.

A contango market is characterized by surplus conditions, which means that as surpluses increase, so does the carry charge increase. Warehouse space is limited in the short term within a warehouse market. The amount of carry-contango in the forward market is indicative of the surplus being carried in the warehouse market.

As warehouse space demand increases, the cost of carry increases to reflect the increasing surplus and the decreasing warehouse space available. This increase has to be reflected in the forward month prices in the futures market. As warehouse space demand decreases, the cost of carry decreases to reflect the decreasing surplus and the increased warehouse space available. This decrease has to be reflected in the forward month price in the futures market.

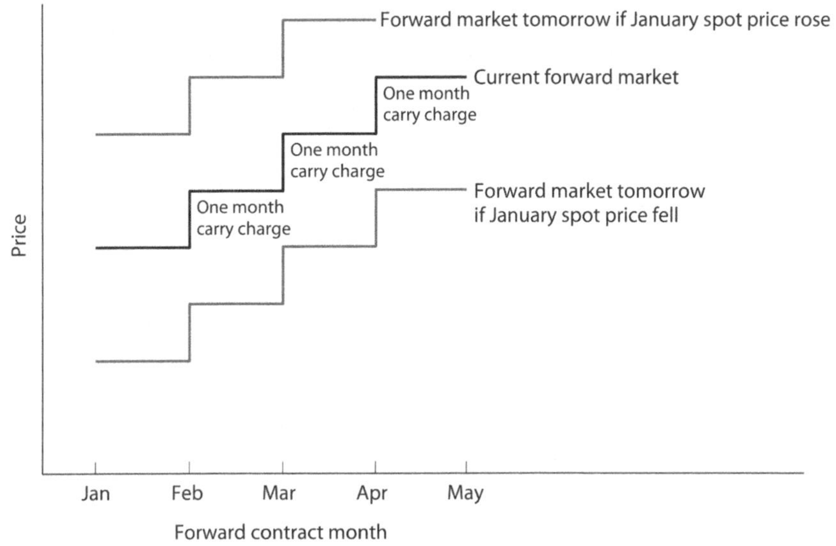

FIGURE 10.1 Parallel movement of forward prices in a contango-surplus market
Source: John Williams

A contango forward market is a characteristic only of a 'trade' market, which means that commodity that is not in the 'trade' supply chain is excluded. Farmers could be hoarding on-farm or storing in an external warehouse or bunker. Merchants could hoard in their own warehouses or bunkers. Both situations result in speculative commodity that is outside the 'trade' market, and therefore this is not reflected in the carry in contango markets, although it could affect overall warehouse space. It means that there could be shortages occurring in the 'trade' market while hoarders such as farmers or merchants are withholding large supplies from the end-user supply chains.

If either actual shortages of supply occur in the food supply chain, or if hoarders withdraw sufficient supply, then end users could increase their spot 'trade' market price to attract more sales from the hoarders. Instead of decreasing price to cause panic amongst the hoarders, the end user frequently increases spot 'trade' prices. This is a panic signal from the end users and merely reinforces the unwillingness of the hoarders to sell. End users could seek product alternatives or substitutes under such circumstances, rather than merely raising the price, which would send the wrong signal to hoarders.

If end users have to choose between shutting down manufacturing plants and standing-down staff with high lay-off costs, or finding substitute food inputs, the latter is often becoming the preferred choice. When wheat prices soared in August 2010 in the aftermath of the Russian wheat crop shortfall, some food manufacturers such as Kraft in the USA increased product substitution. In contrast, Starbucks was forced to raise prices for coffee during the shortage of August 2010 because there were no product substitutes.

Raising spot 'trade' prices by end users can undermine existing merchant relationships within the 'trade' market because of fixed-priced supply agreements. Merchants begin to perceive price opportunities that are denied to them through previous supply agreements, at the same time that hoarders receive psychological price signals to prolong their hoardings. It destroys the motivation for merchants wanting to form strong alliances with end users in the supply chain, while at the same time rewarding hoarders who do not want supply chain relationships.

Merchants react to the higher spot price and decreasing carry in the forward months by bringing selling stocks as spot sales to capture the higher spot price being offered by the end user. Forward storage costs incurred by merchants are not being offset by the carry offered by end users, which explains why merchants will sell stock into the higher price spot market with the hope of buying inverntory back later at a cheaper price. But this strategy is fraught with risk, regardless of risk management tactics.

Wherever there are local futures markets, merchants can sell physical stock and buy distant month futures, in the expectation of taking delivery against the futures contract at a future date to ensure the supply of 'trade' market contractual deliveries to end users. This represents a type of arbitrage that is known as a reverse cash-and-carry strategy.[53]

The spot selling and buying forward month futures contracts by 'trade' merchants will tend to force down the spot price and hold up the distant month prices. This can be referred to as the off-loading phase of an early backwardation market. It can result in temporary surpluses and market prices falling after an initial price rise signal by end users. Also, it is more likely to occur towards the end of the old crop year when supplies are short. Any price rise at that time will often result in hoarders selling, particularly when they only have a limited time to offload before the new crop year. The price fall duration can generally be four to eight months and depend on old-new crop months, the quantity of old crop hoardings to offload, or the amount of merchant stocks to sell in reverse cash-and-carry strategies.

Success in supply chain risk management often depends on future supplies being available. With or without the existence and usage of local futures markets, 'trade' merchants could also have undertaken forward purchase contracts with farmers to ensure forward deliveries to end users. However, farmers might be forced to default on physical delivery against forward contracts with merchants because of adverse biological conditions and production shortfalls. As well and despite futures market guarantees, there may not be enough liquidity in the futures contract delivery period for the physical transaction to occur against the reverse cash-and-carry strategy, thus reverting it to a cash-settled contract. Supply shortages can therefore cause delivery problems regardless of whether forward purchase contracts or reverse cash-and-carry strategies have been used.

In a desperate effort to maintain contractual supply commitments to end users, merchants are forced to raise prices to attract sales from privately hoarded stocks, whilst farmers drive up prices when searching for product to deliver against their forward contractual commitments. Provided that hoarders are prepared to sell late in the old crop year, price rises could be dampened.

Rising spot 'trade' prices to attract spot sales and supplies in a shortage market results in carry decreasing and gradually being eliminated between most or all the forward months. It may start in the nearby months or it may just apply to new crop months. Much depends on new crop production and supply estimates, as well as the expected size of old crop hoarding. Regardless, any expected shortages in the new crop will drive up the nearby month price relative to the distant month prices, and could result in a discount between the old and new crop prices.

If price inversion occurs even between only two months, this inversion of the forward months is known as a backwardation market. Historically, backwardation in 'trade' markets referred to the discount or percentage that a seller had to accept by delaying the delivery of physical stock to a buyer.

The spot month does not necessarily have to be involved for a backwardation market to occur. Backwardation occurs whenever the distant month prices are discounted to the nearby month prices (Table 10.1). It was completely erroneous for Keynes[54] to presume that forward prices are being discounted because of expected adequate futures supplies. 'Trade' market reality suggests that the greater the discount in the forward prices relative to the near prices, the greater is the shortage. Keynes misinterpreted the historical context of 'trade' market backwardation.

There is no regularity of forward prices in a backwardation market. Forward price movement is largely dependent on hedging and speculation in specific months, spread-month trading, or expectations over new crop production. Premiums or discounts can occur between old and new crop years.

A backwardation market will only result in a bullish market if there is a general perception of new crop failure, which would encourage hoarders to continue to withhold supplies. Alternatively, an excellent new crop might push backwardation towards the more distant months and linger accordingly with future new crop production. Only after two to three good new crops would backwardation be expected to be gradually eliminated with a return of increasing carry. Keynes' speculative forecasting only occurs beyond the 12 month 'trade' market agreements and new crop production, where it is usually portrayed merely as a stable price extrapolation.

If there was an actual old crop shortage before the earlier offloading stage, then a potential new crop shortfall will aggravate the shortage and drive spot prices higher. The old–new crop discount usually increases whenever disappointments occur with

TABLE 10.1 Backwardation (inversion) in the forward market

Month	*Price*
March	400
May	380
July	390
September	380
December	360
March (year 2)	370
May (year 2)	380
July (year 2)	370
September (year 2)	360
December (year 2)	360

the new crop expectations, because the new crop physical condition is factored into the old crop spot price.

At this stage of an increasing backwardation market, the disintegration of existing 'trade' markets occurs. Old supply agreements become meaningless because of supply default, new agreements are not finalized because of the uncertainty over supply, and there is general disorder in the supply chain. Merchants, end users, and default farmers compete over limited supply, which often causes a price spike. Backwardation does not always lead to price spikes; however, all price spikes are preceded by periods of backwardation.

Providing that the merchant could accumulate sufficient supplies for the 'trade' market in the longer term, then the 'trade' market might return to some normality, with end users again resuming the payment of forward carry (contango) in supply agreements. This might occur with successive large new crops and potential hoarders being prepared to sell.

Ultimately farmers respond to the high prices by increasing their production unless government intervention occurs. Any intervention-induced price distortion at this stage can act as a disincentive for farmers to produce more. If governments do intervene, such as when the Indian government intervened with price controls in the drought of 2009, then the probability of prolonged food shortages will be high. Farmers need periodic high prices to be profitable, given the high frequency of low prices.

Benefits of strong forward markets

Strong 'trade' markets are essential for forward price discovery. A forward futures market can reflect these 'trade' market forward prices as well as providing liquidity and transparency. Global forward pricing can be benchmarked against a liquid and transparent international futures market.

Without a futures market, transactions are largely reduced to spot physical transactions, with little ability to offset forward price risk. Forward pricing and storage cost compensation within any supply agreement will be difficult to achieve without a futures market and often will be the cause of much conflict. Settlement of both forward pricing and carry cost determination are more likely to be determined amicably if decisions are based on a liquid and transparent futures market benchmark.

Forward month prices of a futures market for a storable agricultural product reflect the supply and demand of the local 'trade' warehouse market. If storage space is short, then carry cost rises, which causes the forward month futures prices to rise. Increasing contango markets therefore are indicative of the increased amount of product being stored. When storage space is excessive, then carry cost falls, which causes the forward month futures prices to fall. Decreasing contango markets are indicative of the decreased amount of product being stored. If the forward month price falls below the previous month price, then price inversion has occurred, and this backwardation indicates that ample warehouse space is available.

The forward pricing structure of a futures market is essential to provide market signals indicating when to store, when to sell, how and when to price. Without a liquid futures market, it is extremely difficult to make such decisions. Paradoxically, increasing carry in forward prices is a signal to a hoarder to sell because of increased supply and decreasing price. Decreasing carry in forward prices is a signal to a hoarder to withhold sales because of decreased supply and increasing price. But rarely would a hoarder follow these carry signals.

Forward futures markets have an important role in offsetting (hedging) physical market position risk for both sellers and buyers. Price risk management can occur against forward futures markets provided that basis risk[55] is not prohibitive. Currency risk may be a problem when attempting to cross hedge between a local market with an international market. However, without a liquid and relevant forward futures market, it is very difficult to manage price risk for buyers and sellers.

Merely having a liquid futures market does not make it relevant for international hedgers. Both basis risk and currency risk can decrease the effectiveness of an international hedge and cause hedge ratios[56] to be lower. Basis risk can be high if there are up-country and port-based differences, or if there are product and contract specification variations. Any local impediment to the formation of local futures markets should be perceived as a major obstacle to achieving efficient local food supply chains and having adequate price risk management.

Role of merchants

The main functions of a domestic merchant include taking ownership of accumulated product, undertaking buying and selling activities, meeting just-in-time deliveries, and absorbing both product and price risk. Merchants also absorb place and timing risk associated with accumulating, handling, and transporting over vast geographical regions and distances, but more importantly, they add time value through storage. To facilitate these functions, they establish supplier contracts with end users. These contracts usually include supply chain storage and carry agreements.

Under regulated supply chain distortions, domestic merchants cannot function effectively because they lack clear price signals, and they often get crowded out by public authorities. International merchants are more likely to be favoured by public authorities under market regulation, in order to sell unwanted commingled commodity internationally.

There are many roles of domestic merchants in a deregulated market, including specific product accumulation, storage, delivery, risk management, financing, and providing supply chain feedback.

Agricultural production is geographically diverse and therefore is characterized by its distance from food manufacturers and consumers. Both transport infrastructure inefficiency and its cost are major impediments to integration between farmers and the food supply chain even in supposedly developed countries.[57] Therefore, there is an important and ever-increasing role for merchants in accumulating from farmers what is being demanded by end users.[58]

The major factor characterizing this accumulation function is its selectivity. Because of product differentiation, merchants need to accumulate only what is required for 'trade' markets. Product rejection can lead to farmer resentment, hostility, and calls for government intervention. However, government intervention cannot make machinery such as flour milling perform more effectively when the input is unsatisfactory. Government intervention cannot change what is being demanded by consumers, such as a particular type of bread, unless there is undue force, which usually results in shortages, queues, black markets, and often riots. Stalin may have murdered the flour miller and sent the consumer to the gulag, but it did not make machinery operate any better or food to be made more edible.

Farmers place unrealistic pressure and expectations on government to undertake the impossible when they demand intervention whenever they produce agricultural output that is not aligned with technological capabilities and consumer demand. The whole food supply chain is made worse if governments respond to this pressure because end users are then forced to close because of machinery and equipment malfunctioning or no consumer demand. Food manufacturers often shift operations to another country. There is a need for political responsibility and good governance behaviour to resist such pressures.[59]

One non-interventionist policy might be for governments to ensure adequate research and development into better agricultural production techniques. Plant varieties might be improved so that they have the dual purpose of satisfying both local production requirements as well as 'trade' market specifications. This is more likely to integrate farmers into the 'trade' market supply chain. The alternative is to allow the required imports despite protests from domestic farmers.

Merchants are more likely to accumulate consistently in both surplus and shortage markets, and deliver continuity of supply to end users, compared to individual farmers. This assumes that there are adequate price signals to base decisions. If price distortions prevent adequate price signals, merchants as well as farmers and end users cannot make the correct decisions, and the domestic merchant is more likely to cease operations or move elsewhere.

There is high risk associated with product accumulation, storage, and delivery because of the biological nature of food products. Merchants are bearers of risk, but to survive they must be risk managers. They will only be prepared to take on risk if there is a risk transfer mechanism either in the food chain through back-to-back contracts or a futures market hedge to offset that risk. Therefore, merchants will only accumulate specified product when there are local end users willing and able to establish strong local 'trade' markets. To do otherwise for merchants without hedging is speculation, and this jeopardizes their financial viability.[60]

Domestic merchants in deregulated supply chains have an important role in storing what has been accumulated for local 'trade' markets, and maybe in some situations for government institutional catering. Storage has to be differentiated according to customer type rather than any producer-driven or government-regulated commingled grading system.

Bulk handling authorities are designed to store bulk commingled commodity from farmers. They are not equipped to store or handle differentiated product storage for end users. This is the function of merchants. Dysfunctionality occurs when regulatory authorities attempt to force the integration of bulk commingled commodity with end-user-based differentiated product.[61] The food supply chain simply collapses.[62]

The domestic merchant is rewarded for creating time value by a carry premium that is added to the price established in the supply chain agreement with the end user in the 'trade' market. It is proportional to time, with a carry charge added for each month stored. This is what causes the stepped forward price barème in contango markets (Figure 10.1).

End users may refuse to pay for storage in times of shortages when product accumulation by the merchant is difficult. The contango-stepped market disappears and is replaced by market price inversion and backwardation. This is a price signal to merchants to deliver any available product immediately and not to store.

It is the function of the domestic merchant to undertake complex logistical operations regarding the maintenance and preparedness of inventory to align in seamless just-in-time delivery operations to the end user.[63] That is the theory which may not be the reality because of the high probability of shortages and disruptions caused by the biological characteristics of the product. Yet merchants are expected to deliver the required product on time every time to the end user under the terms and conditions of the 'trade' market agreement.

There is an added delivery function for international merchants regarding imports and exports. This requires far more specialization compared to the domestic merchant. There is high risk associated with creating international value regarding timing, place, and quantity, with greater skills and networking required.

Because merchants are bearers of risk and yet they must be risk averse to remain in business, there has to be adequate risk transfer mechanisms. The alternative to back-to-back arrangements for merchants is to transfer certain types of risk such as price, currency, and interest rates to a futures market.[64] A liquid risk-transfer mechanism such as a futures market may be required by merchants to reduce short-term risk until longer-term physical back-to-back arrangements are consolidated.

Merchants can consolidate food supply chains through existing 'trade' markets via back-to-back agreements, or they can develop new 'trade' markets by expanding the food supply chain. It was only after the Australian domestic grain market deregulation in 1989 that there was a re-emergence of domestic merchants and clustering of local 'trade' markets, which inexorably led to the reestablishment of lot feeding and an animal feed manufacturing industry.[65]

There is no supply chain security value created when merchants do not transfer their risk in either back-to-back transactions or through the futures market.[66] Operations become reduced to spot market speculation, and the merchant reverts to acting similar to many farmers. No value in the supply chain can be created through speculation, yet there is an increasing proportion of agricultural commodity production being hoarded privately by both merchants and farmers, particularly in countries with price controls.[67]

It is not the function of end users to finance the food supply chain, with most preferring to pay upon delivery. Product accumulation is expensive and drains the merchant of cash flow and working capital, yet merchants must maintain the financial viability of the supply chain if it is to operate effectively. Without reasonable payment terms, farmers will not sell, thus making accumulation extremely difficult. Therefore, there needs to be adequate financing by merchants to pay farmers and to maintain supply chain operations.

Merchants have the advantage of accumulating product in surplus markets because some farmers perceive that they will obtain a higher price if they sell today, relative to a lower price tomorrow. However, farmers would still be unwilling to sell into a falling market if they perceived payment delays and uncertainty over payment.

If there are insufficient cash reserves and liquidity flow, then merchants need access to finance if they are to remain in business. When adequate security exists through product ownership and forward sales, banks are more likely to lend finance to merchants.[68] However, product ownership with no forward sales creates an asset security dilemma for banks to lend finance to merchants. This speculative activity is quite alien to the food supply chain.

This impasse can be sometimes resolved by the bank actually buying the physical product and paying out the merchant in a total return swap when the merchant is unhedged on the futures market, or a total return basis swap when the merchant is hedged on the futures market. The proceeds can be used to pay farmers, whilst the physical product always remains in the merchant's warehouse but owned by the bank. When the merchant receives cash flow from subsequent sales, the merchant buys the physical product back from the bank. Profit for the bank in lieu of an interest return is the difference between the bank's purchase price and the bank's selling price. These swaps are very suitable for countries with anti-usury laws, but unsuitable for cooperatives and pool managers who have unsold product with no title.

Whilst product specification might be beyond the control of the merchant, quality control and assurance is vital for product value maintenance. In highly competitive 'trade' markets, merchants have the responsibility of ensuring specific quality standards, usually under some ISO certification.

Quality control can be extremely difficult to maintain in a food supply chain because of the biological nature of the product and the associated biodiversity. The merchant is quickly replaced by a competitor in the 'trade' market unless there is absolute astuteness in minimizing problems. The alternative is that the particular supply chain collapses due to technical or consumer incompatibilities.

Merchants are vital match-makers in the food supply chain. They usually exist before the food manufacturer actually makes a decision to invest, often preparing the pathway for major manufacturing investment to occur. In some circumstances, there might be new products and varieties discovered for farmers, new inventions in the supply chain, or new methods to transform the product through different end-usage techniques, assuming that there is a consumer focus.

To undertake this product development value function, there needs to be much incentive for motivation and innovation by merchants with sufficient rewards to cover cost and effort. With price distortions, both the incentive and adequate returns and rewards are likely to be missing. The supply chain then begins to lose international competitive advantage with an inability to maintain itself through adequate investment, which ultimately leads to food insecurity.

Stable competitive and innovative food supply chains often lead to integration by merchants with end users through mergers and joint ventures. This can lead to further consolidation of food security in supply chains. Cargill, Conagra-Gavilon, Bunge, Louis Dreyfus, and Sumitomo all began as merchants (sometimes they began as farmers) and ultimately integrated their merchant operations with end-user manufacturing.[69]

There is a very real battle between merchants and hoarders over market intelligence. Market power can swing between merchants and hoarders depending on periods of actual surpluses and perceived shortages. Merchants will attempt every public endeavour to glean information as to on-farm and warehouse stock quantities and grades.

There is a big difference in fundamental intelligence between what is stated by hoarders and what is actually happening, simply because of their vested interest and perceived market power. Sometimes, merchants use governments in an attempt to provide 'intelligence' on international stocks, such as the US Department of Agriculture through foreign trade officials in many countries, or even on domestic market stocks as in Australia. Usually the accuracy in many of these public reports is somewhat limited, and the information often is already known to the 'trade' weeks in advance of the 'news' release.

Despite the weaknesses in fundamental supply and demand 'intelligence', international merchants are more likely to have a better overall knowledge of global stocks than domestic governments or individuals, simply because of their focus and network capabilities. Monitoring of global food stocks does provide trade opportunities for international merchants, and this provides the incentive for some accuracy in market intelligence, but such information is usually kept private.

Factors preventing 'trade' markets

There are many factors that can prevent 'trade' markets being established or cause established 'trade' markets to collapse. Any shortage of supply due to weather-related disturbances can immediately undermine relationships. Distortions in the supply chain pricing mechanism can destroy price signals and good decision making. Good governance must reflect the fragility of domestic 'trade' markets if food security is to be achieved. The impact of large government institutional buying on local 'trade' markets is an important consideration.

'Trade' markets are characterized by high risk, large investment, and intense competition, with such circumstances driving innovation and efficiency.[70] However, if local food supply chains do not function effectively because of bureaucratic regulation or institutional buying, then domestic supply chains will quickly shift to imports unless trade barriers are imposed. Potential exports will be forgone because of the lack of international competitiveness.

People need food, but the consumption possibilities–substitutes complex is likely to be very limited under government domination. Consumers ultimately will eat what they want to eat, and some would rather starve than eat publicly determined food, as many boarding school, institution, and public caterer could testify. It is incongruous to expect government regulation to simultaneously achieve both food availability and diversity. In contrast, the consumption possibilities–substitutes complex generally will tend to expand under competitive efficient 'trade' markets and international trade.[71]

Traditional bulk-handled commingled logistical systems that were rail-port centric destroyed many domestic 'trade' markets during the twentieth century in many countries.[72] Supply chain interventionists were under the misapprehension that this vacuum justified government intervention under supply chain failure circumstances.[73] However, the real answer lay in solving the causal problem of restricted logistics and encouraging 'trade' markets, and not through more government intervention.

Many governments spent most of the twentieth century providing for the foundations of an efficient road transport network. Difficult terrain meant high capital public investment. As road transport networks matured, urban rail-based storages were replaced by just-in-time road deliveries from rural regions in producing countries. The incentive for domestic end-user cluster industries returned, but only in deregulated supply chains.

The rise of supermarkets, consumerism, sealed shipping containers, and better road infrastructure has provided a pathway for reform in importing countries. The question is whether importing governments can overcome ingrained corruption and vested interests to facilitate the reformation of 'trade' markets that so characterized their countries before the advent of exporter-centric bulk handled commingled commodities.

Wherever highly storable products exist, the food chains will be a battlefield between end users and hoarders, thus potentially jeopardizing 'trade' markets. With increased hoarding activity and better bunkering technology, price movement may become more volatile. However, providing that there are adequate risk management structures in existence, the struggle over supply and price can be perceived to be both competitive and efficient, thus driving innovation and economic growth.

However, any policy that actively encourages hoarding, such as government price controls, is an anathema to 'trade' markets. Price controls can increase the hoarding activity of sellers in the expectation of driving any government-controlled price higher. As well, any price distortions supposedly to make the food system more secure can result in corruption and opportunistic behaviour to take advantage of the contrived system. In this highly speculative environment, any wrong decision by either buyers or sellers can result in bankruptcy.

The private hoarder may have the function of ensuring a sufficiently high price to provide farmers with financial sustainability and an incentive for future agricultural production. Without this periodic windfall gain that sustains continued farm production, 'trade' markets would cease to exist, and food insecurity might increase. Paradoxically, the hoarder should be perceived as the friend of those who panic over global food 'shortages'. Stalin shot the hoarder, which resulted in poverty for farmers and decreased food production. Socialists then wonder why food shortages, ration queues, and black markets abound.

'Trade' markets are very fragile from both the supply and the demand perspective. They are very dependent on tailoring supply to delicately align with current demand from consumers in the food supply chain. Sudden shocks caused by international food dumping will shatter the uneasy relationships that exist in a 'trade' market.[74] It can be argued that it has been surplus supply distortions that are the crux of current global food problems. The existence of domestic 'trade' markets has never been the cause of any global food problem,[75] despite the continual rhetoric from some farmers, yet their removal through food 'aid' can be quite devastating for farmers, consumers, domestic merchants, and local food manufacturers.

Non-emergency food 'aid' establishes alternate artificial supply chains that compete with domestic 'trade' markets. Because they are subsidized, generally lack cost and taxation accountabilities, and are usually arbitrary in distribution, these alternate artificial supply chains can cause domestic 'trade' markets to quickly collapse. Without 'trade' markets, farmers become commoditized, vulnerable to the commodity poverty trap, and become quickly replaced by either imports or prolonged food 'aid'. Domestic merchants do not have a function without 'trade' markets, so they either move offshore or cease to exist. Consumers then become vulnerable to food shortages while suffering a lack of product choice.

The precursors to the destruction of 'trade' markets generally include government intervention, price distortions, dysfunctional handling and transport logistical systems, food 'aid', conditional trade, barriers to trade, and a lack of competition. Corruption can thrive at each stage. Controlling corruption must be perceived as a separate issue in the process of deregulation. Quite often the image of privatization and increasing competition becomes damaged through corruption and this then acts as an incentive for re-intervention.[76] This has occurred in Russia.

Factors that integrate farmers into 'trade' markets

The task for farmers entering 'trade' markets can be quite daunting, particularly with geographical isolation in production, transport logistics issues, and a general lack of business negotiating skills and cash flow. They can be prevented from forming supply chain relationships because of costs, economies of scale, the lack of transport infrastructure such as shipping container ports, or the lack of port-user competition.[77]

Farmers primarily are prevented from entering 'trade' markets if they cannot produce the product with the correct specifications. The first prerequisite therefore is to produce what the 'trade' market is demanding, and to do it consistently. Demand for specific products is usually outside the control of the end user, who could be driven by engineering and machinery constraints in food manufacturing capability or due to consumer demand. Farmer exclusion from 'trade' markets should be no justification for resentment towards end users.

Product specificity can refer to the manner in which the product is produced or subsequently handled. Organic grain still has all the innate product characteristics of normal grain, but it avoids the usual chemical contaminant treatments and residues usually associated with traditional bulk handled commingled supply chains. This creates specific product value for a particular food manufacturer and consumer.

The specified product for the 'trade' market requires consistency and commitment in production of the correct type and grade, but with the required quality standards, adequate storage, and delivery procedures. Any need for agronomic rotation of crops therefore disrupts 'trade' market requirements, thus jeopardizing supply contractual obligations unless there are prior buy-in arrangements.

Farm management problems that arise with supply continuity generally involve biological product variability factors and maintaining commitment, particularly when there are labour and time constraints. Such management does favour those farms with extended families or have farm incorporations. Alternatively, the farmer becomes the manager, and delegates responsibilities to either employees, family members, or to contractors.

Whilst the farmer might be producing the correct product specificity, the product could fail due to a lack of quality maintenance. There can be problems with on-farm quality controls due to low investment into adequate infrastructure, inadequate finance, lack of skills, poor commitment, and time constraints. Providing that the farmer can overcome these management issues, then the creation of the correct final product might be possible.

Just-in-time delivery can be a problem for many farmers; however, they can contract this function to professional contractors, presumably those with the skills, experience and hazard controls. Farmers then need to integrate their own on-farm supply chain management with the larger external 'trade' market logistics operations.

The farmer who is more likely to become integrated into 'trade' markets is one who is flexible with high personal management and negotiation skills that probably have been gained through external business pursuits. However, communicative management skills may be acquired through experience, but could be wasted if the required personality trait is largely absent.

Any supply chain integrative approach will fail if the farmer chases market prices. All the requirements for meeting 'trade' market specification may be positively addressed, but if the farmer uses the 'trade' market to manage price, then there will be an extremely high probability of failure.[78]

Price has to be managed or enhanced quite independently of the 'trade' market relationship. If 'trade' markets usually only exist in surplus supply conditions, then farmers must accept low prices from end users. However, if prices rise, then any supply diversion by the farmer in chasing higher prices will result in the collapse of the 'trade' market relationship.

There is an insurance mechanism to enable farmers to enhance their price when they are locked into 'trade' market prices, rather than jeopardizing 'trade' market relationships. The insurance mechanism for farmers is buying call options, whereas the insurance mechanism for end users is buying put options.[79] There is a need for food-based farmers to examine the adoption of such price-enhancing strategies, provided that timing and forward market characteristics are considered, whenever they enter 'trade' market relationships. The cost of buying call options should be perceived as part of the transaction cost associated with managing a 'trade' market relationship, but much depends on the forward market characteristics.

However, this solution is jeopardized wherever such options markets are non-existent, or whenever there is market illiquidity, high currency risk or basis risk.

There is a prerequisite of strong local 'trade' markets leading to liquid futures and options markets before adequate risk management can be properly achieved, which would contribute to food security.

The farmer needs adequate cash reserves and other sources of cash flow to minimize dependence on the periodic cash flows from supply chain agreements. This financial requirement precludes many farmers from 'trade' markets because of the urgency for large immediate cash flows usually at harvest time to pay contractors, merchandisers, and meet bank debt obligations. Therefore adequate funding and financing of farm-logistical operations are important prerequisites for farmers entering 'trade' market negotiations.

Provision can be made whereby the farmer receives a financial incentive for integrating with the 'trade' market, to avoid being disadvantaged through inadequate cash flow. Local merchants may persuade banks to provide adequate advance-finance payments to only those farmers who store the specified product on-farm under 'trade' quality accreditation. This motivates the farmer to produce and store only what is required by the 'trade'. The bank may be willing to lend finance based on product collateral and security in the knowledge that commodity has been shifted to a differentiated product that is aligned with the food supply chain.

This cash-flow provision to farmers and the associated integration with 'trade' markets represents one of the biggest advancements in commodity food supply chains in more than 100 years. Instead of alienating farmers from the food supply chain through traditional bulk handled commingled methods, this cash-flow provision to farmers directly motivates them to move towards 'trade' market integration.

Adequate cash-flow provision and 'trade' market integration removes much of the incentive of farmers to hoard unwanted commodity, provided that they are willing to separate the physical sale from speculative pricing. Farmers in integrated food supply chains still have the opportunity to speculate on higher prices by buying call options, which is an option to take up the underlying buy futures contract whenever prices rise. Using this mechanism is more likely to induce farmers to release physical product into food supply chains while still retaining some control over the final realized price.

Consolidation of 'trade' market integration combined with adequate cash flow provides an incentive for farmers to invest in critical on-farm infrastructure, not just in storage facilities, but also in handling and transport logistics to improve the just-in-time capability of the farmer. Adequate investment can only flow from cash-flow security, which is the biggest weakness incurred through hoarding or pooling commodity. Motivation through increased cash-flow security is required to shift farmers away from speculative hoarding and the price spreading associated with commodity pools, both of which alienate farmers from 'trade' markets.

Notes

1 Dean, William H., 1938, *The Theory of the Geographic Location of Economic Activities*, Edward Brothers Publishers, Ann Arbor, MI.
2 Porter, Michael E., 1998, *Clusters and the New Economics of Competition*, Harvard Business Review, 77:1, November–December, 1–7.

3 North, Douglass C., 1955, *Location Theory and Regional Economic Growth*, The Journal of Political Economy, 63:3, 243–258.
4 Ohlin, Bertil, 1935, *Interregional and International Trade*, Harvard University Press, Cambridge, MA.
5 Haley, B.F., 1936, *Heckscher: Mercantilism*, The Quarterly Journal of Economics, 50:2, 347–354.
6 Keynes, John Maynard 1930, *A Treatise on Money, Volume II: The Applied Theory of Money*, republished in 1971 by Palgrave Macmillan, London, pp. 142–144.
7 Kaldor, Nicholas, 1939, *Speculation and Economic Stability*, The Review of Economic Studies, 7:1, 1–27, especially p. 6.
8 Williams, John Burr, 1936, *Speculation and the Carryover*, The Quarterly Journal of Economics, 50:3, 436–455.
9 Stoker, Thomas M., 1993, *Empirical Approaches to the Problem of Aggregation Over Individuals*, Journal of Economic Literature, 31:4, 1827–1874.
10 Keynes, John Maynard, 1938, *The Policy of Government Storage of Food-Stuffs and Raw Materials*, The Economic Journal, 48:191, 449–460.
11 Alexander, S.S., 1940, *Mr. Keynes and Mr. Marx*, The Review of Economic Studies, 7:2, 123–135.
12 Evgeny Pashukanis introduced commodity form theory based on 'guardians' of commodities. Karl Marx introduced commodity 'fetishism'.
13 Keynes, John Maynard, 1938, *Storage and Security*, The New Statesman and Nation.
14 Keynes was mildly exuberant when his German translations were published 'on paper rather better than usual, and the price not much higher than usual'.
15 Hicks, John R., 1939, *Value and Capital: An Inquiry into Some Fundamental Principles of Economic Theory*, Clarendon Press, Oxford.
16 Working, Holbrook, 1934, *Price relations between July and September wheat futures at Chicago since 1885*, Wheat Studies, 9, 187–237.
17 Davis, Joseph S., 1938, *The Economics of the Ever-Normal Granary*, Journal of Farm Economics, 20:1, 8–21.
18 Dow, J.C.R., 1940, *Addenda to Mr. Kaldor's Note*, The Review of Economic Studies, 7:3, 201–202.
19 Hawtrey, Ralph G., 1940, *Mr. Kaldor on the Forward Market*, The Review of Economic Studies, 7:3, 202–205.
20 Kaldor, Nicholas, 1940, *A Note on the Theory of the Forward Market*, The Review of Economic Studies, 7:3, 196–201.
21 Hoos, Sidney and Holbrook Working, 1940, *Price Relations of Liverpool Wheat Futures, with Special Reference to the December–March Spread*, Wheat Studies, 17, 101–138.
22 Working, Holbrook, 1942, *Quotations on Commodity Futures as Price Forecasts*, Econometrica, 10:1, 39–52.
23 Working, Holbrook, 1948, *Theory of the Inverse Carrying Charge in Futures Markets*, Journal of Farm Economics, 30:1, 1–28.
24 Working, Holbrook, 1949, *The Theory of Price of Storage*, The American Economic Review, 39:6, 1254–1262.
25 Working, Holbrook, 1954, *Whose Markets? Evidence on Some Aspects of Futures Trading*, Journal of Marketing, 19:1, 1–11.
26 Working, Holbrook, 1961, *New Concepts Concerning Futures Markets and Prices*, The American Economic Review, 51:2, 160–163.
27 Brennan, Michael J., 1958, *The Supply of Storage*, The American Economic Review, 48:1, 50–72.
28 Tomek, William G. and Roger W. Gray, 1970, *Temporal Relationships among Prices on Commodity Futures Markets: Their Allocative and Stabilizing Roles*, American Journal of Agricultural Economics, 52:3, 372–380.
29 Kohls, Richard L. and Joseph N. Uhl, 2002, *Marketing of Agricultural Products*, Prentice Hall, NJ, 9th Ed., Ch. 12.

30 McColl, Jim, 1988, *Royal Commission into Grain Storage, Handling, and Transport*, Australian Government Publishing Service, Canberra, p. xxv.
31 Report by the US Federal Trade Commission on the Grain Trade, 1924, Washington, DC.
32 Peck, Anne E., 1985, *The Economic Role of Traditional Commodity Futures Markets*, The American Enterprise Institute for Public Policy Research, Washington, DC.
33 Irwin, Scott H., Garcia, Philip, Good, Darrel L., and Eugene L. Kunda, 2009, *Poor Convergence Performance of CBOT Corn, Soybeans, and Wheat Futures Contracts: Causes and Solutions*, Marketing and Outlook Research Report 2009–02, University of Illinois at Urbana-Champaign.
34 Williams, John and W. Schroder, 1999, *Agricultural Price Risk Management – The Principles of Commodity Trading*, Oxford University Press, Melbourne.
35 Karlson, Nicholas, Anderson, Brad, and Brad Dahl, 1993, *Cash-futures Price Relationships: Guides to Corn Marketing*, University of Minnesota, Department of Agricultural and Applied Economics, Staff Paper 93-1, St Paul.
36 Demick, Barbara, 2010, *Nothing to Envy – Love, Life, and Death in North Korea*, Random House, New York.
37 Eastham, Jane F., Sharples, Liz, and Stephen D. Ball, 2001, *Food Supply Chain Management*, Butterworth Heinemann, Oxford, Ch. 9.
38 Hudson, Richard L. and Benoit B. Mandelbrot, 2004, *The (Mis)Behaviour of Markets: A Fractal View of Risk, Ruin, and Reward*, Basic Books, New York.
39 See Figure 1.3 in Chapter 1.
40 Kohls, Richard L. and Joseph N. Uhl, 2002, *Marketing of Agricultural Products*, 9th ed., Prentice Hall, New York, Ch. 5.
41 Grunwald, H.T. and L. Fortuin, 1992, *Many Steps Towards Zero Inventory*, European Journal of Operations Research, 59, 359–369.
42 Schonberger, Richard J., 1990, *Building a Chain of Customers*, The Free Press, New York.
43 Tansey, Geoff and T. Worsley, 1995, *The Food System: a Guide*, Earthscan, London.
44 Karmarkar, Uday, 1989, *Getting Control of Just-in-time*, Harvard Business Review, 67:5, 122–131.
45 Womack, James P. and Daniel T. Jones, 1996, *Lean Thinking*, Simon and Schuster, New York.
46 Cusumano, M., 1994, *The Limits of Lean*, Sloan Management Review, 35:4, 27–33.
47 Kingsman, Brian G., 1985, *Raw Materials Purchasing: an Operational Research Approach*, Pergamon Press, New York.
48 Brown, Ronald A., Oren, John W., and Jasper S. Lee, 1980, *Physical Distribution in Agribusiness*, McGraw-Hill, New York.
49 Reed, Michael R., 1990, *Wheat Movements in the U.S.: Interregional Flow Patterns and Transportation Requirements in 1985*, Agricultural Experiment Station, Urbana, College of Agriculture, University of Illinois.
50 Larson, Donald W. and Dean E. Baldwin, 1986, *Selected Characteristics of the US Grain Marketing Industry*, Ohio Agricultural Research and Development Centre, Wooster.
51 Spot basis is the difference between the spot 'trade' market price and the spot month futures price.
52 The question arises as to whether a contango can exist in non-agricultural 'trade' markets. The evidence is positive for other commodities, however, gold is one notable commodity exception. Forward gold prices are more effected by forward interest rates and currency movements, government monetary and fiscal policies, speculation, hoarding, and cartel manipulation.
53 Williams, John and W. Schroder, 1999, *Agricultural Price Risk Management – The Principles of Commodity Trading*, Oxford University Press, Melbourne, p. 304.
54 Keynes, John Maynard, 1930, *A Treatise on Money, Volume II: The Applied Theory of Money*, republished in 1971 by Palgrave Macmillan, London, pp. 142–144.
55 Basis is the difference between the local physical price and the futures market price, while basis risk refers to the variability in movement between the local physical price

and the futures market price. It is basis risk that can decrease the effectiveness of a hedge. Williams, John and William Schroder, 1999, *Agricultural Price Risk Management – The Principles of Commodity Trading*, Oxford University Press, Melbourne, Ch. 8.
56 A hedge ratio refers to the quantity hedged relative to the total quantity exposed.
57 Sankaran, Jayaram K., Gore, Amanda, and Baz Coldwell, 2005, *The Impact of Road Traffic Congestion on Supply Chains: insights from Auckland, New Zealand*, International Journal of Logistics Research and Applications, 8:2, 159–180.
58 Chafin, Donald G. and Paul H. Hoepner, 1989, *Commodity Marketing from a Producer's Perspective*, Interstate Publishers, Danville, IL, p. 15.
59 Rafferty, M., 1988, *The History of Flour Milling in Queensland*, Darling Downs Printing Co., Toowoomba.
60 Chicago Board of Trade, 1996, *Buyer's Guide to Managing Price Risk*, General Information Series 3, Chicago.
61 Marriott, Richard S. and Gus McLean, 1991, *Going with the Grain: A History of Queensland's State Wheat Board*, The State Wheat Board, Toowoomba, Ch. 5.
62 Commonwealth Royal Commission, 1934–36, *Report into the Wheat, Flour, and Bread Industries*, Australian Government Publishing Service, Canberra.
63 Towill, Denis R., 1997, *The Seamless Supply Chain*, International Journal of Technical Management, 13:1, 37–56.
64 Chicago Board of Trade, 1996, *Buyer's Guide to Managing Price Risk*, General Information Series 3, Chicago.
65 Refer to the Australian Lot Feeders Association, Sydney.
66 Schaffner, David J., Schroder, William R., and Mary D. Earle, 2003, *Food Marketing: An International Perspective*, 2nd Ed., McGraw-Hill, New York, Ch. 11.
67 Sidhu, D.S., 1979, *Price Policy for Wheat in India*, S. Chand and Co., Ram Nagar, New Delhi.
68 Schap, Keith, 1993, *Commodity Marketing – A Lenders and Producers Guide to Better Risk Management*, American Bankers Association and the Chicago Board of Trade, Chicago.
69 Broehl, Wayne G., 1992, *Cargill: Trading the World's Grain*, University Press of New England, Hanover, NH.
70 Porter, Michael E., 1985, *Competitive Advantage – Creating and Sustaining Superior Performance*, The Free Press – Macmillan, New York.
71 Sachs, Jeffrey D., Warner, Andrew, Åslund, Anders, and Stanley Fischer, 1995, *Economic Reform and the Process of Global Integration*, Brookings Papers on Economic Activity, 1995:1, 1–118.
72 Fielding, Jean P., 1985, *The Golden Grain: A History of Edwin Davey and Sons, Pioneer Flourmillers and Grain Merchants of South Australia*, Hyland House, Melbourne. A sad history of flour-milling demise through regulation.
73 Dunsdorfs, Edgars, 1956, *The Australian Wheat Growing Industry 1788–1948*, Melbourne University Press, Melbourne, Part One, Ch. 6.
74 Friedmann, Harriet, 1990, *Origins of third world food dependence*, in H. Berstein, Ben Crow, Maureen Mackintosh, and Charlotte Martin, *The Food Question: Profits versus People*, Earthscan, London.
75 Annual Seminar series of 'Singapore as a Food Hub', Singapore.
76 This aspect will be discussed further in Chapter 11.
77 This will be discussed further in Chapter 11.
78 Williams, John and William Schroder, 1999, *Agricultural Price Risk Management – The Principles of Commodity Trading*, Oxford University Press, Melbourne.
79 Williams, John, and William Schroder, 1999, *Agricultural Price Risk Management – The Principles of Commodity Trading*, Oxford University Press, Melbourne, Chs. 10–11.

Further readings on 'trade' markets

Carew, Edna, 1993, *Fast Forward – The History of the Sydney Futures Exchange*, Allen and Unwin, Sydney.

Chafin, Donald G. and Paul H. Hoepner, 1989, *Commodity Marketing from a Producer's Perspective*, Interstate Publishers, Danville, IL.

Connor, John M. and William A. Schiek, 1997, *Food Processing: An Industrial Powerhouse in Transition*, 2nd Ed., John Wiley and Sons, New York.

Fornahl, Dirk and Max-Peter Menzel, 2010, *Emerging Clusters*, Edward Elgar Publishing, Cheltenham.

Francois, M. Jacques, 2007, *Even Commodities Have Customers*, Harvard Business Review, 85:5, 110.

Karlsson, Charlie, 2010, *Handbook of Research on Cluster Theory (additional – Innovation and Clusters with Cases and Policies)*, Edward Elgar Publishing, Cheltenham.

Kohls, Richard L. and Joseph N. Uhl, 2002, *Marketing of Agricultural Products*, 9th Ed., Prentice-Hall, New York.

Kuah, Adrian T.H., 2002, *Cluster Theory and Practice: Advantages for the Small Business Locating in a Vibrant Cluster*, Huddersfield University Business School.

Morgan, Kevin, Marsden, Terry and Jonathan Murdoch, 2006, *Worlds of Food: Place, Power, Provenance in the Food Chain*, Oxford University Press, Oxford.

Norwood, F. Bailey and Jayson L. Kusk, 2008, *Agricultural Marketing and Price Analysis*, Pearson Education, Upper Saddle River, NJ.

Sahoo, Pravakar and Rajiv Kumar, 2009, *Efficiency and Futures Trading-price Nexus in Indian Commodity Futures Markets*, Global Business Review, July, 10:2, 187–201.

Schaffner, David J., Schroder, William R. and Mary D. Earle, 2003, *Food Marketing – An International Perspective*, 2nd Ed., McGraw-Hill, New York.

Schrimper, Ronald A., 2001, *Economics of Agricultural Markets*, Prentice-Hall, Upper Saddle River, NJ.

Williams, John, 2009, *Procurement Supply Chains in the Food Manufacturing/Processing Industry – A Study of Behavioural, Management, and Governance Issues*, survey/report by Australian Food Security Institute, Melbourne.

Working, H., 1949, *The Theory of Price of Storage*, The American Economic Review, 39:6, 1254–1262.

11

POST-DEREGULATION ISSUES

There often is the expectation that any shift from regulated to freer trade might be permanent. This might be supported by evidence suggesting that increased competition leads to more efficient markets and supply chains.[1] However, throughout history the forces of re-regulation have been ever present[2] and unless there is constant competition vigilance,[3] the reversion to re-regulation can be swift[4] and may lead to major wars.

The Turgot food supply chain episode in France was an example of deregulatory reforms, followed by re-regulation, a classic consumer-driven revolution in response, and then war to consolidate centralized regulation. This was then depicted in history as a revolt against 'authority', despite Louis XVI being an early Turgot public supporter and perhaps even a later secret admirer. Retribution is often swift, particularly when supporting deregulation. Louis XVI was forced to revert to re-regulation by the end of 1776 because of rising debts associated with the American War of Independence intervention, despite French consumers having languished under centuries of government food controls, with prices being established by local policemen during the eighteenth century.[5]

Turgot had abolished all restrictions on grain sales within France in 1774, but was experiencing resistance from official hoarders, tax collectors, and vested-interest interventionists, combined with weather problems in agricultural production. He was forced to resign by the end of 1776 and French food supply chains immediately returned to government controls under Necker, who conveniently used taxpayer unrest over the gabelle salt tax to justify re-regulating the food supply chains. The 'dreaded' gabelle had been raised to pay for French war debts incurred against the British in North America.

With 3,000 men, women, and children languishing in French prisons because of illegal salt arbitraging, after having brought salt in one town and sold it into another town,[6] the powder keg for revolution had been set. The harsh winter of 1788–89 caused food shortages under Necker's food supply chain controls, which were the last of many straws for the French people who began to revolt. Food chain re-regulation has been often characterized by consumer revolts.

In the ensuing post-revolutionary government period in France, the salt tax was abolished along with many other food supply chain laws. Yet within that short era of government deregulation (1789–94), bread seemingly 'miraculously' became available to urban consumers at 'reasonable' prices.[7]

However, French deregulation ended with the interventionist Thermidoreans who retook control of the food supply chain in 1794 together with the monopolists who had shifted from public to private monopolies in the aftermath of 1789. With the resulting hoarding, profiteering, speculation, and rising food prices during 1795, there were consumer-led bread riots once again, which were increasingly being put down by the military rather than the local police force.[8] The interventionists needed stronger military leadership, and on 17 August 1795, Napoleon received his 'orders' from the Thermidoreans,[9] and the war drums commenced.

What commenced as a suppression of local French consumers inexorably expanded continentally into a revival of the centralized regulated European Empire concept, which was protectionist and generally isolationist. During the nineteenth century, the concept shifted from France to Austria to Prussia, and then to Germany, and still dominates European politics today. Historians were to refer to the Napoleonic period as the commencement of the 'modern era', but it should be called the start of the modern re-regulation period. From a deregulated food supply chain consumer perspective, much of the twentieth century could be described as the 'dark age' that was characterized by protectionism, intervention, nationalism, isolationism, economic depression, and war.

The experiment with food supply chain deregulation thus failed. Napoleon reinstated the dreaded gabelle salt tax in 1804, which lasted in France until 1946.[10] In Britain, David Ricardo argued strongly against re-regulation but to no avail. The English interventionists led by Lord Liverpool passed the 1815 Corn Law legislation that effectively restricted many agricultural imports. This immediately led to consumer riots in London,[11] but this time they were put down by police, in contrast to the military in France. The life of the consumer was lived very precariously under food supply chain regulation. Despite petitions by English merchants in 1820, and by the London Board of Trade in 1821,[12] these food import restrictions continued until 1846.

Deregulation background

Deregulation in the context of the food supply chain may be defined as a shift in government regulation away from actual market intervention and supply chain interference, with the objective of increasing competition, contestability, efficiency, and productivity in the food supply chain.[13] This might be expected to increase international competitiveness.[14]

However, there is still a role for government regarding some aspects of the supply chain.[15] The dilemma becomes one of knowing when to introduce legislation and intervene, but still respect the limits of government regulation and being capable of enforcing these limits.[16]

Regulation is intrinsically linked to creating power for some people who want to control others.[17] Other people want strong leadership through a public medium such as government authority.[18] Some other people want collective direction, either public or private, supposedly to achieve some group objective.[19] Power, domination, and vested interest are individual personal traits that need to be somehow controlled and limited in a functioning society, without having any proper guidelines on how to achieve this.[20] Vagueness on limiting public power with few constitutional checks and balances can quickly revert to authoritarianism, as occurred in Bismarck's Germany in the 1890s.[21]

Money is merely the medium through which power and vested interests dominate in either the private or public domain, which suggests that money is not the root of all evil and it does not all reside in the private sector. Power, political intrigue, and vested interests dominated long before the introduction of money as a means of exchange.[22]

Because power and vested interests can be entrenched through government laws and public intervention, wealth to the few can accrue the fastest under maximum regulation.[23] This has occurred in Soviet Russia, Nazi Germany, Mao's China, North Korea, Zimbabwe, and many other African countries. The corollary is that wealth distribution is likely to be maximized with less government intervention and much less regulation.[24]

However, simple deregulation and privatization of former government monopolies is not the panacea of all ills relating to power and vested interests.[25] These can be merely rearranged under a different flag, as occurred in Russia after several deregulation and privatization attempts during the 1990s.[26] The success of the food supply chain hubs in Singapore derived not just from deregulation, but also from controlling corruption without the interference of food 'aid'.[27]

Government issues

Whilst increasing government intervention involves an inexorable gradualism, government devolvement is a very complex issue.[28] Deregulation is not a simple process, and it is debatable as to whether it can be achieved quickly or slowly.[29] Much depends on the original commitment to deregulation and privatization as in Thatcher's Britain during the 1980s, the degree of corruption in the social hierarchy as was the case in Russia, and the increasing dependency of the people on government as has occurred in France and many Mediterranean countries.[30]

Under the guise of deregulation, market regulation can be shifted to non-market regulation, but this can still have a negative impact on supply chains through increased administrative costs and delays.[31] Merely shifting government regulations between different sectors cannot be referred to as deregulation.[32] At worst, debt-ridden governments can impose new taxes under the guise of deregulatory 'reform'.

This shifting of regulation to other activities or re-regulation under new guises can decrease the international competitiveness and efficiency of the private sector supply chains. An example of this was the Australian government shifting regulation of wheat exporting from the Australian Wheat Board to a licensing wheat export authority. Export initiatives were stifled under some vague regulatory objectives that protected status quo vested interests, particularly when no other Australian export grain industry had such a controlling organization.[33]

If corruption is occurring before deregulation, it is highly likely to continue to exist after deregulation. Corruption cannot be simply eliminated through deregulation, as was the case in Eastern Europe during the 1990s.[34] Privatization can result in previous public monopolies continuing as private monopolies that are supported by government patronage.[35] Largess is largely continued, regardless of the public or private image.

Corruption is more likely to occur whenever industry concentration occurs, regardless of whether it is publicly or privately owned.[36] It is the enforcing of competition laws in a deregulated environment that is more likely to minimize the occurrence of corruption, rather than any deregulation per se.[37] This implies that deregulation must be accompanied by other appropriate laws if it is to be successful, as was the case with anti-corruption and competition laws in Singapore.

Deregulation and privatization are unlikely to decrease the dependency on 'aid'. When the foreign 'aid' industry was privatized during the 2000s after due recognition by donor governments to its detrimental impact on recipient countries,[38] a private global 'aid' industry proliferated, which has increased the dependence on 'aid' in many countries.[39] Banning of such non-emergency 'aid' may be the only solution,[40] but this is difficult when there is widespread corruption and vested interest occurring.

'Aid' can actively undermine the success of deregulation and privatization by preventing or distorting the required price signals for markets and supply chains to function normally, especially given the fragility of 'trade' markets. It can be argued that interventionists can use 'aid' to undermine the deregulation and privatization processes, and when these initiatives fail because of supply chain distortions, vested interests return to government re-regulation.

Import quarantine is often the core of many trade protection debates and can undermine deregulation. Domestic farmers may claim that imports could introduce diseases with the potential to destroy domestic production. For this reason, it can be argued that most food imports can be either banned by government laws, or quarantined so that the high cost and fumigation treatment makes them uncompetitive.

It is a dilemma as to where the quarantine line is drawn. From the importer perspective, it can be perceived as a trade barrier,[41] with domestic consumers being denied access to overseas food that might be produced at varying times of the year. Consumers therefore do not have the opportunity of annual continuity of supply, and perhaps access to even better quality than domestic food products. From a domestic food manufacturer perspective, import quarantine can destroy local food supply chains whenever domestic production fails.

As well, it can be argued that if the origin-country can survive the ravages of disease and thrive with exports, then the destination-country should not be unduly protected from imports. From a disease–genetic perspective, the complete lack of resistance is more devastating in the longer term, compared with gradual exposure and slowly building resistance. Many plant species disappeared quickly in the Americas in the sixteenth century after thousands of years of being isolated from Asia, Africa, and Europe.[42] There was simply no resistance bred into the species.

Quite often, import quarantine laws disguise a lack of research and development to improve disease resistance in domestic plant varieties and animal breeds. The greater the vulnerability of a country to a particular food or animal pathogen, the more resources are required to decrease that vulnerability. Quite often, import quarantine is a substitute for not making the required changes and investment to develop more resistant varieties and breeds. Merely increasing trade barriers may be the short-term cheap solution to a potential long-term expensive disaster in food supply. The alternative to natural resistance is increased vulnerability. Therefore, there is a need for time limits on import quarantine protection.

Export restrictions may be perceived as an oxymoron, but they regularly occur. During deregulation, Wheat Exports Australia was established to approve or reject bulk wheat export licences under an export quarantine policy.[43] This government authority acted as a de-facto global end user or consumer, protecting a few preferred exporters, with little understanding of what was really being demanded in international specifications. It implemented global specifications by decree and was perceived as re-regulating the supply chain under the guise of a 'deregulatory' environment.[44]

Some global customers want low-quality products for mixing, blending, or using them as bulk-filler ingredients in food manufacturing or animal feed. There are no reasons why trade should be discriminated on quality. If the exporter fails to deliver the required trade specifications, the importer is highly unlikely to undertake trade with that exporter again. That is the normal market mechanism.

There should be no need for government intervention to enforce quality control in a deregulated market. That is not the function of government. ISO standards and certification are an adequate incentive for private quality control. If private quality systems have failed, it is highly unlikely that government systems could do better.[45]

It can be argued that export restrictions prevent importers from accessing any required low-quality product from the exporting country. As well, it forces low quality back onto the domestic home market and suppresses prices, which is unlikely to assist domestic farmers unless low-quality buyers are available locally.

There is no function for a government to monitor private inventories in deregulated supply chains.[46] Government spying on private inventories on behalf of vested interests under the guise of 'orderly marketing' arrangements undermines competition and is likely to encourage the re-emergence of oligopolies.

Problems with deregulation

Intervention did not stop with Richard Cobden[47] and the repeal of the Corn Laws.[48] Throughout much of the second half of the nineteenth century, there were polarized policy differences in Britain between the governments of the deregulatory freer-market Gladstonian Whig-Liberals and the regulatory interventionist Disraeli Conservatives.[49] The main problem with deregulation and privatization was the lack of commitment by bureaucracy, institutions, political parties, and corporations to deregulation,[50] aided by vested interests. This reluctance to commit leads to re-regulation pressures.

Centralization is not the answer to deregulatory problems, especially if the policy lacks commitment to enforcing competition and contestability.[51] However, the real danger in centralization is the threat of hastening re-regulation, protection, and nationalism.[52]

Partial deregulation can lead to privatization without competition such as a private monopoly replacing a public monopoly, as occurred in many former USSR countries,[53] with vested interests taking control again under a different format.[54] If deregulation and privatization occurs too quickly, there is too little time for the development of normal supply chain mechanisms and domestic 'trade' relationship consolidation. There is a need for an adjustment period, otherwise corruption could proliferate in lieu of adequate price signals just to accomplish operations.

It can be argued that deregulation in the food supply chain must be undertaken gradually, but there is a very real danger that this slow process can induce pressures for re-regulation, as has occurred on a number of occasions in Europe.[55] As an example of the lead time for 'trade' market consolidation, the Australian lot-feeding industry took ten years after domestic grain deregulation before it was large enough to be recognized as a 'trade' market with cluster industry support.

There may be confusion over roles particularly before 'trade' markets become established. Food surpluses need to regularly occur before merchants and end users undertake supply agreements and consolidate 'trade' markets, with investment taking many years.[56] Cluster industries need to be established, combined with adequate support and service industries.[57]

Governments have a vested interest in supply chain intervention because it can minimize infrastructure expenditure, particularly for those governments with a small tax-revenue base or huge debt.[58] Food commodities may be forced to be transported by existing railways between rural–urban centres and ports by necessity because of inadequate roads. 'Trade' markets thus can be prevented by infrastructural constraints, regardless of any government deregulation of the food supply chain.

The problem with deregulation that is linked to just-in-time deliveries is its dependency on vast road transport networks, particularly when such logistics systems are demanded by end users, wholesalers, distributors, and retailers. Unless there is an adequate road infrastructure, the full benefits of deregulation will not be achieved.[59]

From a historical perspective, it may have been the lack of adequate infrastructure that made commodity chains so dependent on government control. It can be strongly argued that it has only been since adequate road networks have been established that effective private food chains have been made possible. 'Trade' markets have a long history, but they were usually restricted mostly to ports because of inland transport inadequacies. Chicago really only developed after the completion of the Illinois–Michigan canal in 1847 that linked Lake Michigan to the Mississippi River basin.

During the second half of the twentieth century, ports were designed for bulk handling large volumes of undifferentiated commodity between railhead and ship.

These ports were never designed for small deliveries of segregated differentiated product arriving by road transport. Therefore, there can be major port and shipping issues with deregulation.

Privatization of previous public bulk handling monopolies merely causes competition problems at ports.[60] The devolvement of public assets to a private monopoly company, especially when no asset purchase payment occurred, can create enormous monopoly advantages for the incumbent company to the exclusion of possible competitors.[61] Under such circumstances, there are formidable entry barriers for potential competitors to construct new port facilities.

The alternative is shipping containerization, but the growth in the general container trade through ports has caused congestion and delays.[62] Containerized regional ports that are isolated from main shipping routes are expensive and may incur high annual harbour and channel maintenance. Specialized food containers can be in short supply and depend on back loading. Port constraints and inefficiency can restrict export trade and lower international competitive advantage.[63] However, containerization does have the advantage of just-in-time delivery presuming no mishap, improved quality control particularly when sealed, and product can be tailored specific to a customer's demand.

As a solution, there needs to be greater provision of private ports with the flexibility for containerization. Ironically, private ports and lot cargoes existed 100 years ago in many countries before the nationalization of railways. Successful deregulation therefore requires the devolvement of much inefficient infrastructure that evolved over the past century and replacement with more relevant structures. But that requires capital investment commitment from governments who are generally already burdened with social debt.

Effective food supply chains require huge financing requirements for storable products because of the seasonal production and the need to fund farmers' cash flow. Adequate financing of the supply chain may mean that a few large merchants dominate domestic market food supply chains rather than many small merchants.[64] This should not be perceived as a lack of competition, but rather the inability of smaller merchants to adequately finance the supply chain functions and operations.

There needs to be sufficient competition law enforcement under deregulation to allow the small merchants to succeed and the large merchants to fail. Small merchants should not be entirely excluded from adequate financing arrangements. Such new entrant competition with existing larger merchants and low barriers for entry and exit should maximize competitiveness, efficiency, and ultimately provide export opportunities.[65]

Although operational size may be important to achieve economies of scale, adequate financing, and food chain integration, there is a need to keep barriers to entry relatively low in the food supply chain. Competition is required in deregulated markets to drive efficiency, productivity, and innovativeness, otherwise it will stagnate. Governments are required to enforce competition in the food supply chain and prevent oligopolistic behaviour, including those arising from farmer-selling cartels, because of the impact on price signals in 'trade' markets.

Periods of real or artificial product shortages are not conducive to orderly product flow through 'trade' markets. Protagonists have used high price and the associated problems with 'trade' markets as reasons for re-regulation, so as to induce complexities such as food price stability, inflation control, reduction of crime and vagrancy, and improvement in health.[66] The paradox is that without periodic price-reward opportunities for farmers, prolonged surpluses and low prices are not conducive to the financial sustainability for farmers, which can jeopardize longer-term food security.

Provided that there are clear price signals to farmers and adequate farm resources, the increased farm production response to higher prices should be swift. High prices in deregulated markets should be perceived as an opportunity for farmers to remain financially viable. From an end-user perspective, periodic shortages and high prices need to be perceived as the sacrifice to ensure production sustainability and future growth.

Food supply chain deregulation has also been linked to destabilizing monetary policy through price and cost inflation, particularly in the EU.[67] If government survival depends on artificial food surpluses and subsidies to maintain low inflation, then it could be argued that the government deserves to fail because they are undermining their food production base and destroying domestic farmer viability, particularly in long-term farm capital investment. The end result is likely to be dependence on government, or food imports, or vulnerability to food shortages.

The formation of 'trade' markets under deregulation and the increased adoption of just-in-time deliveries at a time of improved bunker-storage technology have shifted the power of withholding supply back towards the farmer and first-stage merchant. To counter hoarding and periodic artificial shortages, there needs to be sufficient further deregulation to permit domestic food production diversity, facilitate competing food supply chains including between supermarket chains, and allow food imports. The greater the source diversity of food supply and the availability of food substitutes, the lesser will be the probability of overall food shortages, and the lower will be the impact of any hoarding. This was the lesson learnt in Singapore, but it was also the lesson earlier learnt after Stalin's and Hitler's gross errors of judgement in the food chain.[68]

Deregulation is frequently criticized from a product quality perspective. Arguments over quality in deregulated supply chains assume perfect quality standards under regulation, which was never the case.[69] Under market regulation, commingled bulk handled commodity pooling disguised many quality problems that were due to the biological nature of production. Absolute quality control may be virtually impossible when agricultural production is exposed to weather, temperatures, humidity, pests, and diseases. Bulk handling monopoly authorities could blend poorer quality product or it could be regionally isolated, but the problem still existed.[70]

Supply chain regulation meant that quality export markets could be serviced through sourcing from a few selectively known quality storages. End users could be deluded into believing that all agricultural production from that country represented that quality standard. What was often poorly recognized was that the remaining low quality slow-selling commodity was eventually dumped onto the domestic or international supply chains whenever annual storage maintenance had to occur. The quality problem was always there, but it was carefully hidden by regulation authorities.

Market deregulation can often expose the underlying quality problem, to the aghast of end users. The issue can be exacerbated by poor private hoarding storages, inappropriate private mixing and blending, and inadequate quality and fumigation maintenance. Quality problems are exacerbated in countries with marginal production linked to marginal regions producing marginal quality.

Innate quality problems should not be perceived as the fault of deregulation, but rather the exposure of agricultural production to natural external occurrences in marginal growing regions. The blame in countries such as Australia could lie in government soldier resettlement farm schemes that inappropriately subdivided marginal pastoral land and induced marginal production-quality problems after both world wars.[71]

Improvement in quality should derive from the pricing mechanism, not through government re-regulation.[72] If farmers continually produce quality that is not demanded in food supply chains, then they will receive continual low prices, which should act as an incentive for change. Either the farmer can continue to realize low prices and go bankrupt, or change to other more profitable farm pursuits that have higher financial return.

In contrast, government re-regulation sends the wrong signal to farmers, and they are more likely to continue producing a poor quality product. There is a perception of reward for mediocre performance when all product ills can be pooled and mixed into average quality. It penalizes those farmers with good quality. This seriously distorts the efficient allocation of the country's resources, decreases international competitiveness for farm outputs, and removes the potential for sustainable exports.

If farmers can only produce a particular plant variety, but this plant variety is not suitable for end usage, then the problem lies with production, and not with the deregulated supply chain. Either there has to be more plant breeding to develop multipurpose varieties suitable for both production and end usage, or the farmer needs to shift to alternate farm enterprises. Ironically, trees may have more relevance to the food supply chain through cardboard packaging than some varieties of wheat or rice.

After many decades of mandated bulk handled commingling, many farmers have difficulty in perceiving that what they produce is differentiated product rather than undifferentiated commodity. Many farmers have never known a delivery system other than for bulk handled commingled commodity. It requires a different mindset to market differentiated product in a deregulated supply chain compared to merely delivering to a commingled commodity depot.[73]

Alternatively, there is also the biological problem for farmers of not consistently producing what end users want, assuming that they were able to do so occasionally. It is easy for farmers to get despondent over biological inconsistencies and perceive their product as commodity and unwanted by the deregulated 'trade' market. However, rejection from 'trade' markets may be due to many logistics and supply factors, and not just product inconsistencies.

If farmers cannot differentiate and develop competitive advantage through their product, then they really have no supply chain advantage in producing what they do. There is no amount of government intervention, other than product research and development, which is going to shift these farmers out of the commodity poverty trap of falling real prices and rising costs. Increasing productivity will only result in producing more unwanted product and increasing poverty. The blame cannot be passed onto deregulation, and re-regulation is certainly not the solution.

Product grade standards under regulation were usually production driven, and not end user or consumer driven. There is a need for signals from 'trade' markets as to the required product differentiation, otherwise product grades can be artificial and meaningless to the 'trade'.

In deregulated 'trade' markets, the role of the accumulator becomes dominant, much to the chagrin of the farmer. The accumulating merchant needs to undertake objective product testing, not just for quality but for the innate product characteristics that ensure a quality final product can be achieved after food manufacturing has occurred, and one that is aligned to consumer demand. This objective testing determines whether the farmer's output is either accepted as differentiated product or rejected as non-demanded commodity. Mere subjective observations can lead to farmer perceptions of bias, without adequate feedback.

Individual end users can still issue particular specifications in supply contracts, irrespective of the general supply chain consensus on product differentiation. This means that there cannot be one industry standard product-grade classification, much to the disappointment of farmers, government regulators, and plant breeders. Deregulation is the antithesis of standardization.

There may be some industry sectors such as finance and currency that respond quickly and positively to deregulation. The slow response to deregulation from the agricultural and food sectors should not be perceived as an industry indictment and a cause for re-regulation.[74]

Much of the private food supply chain requires large organizational, logistical, operations, technology, and infrastructural investment, not just for the primary business but also for all the secondary input suppliers and relevant cluster support industries that need synergizing with each other. It can be contrasted to the starkness of solitary government-owned canned food manufacturing sites.

The sensitivity of 'trade' markets to collapse whenever there are shortages suggests that time and patience are required to develop 'trade' relationships. It may be a case of experimenting with flexible suppliers from different regions. Each different stage or experiment requires large resources and expenditures. This is unlikely to occur quickly. The longer the time interval, the more likely will the correct decisions be made and the food supply chain made more secure.

There are some expectations that forward markets will develop as soon as deregulation occurs. This can be a very false presumption. Forward markets are based on underlying 'trade' markets, and if these 'trade' markets take a long time to develop and consolidate, then the development of a forward market will reflect this. It took Chicago 40 years during the second half of the nineteenth century to develop its agricultural forward markets and achieve liquidity in all contracts out to 12 months. In Australia, wheat futures markets occurred within six years of domestic deregulation, but did not gain adequate liquidity until export market deregulation occurred almost 20 years later.[75] In the meantime, there was much enforcement of competition laws as well as secondary deregulation by government.

Characteristics of deregulation

In countries such as Australia, supply chain deregulation effectively destroyed the cozy relationships between statutory export authorities on the selling side, and international merchants and government import agencies on the buying side. No longer was the leased air-conditioned urban single office of the buyer suitable for their transactions. There was now a requirement for international merchants to venture into rural areas and discuss, negotiate, invest locally, and provide supply chain feedback to farmers, domestic merchants, and local end users. The days of cognac and cigars in city offices were effectively over, as was a government agency accumulating product for international merchants.

It is usual for local merchants to reappear quickly under deregulation because of the need to effectively link the farmer to diverse end-user segments. However, it could take ten years before there is a consolidation of domestic merchant integration into 'trade' markets and the effective performance of their normal food supply chain functions of accumulation, storage, and just-in-time delivery.

Farmers may be resentful of supply chain intermediaries such as merchants, particularly when the latter are very selective in their product accumulation strategies. However, it took much of the twentieth century through supply chain regulation to prove that farmers could not undertake the role of the domestic merchant. Farmers are ill-equipped for supply chain functions, despite regulatory authorities and directives. Any such supply chain interference is likely to destroy the delicate balance that characterizes 'trade' markets. This then forces many farmers onto the low price global commodity market, thus becoming increasingly dependent on international merchants.

Some local merchants succeed while others fail, but that reflects normal commercial outcomes under a high-risk food supply chain environment. It is not the function of governments to pick winners and losers or to assist ailing merchants. However, it is the function of government to ensure adequate competition amongst merchants, which is likely to induce efficient food supply chains. There were occasions in the first half of the twentieth century where previous government intervention had caused domestic merchants to fail, but then the government used the pretext of assisting ailing merchants to further intervene in the domestic food supply chain.[76]

Supply chain interventionists have strongly argued that deregulation would result in the domination of international merchants. The paradox was that international merchants rose and held their market dominance during the decades of government intervention.[77] Even after 50 years of statutory export selling and supposedly immense global trading experience, at least 40 per cent of the Australian Wheat Board's export sales were done through international merchants,[78] with most of the remainder sold to government import agencies, usually through the tender system.

The reality was that international merchants were threatened by deregulation. Competitive margins were always tight in contradiction to the belief of many farmers, which meant that all of the larger international merchants had significant diversification into secondary industries by the time that grain market deregulation occurred in Canada, Australia, Eastern Europe, and Western Asia. Many of these diversifications involved non-food industries, which belied the proposition that international merchants were skimming profits from the food supply chain that were not commensurate with the risks involved.

However, once domestic deregulation occurred, many international merchants were quick to adapt to the new food chain environment. They had the capital, experience, knowledge, and financing capability that had been gained from other industry supply chains, and generally used this expertise to develop food supply chains. International merchant experience was suddenly injected into the local learning curve aided by appropriate funding and investment. Without this capital and experience, local merchants and potential domestic end users would have struggled to provide sufficient funding and expertise in the early deregulatory stages.

International merchants did not monopolize Australia's newly deregulated wheat export market after 2008. Instead international merchants competed with local entrepreneurs, particularly in the container trade, and with other exporters who found supply chain synergies with wheat. Some of these other exporters included both local and international food supply chain specialists who were already linked with buyers representing overseas end users. This export supply chain polyglot was hardly the predicted catastrophe that alarmist market interventionists had predicted. The market was merely behaving with efficient decision making and investment as it should have done for the previous 70 years under regulation.[79]

International merchants integrated deeply into the food supply chain once deregulation occurred, focusing on differentiated product and often creating whole new food supply chains. Important linkages between international merchants and local merchants usually generated mutual benefit. The international merchant gained access to farmers, while local merchants gained access to international buyers. Farmers could now receive international supply chain feedback, which could be factored into important farm production and marketing decisions.

This contrasted with privatized entities that previously were either former regulatory authorities or farmer cooperatives, but which remained largely focused on selling undifferentiated product. Whereas merchant integration was more likely to be sustainable and lead to food security, privatized previous entities were more likely to result in the commodity poverty trap for farmers and lead to longer-term food insecurity.

Supply chain regulation during the twentieth century was characterized by a void of cluster support industries to the food supply chain. Governments were often forced to further intervene in food manufacturing after previous public intervention had either removed or bankrupted supply chain intermediaries, end users, cluster industries, and support organizations. Under deregulation, solitary silos silhouetted against the sky suddenly appeared outdated, and a stark reminder of the regulatory period.

Deregulation provides the environment for the re-emergence of both horizontal and vertical cluster industries and support organizations in the food supply chain. However, the formation of these associated industries can take decades to achieve, and much depends on appropriate government monetary inflation and investment policies as well as fiscal taxation policies. Generally, it was government policies and intervention that quickly destroyed these high-risk entrepreneurial industries in the first instance, so they would be expected to take time to become reestablished even in a favourable economic environment.[80]

Benefits of support organizations and cluster industries include competition within the food supply chain, which drives efficiency, productivity, innovation, food security, research, and the potential to develop sustainable export industries.[81] The greater the diversification of the food supply chain, the higher the probability of domestic food security and exports. Under such circumstances, there is more likely to be stability in the food supply chain and maximum choice for consumers, although this also depends on government food import policies. Regional food hubs are more likely to develop out of an environment of domestic deregulation combined with import competition.

The continual merging and demerging of companies with joint ventures and takeovers, as well as cooperative demutualizing, are characteristics of deregulated food supply chains and associated cluster industries.[82] Whilst some have perceived this to be supply chain instability, much of this organizational change involves attempts to establish synergistic alignment in a dynamic food chain. It can result in greater niche relevancy and promote growth expansion in the food supply chain, and usually involves the injection or consolidation of capital to sustain the food supply chain through new investment stages, thus consolidating food security.

The injection of global capital should not be perceived as a foreign takeover of the domestic food chain. It usually brings much needed capital funding for productivity gains and technological improvement. As well, it usually introduces new ideas, innovative processes, and strengthens the learning, knowledge, and experience pathway. To reject all of these on the grounds of losing national sovereignty over food is to jeopardize the efficiency and growth of the domestic food supply chains, which could ultimately decrease the food security of the country and lower the export potential.

In many cases, it is extremely doubtful whether local businesses have sufficient access to required capital funding sources, and whether they have adequate skills and knowledge from a global perspective. Usually there are major benefits for domestic food supply chains from overseas mergers and joint ventures. At best, it creates important synergies and linkages with international food supply chains, which could be used to develop food exports. At worst, the foreign venture fails, leaving behind capital, resources, learning, and experience for the host country.

For many consumers and even end users, food security translates into an ability to trace the food product back to its origin, particularly when there are audited records at all stages along the complete supply chain.[83] Alternatively, food insecurity arises for many end users and consumers when bulk commingling masks a multitude of unknown origins, chemical treatments, and exposure to potential contaminants, as well as lacking quality assurances to enable buying.

Traceability and identity preservation are more likely to occur when product differentiation is segregated in a deregulated food supply chain. This contrasts with the inability of traceability and identity preservation under traditional bulk commingled commodity systems, which are more likely to occur under government regulation.

To sell undifferentiated bulk commingled commodity internationally is extremely difficult and usually dependent on import government tenders that only purchase at the lowest possible price. Foreign policy officials from the exporting country often have to intervene merely to dispose of much surplus commodity into a host country, sometimes as 'aid' programmes. Commodity sale agendas can then become interwoven into much foreign policy and intrigue.

Deregulation unshackles the food supply chain from government foreign policy dictates. Usually there are no discriminatory trade distortions under deregulation, which might occur when foreign policy determines trade destinations. However, there might still be a role for foreign trade officials when food exporters are not paid by importers in the desired currency, such as occurs in countertrade, but this is more likely to occur with undifferentiated commodity in an opportunistic sale within a non-integrated food supply chain.

Under regulation, there was an immense barrier between farmers, end users, and consumers caused by a commingled pool system, prolonged storage and selling periods, a lack of local merchants, and distance to markets especially when local end-user industries had been removed under regulation. Price signals between farmers, end users, and consumers were virtually non-existent in commodity pools where prices were averaged over a major time period. Farm decisions were mostly made in a vacuum of knowledge relating to exactly who were the actual end users and consumers, what product and services did these end users and consumers actually want, and how much were they prepared to pay. Under deregulation, local merchants become the conduit for supply chain intelligence and the provision of price signals, particularly between end users and farmers.

Arguments against re-regulation

Without strong enforcement of competitive practices and unrelenting commitment to deregulation, the forces of regulation will reassert themselves through complacency and vested interests. Some farmers perceive regulation as a means to either control the price of their output, or transfer their losses to taxpayers, or as an egotistical power-base. They ignore the paradox that government intervention reverts farm output into a price-taking undifferentiated product in a global commodity market, with a high probability of low prices.

Regulation of food supply chains and the enforcement of traditional bulk commingled commodity storage-handling systems usually necessitated pooling, which then led to pool price averaging over a 12- to 18-month selling period. This price is not indicative of either the spot or forward market price, occurs post-harvest for old crop, which means that it has no relevance for new crop decision making, makes it very difficult to manage price risk, and it distorts vital signals in the supply chain and 'trade' markets.[84]

Farmers need to be focused on price signals before, during, and immediately after the new crop to make decisions regarding acreage, discretionary farm input usage, and whether to sell or store. Decisions need to be made as to what to produce, how much to plant, when to plant, how much farm inputs to use, when to price, and when to sell. If farmers are dependent on a time-lagged old-crop pool price for income, then this will distort major decision making regarding the new crop. Discretionary expenditure on added nitrogen fertilizer should be dependent on grade spread price differentials such as between 9 per cent protein wheat at Chicago and the 11.5 per cent protein wheat price at Kansas, rather than attempting to predict a forward commodity pool price 18 months from the current spot price.

If forward 'trade' prices that reflect surpluses and shortages are the essence of 'trade' markets, then these price signals for domestic merchants to accumulate, store, and deliver in the food supply chain will quickly collapse with any government re-regulation of the food supply chain. As well, if a 'trade' market is essential for product differentiation in the food supply chain, then any re-regulation will quickly cause the reversion of farmers' output to unwanted commodity and create dependence on global commodity prices that are heavily skewed against the interests of farmers. This acts as a disincentive to produce, which causes farmers to pursue other enterprises or simply move out of farming. If the objective of re-regulation was food security, then the government is highly likely to achieve the paradox of food insecurity.

Re-regulation is highly likely to distort and destroy domestic 'trade' markets, thus leading to re-commoditization of the food chain and making it more dependent on traditional bulk commingled exports and imports. This is the antithesis of synergizing with global food supply chains. Product is likely to be unwanted, leading to an exodus of domestic farmers, merchants, and end users.

Any re-regulation policy assumes that the government has the institutional capability and collective commitment to achieve food security through an efficient food supply chain mechanism. It also assumes that government interventionist bureaucracy is the best method to achieve maximum welfare at least cost, presumably balancing the welfare of farmers, merchants, end users, and consumers equally, and presumably not favouring one vested-interest group over another.[85]

If much of the justification for initial government intervention into food supply chains arose because of debt and associated transport and infrastructure limitations, as well as a need to control exports and imports because of trade deficits and imbalances, then the whole rationale of re-regulation by government may be seriously questioned when the financial and currency markets have been deregulated and infrastructure improved. Alternatively, mistakes in macro-economic policies should never have been made to impinge on micro-economic efficiency, and thus jeopardize food security.

There is little past evidence to prove that government intervention in the food supply chain had any real micro-economic benefits that might justify re-regulation in an efficient economy. It was government regulation that forced farmers into a commodity poverty trap, consumers suffered through a lack of product choice and frequent food shortages, whilst it almost completely destroyed domestic merchants, local cluster industries, and food manufacturers. The macroeconomic beneficiaries were possibly taxpayers and convenience-driven politicians who did not want to make the hard decisions regarding debt, corruption, microeconomic reform, and better governance.[86]

Notes

1 Kagami, Mitsuhiro and Tsuji Masatsugu, 2000, *Privatization, Deregulation, and Economic Efficiency: A Comparative Analysis of Asia, Europe, and the Americas*, Edward Elgar Publishers, Cheltenham.
2 Christensen, Jorgen Gronnegard, 1987, *Deregulation: Recent Danish Experience*, Paper delivered at APSA Annual Meeting, Chicago, September.
3 Gart, Alan, 1994, *Regulation, Deregulation, Reregulation Future of the Banking, Insurance, and Securities Industries*, John Wiley and Sons, New York.
4 Christensen, Jorgen Gronnegard, 1989, *Regulation, Deregulation, and Public Bureaucracy*, European Journal of Political Research, 17, 223–239.
5 Miller, Judith A., 1998, *Mastering the Market: The State and the Grain Trade in Northern France, 1700–1860*, Cambridge University Press, Cambridge, Ch. 1.
6 Kurlansky, Mark, 2003, *Salt – A World History*, Random House, London, p. 233.
7 McLynn, Frank, 1998, *Napoleon – A Biography*, Random House, London, p. 97.
8 McLynn, Frank, 1998, *Napoleon – A Biography*, Random House, London, p. 81.
9 McLynn, Frank, 1998, *Napoleon – A Biography*, Random House, London, p. 92.
10 Kurlansky, Mark, 2003, *Salt – A World History*, Random House, London, pp. 233–234.
11 Hirst, F.W., 1925, *From Adam Smith to Philip Snowden – A History of Free Trade in Great Britain*, T. Fisher Unwin, London, pp. 15–16.
12 Schonhardt-Bailey, Cheryl, 2006, *From the Corn Laws to Free Trade: Interests, Ideas, and Institutions in Historical Perspective*, The MIT Press, London, p. 9.
13 McColl, Jim, 1988, *Royal Commission into Grain Storage, Handling, and Transport*, Australian Government Publishing Service, Canberra, p. xxv.
14 Australian Competition and Consumer Commission, 2008, *Report of the ACCC Inquiry into the Competitiveness of Retail Prices for Standard Groceries, July 2008*, ACCC, Canberra, 5 August.
15 These were described in Chapter 3.
16 Christensen, Jorgen Gronnegard, 1989, *Regulation, Deregulation, and Public Bureaucracy*, European Journal of Political Research, 17, 223–239.
17 Olsen, Mancur, 1993, *Dictatorship, Democracy, and Development*, American Political Science Review, 87:3, 567–576.
18 Meier, Kenneth J., 1985, *Regulation, Politics, Bureaucracy, and Economics*, St. Martin's Press, New York.
19 Kingdom, John W., 1984, *Agendas, Alternatives, and Public Policies*, Little, Brown and Co., Boston, MA.
20 Olsen, Mancur, 1993, *Dictatorship, Democracy, and Development*, American Political Science Review, 87:3, 567–576.
21 Clark, Christopher, 2007, *Iron Kingdom: The Rise and Fall of Prussia 1600–1917*, Penguin, London, Chs. 12–17.
22 Goodin, Robert E., 1976, *The Politics of Rational Man*, John Wiley and Sons, New York.
23 Wilson, James Q., Ed., 1980, *The Politics of Regulation*, Basic Books, New York.
24 ACIL Tasman, 2006, *Marketing Australian Wheat: Competition and Choice in the Australian Export Wheat Market – Increasing Growers' Net Returns*, November, Melbourne.
25 Debande, Olivier, 2001, *Deregulation and Privatizing Statutory Monopolies*, Journal of Economics and Business, 53: 2-3, 111–137.
26 Anderson, Kathryn and Richard Pomfret, 2003, *Consequences of Creating a Market Economy: Evidence from Household Surveys in Central Asia*, Edward Elgar, Cheltenham.
27 Quah, Jon S.T., 1997, *Singapore's Experience in Curbing Corruption*, in Arnold J. Heidenheimer, Michael Johnston and Victor T. Le Vine, Eds., *Political Corruption: A Handbook*, 4th Ed., Transaction Publishers, New Brunswick, NJ.

28 Richardson, Megan, Ed., 1996, *Deregulation of Public Utilities: Current Issues and Perspectives*, Faculty of Law, University of Melbourne.
29 Bromley, Daniel W., 1993, *Creating Market Economies from Command Economies*, Economic Issues, Department of Agricultural Economics, University of Wisconsin-Madison, No. 121.
30 Yergin, Daniel and Joseph Stanislaw, 2002, *The Commanding Heights – The Battle for the World Economy*, Simon and Schuster, New York.
31 Beesley, Michael E., 1992, *Privatization, Regulation and Deregulation*, Routledge in association with the Institute of Economic Affairs, New York.
32 Derthick, Martha and Paul J. Quirk, 1985, *The Politics of Deregulation*, Brookings Institute, Washington, DC.
33 ACIL Tasman, 2009, *A Review of the Wheat Export Marketing Act 2008*, Report prepared for the Co-operative Bulk Handling Ltd., Melbourne.
34 Kornai, Janos, 1980, *The Road to a Free Economy, Shifting from a Socialist System: The Example of Hungary*, W.W. Norton and Co., New York.
35 Schmidt, Klaus, 1996, *The Costs and Benefits of Privatization: an Incomplete Contract Approach*, Journal of Law, Economics and Organization, 12:1, 1–24.
36 Tullock, Gordon, 1967, *The welfare costs of tariffs, monopolies, and theft*, Western Economic Journal, 5:3, 224–232.
37 Ades, Alberto and Rafael Di Tella, 1999, *Rents, Competition, and Corruption*, American Economic Review, 89:4, 982–993.
38 Reusse, Eberhard, 2002, *The Ills of Aid – An Analysis of Third World Development Policies*, The University of Chicago Press, Chicago, p. 112.
39 Maren, Michael, 1997, *The Road to Hell: The Ravaging Effects of Foreign Aid and International Charity*, The Free Press, New York.
40 Glennie, Jonathan, 2008, *The Trouble With Aid: Why Less Could Mean More for Africa*, Zed Books, London.
41 Tubiana, Laurence, 1989, *World Trade in Agricultural Products from Global Regulation to Market Fragmentation*, in David Goodman and Michael Redclift, Eds., *International Farm Crisis*, Macmillan, London.
42 National Research Council, 1989, *Lost Crops of the Incas*, National Academy Press, Washington, DC.
43 Australian Government Productivity Commission, 2010, *Wheat Export Marketing Arrangements Inquiry Report*, Canberra, 28 October.
44 Bailey, Elizabeth and Janet Pack, Eds., 1995, *The Political Economy of Privatization and Deregulation*, Edward Elgar, Cheltenham.
45 Whitwell, Greg and Diane Sydenham, 1991, *A Shared Harvest – The Australian Wheat Industry 1939–89*, Macmillan Education Australia, Melbourne, pp. 110–114.
46 Australian Government Productivity Commission, 2010, *Wheat Export Marketing Arrangements Inquiry Report*, Canberra, 28 October, Ch. 7.
47 Hinde, Wendy, 1987, *Richard Cobden*, Yale University Press, New Haven, CT.
48 Lusztig, Michael, 1995, *Solving Peel's Puzzle: Repeal of the Corn Laws and Institutional Preservation*, Comparative Politics, 27:4, 393–408.
49 Willis, Michael, 1989, *Gladstone and Disraeli*, Cambridge University Press, Cambridge.
50 Christensen, Jorgen Gronnegard, 1987, *Deregulation: Recent Danish Experience*, Paper delivered at APSA Annual Meeting, Chicago, September.
51 Head, Brian and Elaine McCoy, 1991, *Deregulation or Better Regulation: Issues for the Public Sector*, Macmillan, Melbourne.
52 Gerwath, Robert, 2007, *The Bismarck Myth: Weimar Germany and the Legacy of the Iron Chancellor*, Oxford University Press, New York.
53 Keren, Macours, 1990, *Privatization in Eastern Europe and the Soviet Union*, Jahrbuch der Wirtschaft Osteuropas, 14, 159–186.
54 Cox, Andrew W., Ireland, Paul, Lonsdale, Chris, Sanderson, Joe and Glyn Watson, 2001, *Supply Chains, Markets and Power*, Routledge, London.

55 Bös, D., 1993, *Privatization in Europe: a Comparison of Approaches*, Review of Economic Policy, 9, 95–111.
56 Sidhu, Surjit S. and Mohinder S. Mudahar, 1999, *Privatization and Deregulation*, Kluwer Academic Publishers, Norwell, MA.
57 Kuchik, Akifumi and Masatsugu Tsuji, 2011, *Industrial Clusters: Upgrading and Innovation in East Asia*, Edward Elgar, Cheltenham.
58 Whiteley, Paul J., 1986, *Political Control of the Macro-Economy*, Sage, London.
59 Bradshaw, William and Helen Lawton-Smith, Eds., 1999, *Privatization and Deregulation of Transport*, Macmillan, Basingstoke.
60 Co-operative Bulk Handling Ltd., 2010, *Port Capacity Allocation System and Port Terminal Rules*, Discussion Paper, Perth.
61 Gayle, Dennis and Jonathan N. Goodrich, 1990, *Privatization and Deregulation in Global Perspective*, Pinter, London.
62 Morrison, Gerard, 2011, *Equipment Availability for Containerised Grain Exports*, paper presented to the Australian Grain Industry Conference, 28 July, Melbourne.
63 Thompson, Stephen, 2011, *Trends in the Freight Market and Exporting Grain from Australia*, paper presented to the Australian Grain Industry Conference, 28 July, Melbourne.
64 Australian Government Productivity Commission, 1999, *Impact of Competition Policy Reforms on Rural and Regional Australia*, AusInfo, Canberra.
65 Porter, Michael E., 1985, *Competitive Advantage – Creating and Sustaining Superior Performance*, The Free Press – Macmillan, New York.
66 This advocacy was made by Persson, Karl Gunnar, 1999, *Grain Markets in Europe, 1500–1900: Integration and Deregulation*, Cambridge University Press, Cambridge.
67 Whiteley, Paul J., 1986, *Political Control of the Macro-Economy*, Sage, London.
68 Refer to case studies in Chapters 6 and 7.
69 Whitwell, Greg and Diane Sydenham, 1991, *A Shared Harvest – The Australian Wheat Industry 1939–89*, Macmillan Education Australia, Melbourne, Ch. 13.
70 McColl, Jim, 1988, *Royal Commission into Grain Storage, Handling, and Transport*, Australian Government Publishing Service, Canberra, p. xxv.
71 Whitwell, Greg and Diane Sydenham, 1991, *A Shared Harvest – The Australian Wheat Industry 1939–89*, Macmillan Education Australia, Melbourne, pp. 26–27.
72 Bartholmaeus, Malcolm, 2011, *Wheat Quality Rewards Good for Pools*, Australian Farm Journal, 21:7, 10–12.
73 Francois, M. Jacques, 2007, *Even Commodities have Customers*, Harvard Business Review, 85:5, 110.
74 Richardson, Jeremy J., Ed., 1990, *Privatisation and Deregulation in Canada and Britain*, Gower Publishing, Aldershot.
75 Australian domestic wheat deregulation occurred in 1989, the first domestic wheat futures market appeared in 1995, but a sustainable futures market with adequate liquidity was not achieved until export market deregulation occurred in 2008.
76 Marriott, Richard S. and Gus McLean, 1991, *Going with the Grain: A History of Queensland's State Wheat Board*, published by the State Wheat Board, Toowoomba, Ch. 5.
77 Morgan, Dan, 1979, *The Merchants of Grain*, The Viking Press, New York.
78 Whitwell, Greg and Diane Sydenham, 1991, *A Shared Harvest – The Australian Wheat Industry 1939–89*, Macmillan Education Australia, Melbourne, p. 2.
79 The Australian export single desk under the Australian Wheat Board and AWB Ltd. existed from 1939 to 2008.
80 Karlsson, Charlie, Ed., 2008, *Handbook of Research on Innovation and Clusters – Cases and Policies*, Edward Elgar, Cheltenham.
81 Karlsson, Charlie, Ed., 2008, *Handbook of Research on Cluster Theory*, Edward Elgar, Cheltenham.
82 Porter, Michael, 1998, *Clusters and the New Economics of Competition*, Harvard Business Review, 77, 1–7.

83 Harvey, Katherine, 2004, *Supply Chain Analysis of Traceability Systems within the Grain Industry*, Department of Agricultural Sciences, La Trobe University, Melbourne.
84 McColl, Jim, 1988, *Royal Commission into Grain Storage, Handling, and Transport*, Australian Government Publishing Service, Canberra, p. xxv.
85 Christensen, Jorgen Gronnegard, 1989, *Regulation, Deregulation, and Public Bureaucracy*, European Journal of Political Research, 17, 223–239.
86 Greaves, Bettina Bien, 1998, *Interventionism: An Economic Analysis by Ludwig von Mises*, Foundation for Economic Education, Irvington-on-Hudson, New York.

Further readings on deregulation

ACIL, 2004, *The Effect of Grain Market Deregulation under National Competition Policy: A Report Prepared for the National Competition Council*, Melbourne, October.

Barkema, A., Drabenstott, M. and K. Skold, 1992, *Agriculture in the Former Soviet Union: the Long Road Ahead*, Economic Review of the Federal Reserve Bank of Kansas, 77:4, 79.

Bradshaw, William and Helen Lawton Smith, Eds., 1999, *Privatization and Deregulation of Transport*, Macmillan, Basingstoke.

Essinger, Stacy M., Hill, Lowell D. and Alfonso Rulz-Guinazu, 1996, *The Argentine Grain Industry: the Start of Privatization*, Department of Agricultural and Consumer Economics, University of Illinois at Urbana-Champaign, AE-4716, November.

Essinger, Stacy M., Hill, Lowell D. and Jacobus M. Laubscher, 1998, *Privatization Progress in the South African Maize Industry*, Department of Agricultural and Consumer Economics, University of Illinois at Urbana-Champaign, AE-4723, June.

McCorriston, Steve, and Donald MacLaren, 2007, *Deregulation as (Welfare Reducing) Trade Reform: the Case of the Australian Wheat Board*, American Journal of Agricultural Economics, 89:3, 637–650.

Nielsen, Randall, 2001, *Grain Markets in Europe, 1500–1900: Integration and Deregulation*, The Journal of Economic History, 61:1, 260–262.

Ohmae, K., 1995, *The End of the Nation State: The Rise of Regional Economies*, The Free Press, New York.

Park, Albert, Jin, Hehui, Rozelle, Scott and Jikun Huang, 2002, *Market Emergence and Transition: Arbitrage, Transaction Costs, and Autarky in China's Grain Markets*, American Journal of Agricultural Economics, 84:1, 67–82.

Tubiana, Laurence, 1989, *World Trade in Agricultural Products: From Global Regulation to Market Fragmentation*, in David Goodman and Michael Redclift, Eds., *International Farm Crisis*, Macmillan, London.

Watson, Alistair S., 1999, *Grain marketing and National Competition Policy: reform or reaction?* Australian Journal of Agricultural and Resource Economics, 43:4, 429–455.

Whitwell, Greg, 1993, *Regulation and Deregulation of the Australian Wheat Industry: the 'Great Debates' in Historical Perspective*, Australian Economic History Review, XXXIII, March, 22–48.

Wilson, William W. and Alexander Belozertsev, 1995, *Russian Grain Trading and Marketing: Evolution and Struggles*, Agribusiness, 11:1, 1–12.

Yergin, Daniel and Joseph Stanislaw, 2002, *The Commanding Heights – The Battle for the World Economy*, Simon and Schuster, New York.

12

FOOD SUPPLY CHAIN EFFICIENCY

Failure and blame are seemingly forever associated with markets and supply chains,[1] despite global consumers and farmers constituting the food market, and despite the efforts of global agricultural supply chain intermediaries to feed, clothe, and house people. This suggests that there is some degree of efficiency despite the huge disparity between the decision making of farmers and consumers who generally tend to ignore each other. The question arises as to where does this perceived 'failure' occur, and is blame more of a psychosomatic symptom of a well-fed region. The corollary may offer the answer in that under-fed people always seem to blame the government, war, or anarchy, but never the market and the food supply chain.

The problem with the concept of efficiency is that it is mostly indefinable even within an organization, but extremely difficult to calculate from an external supply chain perspective.[2] It is interwoven with supply chain effectiveness, product flow, value creation, value stream management, and vague notions of performance. Economists usually resort to a vague incalculable social 'welfare' to define external efficiency.[3]

One of the biggest consequences from food supply chain efficiency is that it has led to alienation of the consumer from the farmer, which has bred many myths and conspiratorial theories. The twentieth century led to a decimation of much of the previous rural workforce, a massive shift from rural living to urbanization, and a transfer of income, wealth, and politics from farmers to city dwellers.[4] This has created an unusual type of information asymmetry that is not readily solvable.[5] Much urban leisure time is utilized in pursuing a rustic delusion that may be a relic of some simplistic agrarian archetype, which makes it difficult to separate myth from fact.

There are many perceived 'failures' that seemingly focus on 'inefficiency' that might have more relevance to an archetypal dream, such as that of Richard Darré. Even a perception of inefficiency can result in supply chain intervention, despite the perception being restricted to vested interests such as sectionalized government bureaucrats and politicians, or a small group of farmers or consumers. Darré's dream of 'blood and soil' ended with famine, war, and bloodshed.

Capitalism is often suggested as being the cause of perceived inefficiency, which is probably very correct; however, the ills of capitalism historically have mostly derived from the effects of oligopolies and monopolies and not from competitive markets and supply chains.[6] If most food supply chain inefficiencies are caused by a lack of competition or by direct intervention, then intervention policies to correct 'capitalist ills' will more likely result in monopolies and intensify the non-competitive supply chain problems and inefficiencies.[7]

The greater the success of food supply chains, the more choice is provided to consumers. However, if consumers have an array of choices and they seemingly abuse this choice through obesity, ill-health, and disposal wastage,[8] does this constitute warranted criticism and blame against the supply chain mechanisms and efficiencies? The paradoxical lesson quickly learnt in public catering is that any restriction solely towards providing healthy food choices usually results in low profitability and quick bankruptcy. If consumers demand a certain food type, then food supply chains and caterers are forced to meet that demand, simply because that is where profitability and sustainability exists. To do otherwise would soon result in a closure of business operations. Advertising is only successful when it reaches an appreciative audience.

Another issue is that what may be efficient to the food supply chain in meeting consumer demand may not be perceived as efficient for a government health provider, such as hospitals or medical insurance. If the consumer has free choice over what is eaten, blame can hardly be placed upon the supplier who is merely the deliverer of what is being demanded.

Government dictates over public health efficiencies assume that it has the right to be the ultimate decider over what food chain is good or bad for its citizens, despite many contradictions between government departments such as tourist promotion, agriculture, and public institution providers. It presumes that people cannot learn quickly from their experiences and make the required adjustments as part of life's learning experience. The intrinsic problem is that people develop dependency on government, do not learn from their own experiences, and expect the government and taxpayer to be the ultimate provedore whenever health difficulties arise. If governments realized the ills of 'aid' by the 1980s, then they need to recognize the same dependency problems in current public health policies, which are a relic of Bismarck's Sozialstaat (Welfare State) in the 1870s[9] and its disastrous dependency aftermath[10] that craved for simplistic authoritarian leadership to provide for socially desirable outcomes.

Deregulation, supply chain efficiency, and consumer choice might be considered the luxuries that can only occur 40–50 years after a major war, once war-debt repatriations have ended and infrastructure repaid. There could be trade-offs between war-debt repatriation and supply chain efficiency for many countries. Price controls, trade protection, and government supply chain intervention may be forced upon some governments in order to generate foreign currency that can repay war debt. War or civil disruption may therefore have a serious impact on domestic supply chain efficiency many decades after such events have ceased.

Efficient food supply chains do not preclude governments, quite the contrary. There is a need for government to enforce competition and contract law, defend trade routes, impose law and order, prevent food contamination, ensure food safety, and issue temporary quarantine restrictions.[11] Some government actions can actively reinforce the efficiency environment,[12] but other policies such as taxes and intervention can destroy efficiency.

Despite being largely undefinable, efficiency is what underlies the success of food supply chains. It exists in a very dynamic environment with many causes, impacts, paradoxes, and variables. Without adequate robust operations management systems being implemented to induce efficiency, apathy can lead private food chains to depend on government intervention, which historically has evolved into much famine and death.

The ultimate efficiency criterion

If power in the food supply chain lies in the choice of the consumer to either buy or not to buy, then any measure of efficiency must account for customer satisfaction.[13] The problem with measuring the efficiency of achieving this satisfaction is compounded by individuality and fickleness of the consumer.[14] What may be efficiently provided to one consumer may be contrary to another. Alternatively, something that may satisfy a customer today may not satisfy tomorrow.

Efficiency criteria in the food supply chain must therefore be linked to current product and service flexibility, and the ability to change over time.[15] Product and service flexibility might be expected to increase during product surpluses and decrease during shortages. Any elimination of flexibility under shortage conditions, such as the ability to import, could lead to famine. The prerogative to import therefore should lie with the merchant or end user and not be determined by vested sectional interests under the guise of 'protection' or unwarranted quarantine laws.

The science of politics should be linked to the psychology of consumer behaviour because political stability is more likely to occur when consumers are satisfied. However, politicians cannot control the biological variations of food supply, despite the egotistical delusions of many dictators.[16] The standard practice under supply variation is to intervene and disrupt the very mechanisms that can cause consumers to be satisfied, thus creating political instability, which is self-destructing to politicians, bureaucracy, and supply chain intermediaries. A life of a dictator was never meant to be easy.[17]

Performance in meeting consumer expectations should therefore be the prime measure of efficiency in the supply chain.[18] The corollary suggests that a dysfunctional supply chain will have large inefficiencies and not meet the expectations of most consumers. Most food supply chain dysfunction lies with an inability to satisfy the final customer.

Consumer satisfaction transcends mere cost efficiencies and in most countries it must include the provision of quality, consistency, availability, continuity, suitability, safety, and more frequently include traceability, ethicality, and sustainability. Supply chain efficiency increasingly needs to be measured against meeting these criteria for both product and service provision. However, efficiency measurement benchmarking is extremely difficult when individuals have different value judgements on these criteria.

Another paradox is that quality, consistency, availability, continuity, suitability, safety, traceability, ethicality, and sustainability only become real issues in deregulated supply chains. It is deregulation that has allowed these issues to be initiated and openly discussed with a possibility for further improvement, which may encourage economic growth and development. Under government invention, these issues are frequently subsumed under a plethora of major problems through induced food shortages, lack of choice, price distorting signals, subsidies, 'equilibrium' schemes, hoarding, corruption, trade barriers, and ultimately leading to food 'aid' problems. Advocacy for government intervention to achieve traceability, ethicality, and sustainability is therefore highly incongruous.

Attempts to usurp supply chain power from consumers under the banner of 'efficiency' possibly extend over thousands of years. Usurpation frequently assumes that the efficiency in meeting consumer needs in the supply chain either can be ignored or treated with contempt and irrelevancy. Governments have used the excuse of sovereignty, defence, taxation, and law and order to usurp this supply chain power. Farmer organizations and minority consumer groups have often used political pressure to coerce governments into usurping the power of general consumers.[19]

If the efficiency associated with meeting final consumer satisfaction is ignored, then each participant along the supply chain is reduced to measuring their own efficiency, which is highly likely to have different internal and external perspectives of efficiency. Under such restrictions, the problem is that what is efficient to one supply chain participant is not efficient to another supply chain participant. Benchmarking performance is very difficult under different value judgements that are not consumer based.

Efficiency measurement

Efficiency is very difficult to quantitatively measure,[20] despite lavish internal subjective tools such as key performance indicators, key success factors, key core competencies, and factors creating competitive advantages.[21] External constraints and subjectivity can reduce objective efficiency performance measurement to competitive benchmarking, but even that requires competitors and assumes that there is sufficient competition to drive motivation, innovation, investment, and growth.[22]

Measures of supply chain performance must somehow relate to the ability to meet product and service specification and satisfy the final consumer. The measurement of supply chain efficiency must go beyond mere cost of production relativities and economies of scale to the behaviour of individual participants.[23] However, behavioural measurement criteria are weak because of subjective individual value judgements that vary over time under changed circumstances.

Micro-economic theory is ill-equipped to measure consumer behavioural–satisfaction criteria. John Stuart Mill introduced the theoretical concept of 'utility' during the first half of the nineteenth century to help explain psychological consumer behaviour, but it has always remained very difficult to actually measure.[24] Theoretical general utility curves with many assumptions are no substitute for actual behavioural measurement of individual consumers over time. A huge marketing and advertising industry is entirely devoted to this measurement identification aspect.

Measuring competition might be a better criterion of performance and efficiency in the supply chain than attempting to gauge customer satisfaction.[25] The number and size of supply chain intermediaries can be quantitatively measured and managed by government through enforcing competitive practices, improving infrastructural bottlenecks and lowering industry entry barriers.[26] However, there still remains subjectivity in establishing cause and effect relationships, and there needs to be robust commitment by government, efficiency within the bureaucracy, with low corruption occurrences between industry and government.

If efficiency is measured by food security, then generally there are many paradoxes and contradictions. Ironically, food security is hardly an issue in deregulated food supply chains with few import controls, even for countries with an extremely small domestic agricultural base such as Singapore. In contrast, food security is an issue in those countries such as Africa, the Middle East, and the Asian Subcontinent, which have burdensome government domestic regulations and import controls.[27] As well, food insecurity has risen to prominence since the Second World War, which has been an era characterized by highly regulated food chains.[28] This insecurity and regulation association suggests that food security is more of a measure of food chain inefficiency than supply chain efficiency. Therein lays a possible answer to the food dilemma.[29]

Efficiency of supply chains might be measured by the existence and performance of the forward market. There is no evidence in any country of price controls coexisting with strong liquid forward markets. Supply chains with no forward market lack the 'trade' clustering sophistication that would normally lead to futures markets.[30] A supply chain that was considered inefficient might be expected to have little forward market functionality and therefore limited ability to manage risks.

Similarly, if futures markets accurately reflect the conditions of the underlying physical supply chain, then local futures markets might be a barometer of the efficiency of the domestic supply chain. Any weaknesses in the futures market will usually reflect serious underlying supply chain problems or impediments, more so than any weakness in the futures market per se.[31] It is erroneous to blame the futures market when it is merely reflecting the problems in the underlying physical supply chain.[32]

Alternatively, balanced strengths in the futures market will reflect underlying efficiency in the physical supply chain. Balanced strength of a futures market can be measured by how well the speculators are balanced by the hedgers. Efficiency of a futures market should be measured by this balanced strength, otherwise high liquidity could merely reflect much speculative activity with little hedging activity.[33] While speculators are important to add trading volumes to futures markets, it is the volume of hedgers relative to speculators that really reflects the efficiency in the underlying physical supply chain.[34]

If the lack of food choice is so characteristic of regulated supply chains, then the corollary suggests that measuring the range of food choices for consumers might be an efficiency performance criterion.[35] Managing and expanding the food consumption possibilities–substitutes complex is the role of private food supply chains, and an increasing function of supermarket chains in more recent times.[36] The adoption and transformations occurring with these supermarket supply chains could reflect efficiencies in domestic food supply chains because of the competitiveness of global supermarkets.[37] This has implications for those countries wanting to unshackle themselves from debilitating regulation and food aid dependence.[38]

Any stakeholder approach to efficiency is difficult because of the many conflicting and often contradictory attitudes and behaviours of different vested interest groups. Proactiveness with one group may result in reaction by other groups.[39] Consumers are often subsumed into ill-definable or irrelevant groupings as stakeholders, so clear focus on the primary objective is often lost. There are legal obligations to many stakeholder groups, which often leads to the stakeholder approach being reduced to efficiency in satisfying minimal legal obligations.

The systems approach to efficiency may have some desirable attributes, despite its Bogdanov origins, because it analyses the supply chain as a whole and not segments of the supply chain independently. Just-in-time deliveries may be perceived as a systems approach, but such a holistic approach does not guarantee win-win outcomes for all supply chain participants.[40] Conceptualization does not always deliver efficiency in a dynamic market with many conflicting stakeholders, and it is frequently limited by the subjectivity of the designer. Lange could never make his socialist production concept workable, with failed results being explained through vague future inexorable inevitability.

Efficiency as measured by the competing values method examines the trade-offs within the supply chain. There may be a cost–value trade-off, beyond which the consumer perceives no extra value for money. The benefits of economies of scale through rail transport often preclude the required product segregation, traceability, and identity preservation. Certain consumer demands for services incur a high logistics cost that might be beyond the willingness of consumers to pay. Deregulation may squeeze profit margins in the supply chain through competition, and this can reduce the number of participants and the services and functions provided. The competing values approach has relevancy to the food supply chain, but its contribution as an efficiency measure is limited.

The ultimate barometer for inefficiency

If measurement of efficiency is difficult, then perhaps it is easier to identify inefficiency. Whenever the supply chain fails, there is an increased opportunity for speculation, either in the physical market with hoarding or pooling activities, or in futures markets that have been isolated from the underlying 'trade' markets[41] and become dysfunctional. If integration can be a measure of supply chain efficiency,[42] speculation resulting from disintegration therefore may become a barometer for inefficiency.

Wherever futures markets exist, inefficient underlying supply chains can result either in too little liquidity or excessive futures market speculation (excessive being gauged by speculator volume overwhelming hedger volume).[43] Alternatively, if futures markets do not exist, physical speculation through hoarding and frequent pooling reflect supply chain inefficiency. Hoarding, pooling, illiquid futures markets, excessive speculation, and general supply chain inefficiency that induces food shortages frequently occur concurrently in countries such as India, which has government price controls, monopolies, distorted supply chains and corruption.[44]

Underlying inefficiency is largely ignored when governments conveniently blame futures market speculation or private hoarding rather than distortions in the physical spot price. The formation of 'trade' markets and forward markets is quite complex and frequently misunderstood.[45] Blame is easily apportioned to the effect rather than the cause under such circumstances.

Excessive speculation associated with price distortions and supply chain dysfunction undermines risk management,[46] thus leaving 'trade' markets vulnerable to financial unsustainability, particularly when combined with weather-driven supply uncertainties.[47] Alternatively, an inability to manage risk within the supply chain necessitates speculation. Any exposure to physical price or currency risk is speculation. Price distortions can therefore lead to both external and internal speculation.

Subsidies can undermine supply chain efficiencies despite their seemingly benign largess from taxpayers. Futures markets should reflect the underlying 'trade' market conditions, but prices become distorted when subsidies are given to farmers, end users, or even supply chain intermediaries. Up-country markets usually behave quite differently to port-based markets because of local supply and demand differences, but this is likely to become accentuated with domestic subsidies and could make an up-country futures market price quite irrelevant to the port-based price, despite the predominance of exports.[48] This price distortion then creates big opportunities for excessive speculation.[49]

Having identified supply chain inefficiency, governments then need to isolate the cause and rectify it. The solution does not involve more supply chain direct intervention, rather it can involve deregulation, privatization, improved competitive practices, removal of subsidies[50] and trade barriers, minimizing corruption and the impact of vested-interest pressure groups, and more infrastructural investment. Focus needs to be on supply chain signals that are efficiently transmitted with transparency in a physical environment that facilitates the flow of trade, ideas, information, and investment.

Identifying private inefficiency does not imply that there are sufficient organizational and operational change mechanisms to rectify or solve the problem. The skills and experience of operations management become important if change is to be implemented.[51] As well, the identification of inefficiency also does not imply that today's efficiency measures will continue into the future. There is a time element associated with efficiency that needs to be monitored through change management strategies.[52]

Efficiency trade-offs

Perceived economies of scale whereby unit costs decrease with increased quantity may not lead to supply chain efficiency. Commoditization of commingled product through traditional bulk handling may have achieved perceived economies of scale efficiencies from a bulk handler perspective. It removed the necessity for segregated differentiation of product, thus removing a major cost for bulk handlers, which had all the hallmarks of great costs savings in the supply chain. Because merchants were integral to accumulation of differentiated product and storage of this segregation, they were perceived to be redundant. A seamless transition of product from farmer to consumer through a bulk handler was considered possible.

However, from an end-user perspective, commingled commoditization was an unmitigated 'dysfunctional' disaster from the perspectives of efficiency, technology, logistics, inventory finance and management, and consumer demand.[53] It did not provide the necessary product specification either being demanded by consumers or required for manufacturing transformation. It resulted in an unwanted product that was aligned and made dependent on an increasingly inefficient rail freight system. What was seemingly efficient on the production side of the supply chain was inefficient from the consumption side of the supply chain.

Seemingly efficient outsourcing and just-in-time strategies by food manufacturers can merely transfer inventory and logistics inefficiencies and costs back in the supply chain to suppliers such as farmers and merchant accumulators. Efficiency can therefore have trade-offs in such win-lose situations.[54] Joint value creation between an end user and an input supplier that might result in perceived improvements in overall supply chain efficiency could be very difficult to achieve.[55]

There is no automatic efficiency achieved in the supply chain just by the adoption of just-in-time principles.[56] What is perceived conceptually from an end-user perspective that might lead to efficient rationalization amongst competing suppliers, could also lead to poor performance of suppliers, inadequate investment in the supply chain, and ultimately supplier closures that could have serious impact on the financial viability and economic sustainability of end users, with an inability to source specified product. Just-in-time delivery operates best when there is a robust deregulated warehouse market, when there is adequate carry compensation in the forward market, and only when there are surpluses of biologically dependent product.

Surplus conditions would be expected to lead to greater certainty, less risk, and more efficient outcomes in the food supply chain. End-user activity and cluster industries would be expected to develop under supply chain deregulation whenever prolonged surplus conditions existed. However, the downside is the jeopardizing of longer-term supply because of decreasing real prices and returns to farmers.[57]

Alternatively, frequent supply shortage conditions would be expected to lead to greater uncertainty, higher risk, and less efficient outcomes in the food supply chain, which could even collapse under supplier default circumstances with bankruptcy amongst supply chain intermediaries. Increasing real prices, lower margin returns to end users, and an inability to support any warehousing or merchant activity can undermine 'trade' markets. However, the resultant periodic higher prices to farmers would motivate them to produce more for the future.[58] Food supply chains must be robust and flexible enough to withstand such shocks in the short to medium term.

There is a continual struggle between farmers and end users for sustainability or even survival, which frequently subsumes any goal to achieve supply chain efficiency.[59] The role of supply chain efficiency might be diminished whenever this struggle is the greatest.[60] Alternatively, supply chain efficiency could arise from a more balanced struggle between farmers and end users, with occasional swings between risk and rewards, but this is entirely theoretical because supply is weather dependent.

A question arises as to whether degrees of production and supply uncertainty will lead to greater or lesser supply chain efficiency. Therein lies another paradox. The countries that have the greatest production certainty, such as the four global food bowls of the Asian Subcontinent, China and Indo-China, Continental Europe, and the USA, all have the highest government intervention in supply chains either through price controls (India), 'buffer' stocks (China), or subsidies and tariffs (USA and EU). An example of US subsidy–tariff supply chain inefficiency occurred in August 2011 when 52.4 million gallons of ethanol were exported from corn-derived sources at the same time that 45.4 million gallons of ethanol were imported, which was derived from Central and South American sugar. Inefficiency caused by supply chain distortions in food bowl countries can explain why the world is in a dilemma over food supply, at a time of increasing global demand through population increases and rising income. Production and supply certainty is therefore no measure of supply chain efficiency.

This contrasts with some of the larger marginal agricultural producing and exporting countries that have the least production certainty, such as Australia and Brazil, but which tend to have the lowest government intervention and maybe more efficiency in food supply chains whenever surpluses occur, albeit with transport infrastructural inefficiencies. The trade-off in marginal producing and exporting countries such as Australia can be between deregulation and the lack of transport infrastructural investment by government. However, from a government perspective, there is a trade-off between marginality of production and transport infrastructural investment. Efficiency trade-offs can therefore depend on different stakeholder perspectives.

The higher the production marginality and uncertainty, there is more probability that risks could undermine any supply chain efficiency, especially for 'trade' markets when there are import barriers. If global food demand is increasingly being made dependent on supplies and exports from high production risk countries with government supply chain intervention, at the same time that there are subsidies, tariffs, 'buffer' stocks, and price and export controls in food bowl countries, this can explain why global food insecurity is currently perceived as a problem. However, the paradoxical solution of more government intervention will only worsen supply chain insecurity and inefficiency.

If prices are allowed to rise, farmers normally will respond in the medium term with more production. Therein lies more efficiency trade-offs. The problem is that inefficient food supply chains can prevent increased food production being aligned with food shortage regions because of inadequate price signalling and merchant arbitraging, whilst much food production can be alien to many food supply chains because of non-specific commodity characteristics and quality. Merely increasing agricultural production is no surety of security if there are inefficient supply chains, especially when storage technology has improved hoarding efficiency.[61]

As well, there is high food wastage in most countries,[62] which suggests that increased food production and supply chain efficiency alone may not remove food shortages. The efficiency of food usage in public catering is also important. Wastage of processed food may be relatively low for supermarkets[63] compared to the high wastage from private restaurants, takeaway food industries, and institutional catering.[64] This is despite such non-interventionist food preparation being perceived as efficient in the context of global supply chains.[65]

Questions arise as to whether there is trade-off between food wastage and supply chain efficiency, and whether such food wastage can lead to unsustainability and food insecurity.[66] Problems exist in both developed and less-developed countries,[67] which suggest that it is not income dependent. The corollary suggests that wherever supply chain inefficiency is maximized, food wastage is minimal with famine food rationing. Whilst this supports the association between food wastage and supply chain efficiency, the answer is more likely to lie in improved public catering efficiency, but that assumes the consumers will become less fickle in their tastes and less influenced by weather and seasons, and that caterers can become better at food demand forecasting. All of which is highly unlikely.

Any post-harvest storage to improve logistical efficiency can result in high losses and wastage.[68] Stored products can deteriorate through quality losses, chemical contamination, insect and fungal infestations, moisture-shrinkage loss, and humidity and temperature problems. Therefore, storage and efficiency can be considered trade-offs, despite major quality control management.

Partial deregulation problems can occur when some industries such as supermarkets are deregulated whilst other activities such as labour laws are not. Without sufficient domestic labour market reforms and human resource mobility and flexibility, import supply chains will be preferred by supermarkets because of their lower cost in contrast to their high-cost domestic supply chain counterparties. There can be major trade-off problems between labour regulation laws and domestic food supply chain efficiency. Supermarkets will tend to become increasingly dependent on food imports unless domestic labour deregulation and cost reduction occurs.

Efficiency and integration

There could be an inverse relationship between supply chain linkages and efficiency. Corruption could strengthen supply chain illegal relationships, but decrease efficient outcomes.[69] Alternatively, anti-corruption laws might weaken these perverse linkages and improve the more normal relationships that are required for supply chain efficiency.

Some organizations perceive specialization or vertical integration as the ultimate efficiency measure,[70] but ignore the many trade-offs such as increased supply chain risk, especially supply vulnerability and demand fickleness.[71] Without adequate risk management opportunities, diversification or horizontal integration might be preferred, simply to manage supply chain risks despite the possibility of associated increased inefficiencies. There is an efficiency–risk trade-off in many situations, and much depends on the ability to manage that risk.[72] The need to manage or minimize risk might overwhelm the desire for vertical integration and efficiency.

Forward contractual integration is often portrayed as a good measure of supply chain efficiency.[73] However, in weather-dependent food supply chains, this association cannot be easily resolved without firstly considering production and quality risk in the particular region or country.[74] High supply risk can undermine contractual integration and eliminate efficient supply chain mechanisms.

Uncertainty creates risk. If uncertainty is high, risks may be identifiable and measured, but extremely difficult to manage.[75] Regardless of the amount of integrative forward contracting and the sophistication relating to whether the contracts are vertical or horizontal in the supply chain, performance and efficiency depend on contract assurance.[76] The greater the production and quality risk, the higher will be the probability and frequency of supply chain collapse, irrespective of the amount of integration. This will lead to food insecurity. Alternatively, low production and quality risk could result in greater food security, irrespective of the amount of supply chain integration. Therefore, contractual supply chain integration may be quite independent of food security.[77]

However, supply chain integration is integral to food security through the formation of 'trade' markets and by the ability to offset risks through forward markets. An ability to establish both 'trade' markets and forward markets is more likely to lead to long-term sustainability of food security because of their risk transfer mechanisms. The corollary suggests that less integrative food supply chains would be expected to be less sustainable due to high risk and more vulnerable to supply chain intervention by either governments or monopolies.[78]

Impact of infrastructure and logistics

Infrastructure, logistics, and freight transport networks are vital links if supply chain efficiency is to be achieved.[79] Inadequate infrastructure, inefficient transport and delivery operations, bottlenecks, and some logistical complexities involved with containerization can destroy supply chain efficiency, despite the implementation of deregulation and just-in-time delivery policies.[80] Huge costs can be quickly added to a supply chain whenever there are failures in freight transport and inventory systems.

Alternatively, logistics and freight efficiency systems could fail, and infrastructure could be underutilized, if supply fluctuates widely.[81] Much infrastructure investment could be wasted in low production years, whilst low returns during these years discourages future investment. During the 2001–09 drought in eastern Australia, much railway rolling stock was transferred from agricultural to mining industries, and never returned in the good seasons.[82]

Capital infrastructural investment tends to flow only to those industries that have a much higher probability of long-term investment returns and positive multiplier benefits relative to other industries. Fiscal rectitude forces governments to comply with similar investment expenditure conditions as private enterprise. Security of supply is therefore vital to assure adequate average returns and cash-flow projections in large long-term capital infrastructural investment.

The insecurity of agricultural supply thereby weakens the justification for infrastructural and logistics investment, but this only undermines long-term food supply chain efficiency. Inadequate food logistics infrastructure tends to exacerbate any global food shortage conditions and worsen the food insecurity such as was experienced in 2008. There can be adequate production, but if geographic supply is not aligned to meeting specific location demand through efficient logistical systems, then food shortages may occur.

Inadequate supply response mechanisms and the existence of trade barriers offer potential windfall gains for hoarders. The propensity to hoard is more likely when there are supply restrictions, caused either by infrastructural logistics deficiencies or import barriers. Any improvement to infrastructural logistics and a decrease in trade barriers is likely to lessen the incentive for hoarding whenever acute supply pressures occur.

Complex logistics conversions in handling and transport systems are unlikely to lead to win-win efficiency outcomes. What is perceived as efficiency improvement from one perspective is not viewed the same from other perspectives. The conversion of agricultural output handling to rail transport was perceived to bring much supply chain efficiency by the early twentieth century. Similarly, the conversion from product segregation to bulk commingling was perceived to bring much supply chain efficiency by mid-twentieth century. Yet both logistic system conversions were to be disastrous in terms of value creation and product transformation in food supply chains,[83] and destroyed the ability of both manufacturers and consumers to trace product flows, preserve product identity, and maintain total quality control management.

Both the rail and bulk handling system conversions originated from perceived fiscal propriety for those governments who were made responsible for debts incurred after road transport increasingly made private rail freight unprofitable. The trade-off for this seemingly fiscal efficient policy was food supply chain inefficiency. Whether governments should have assumed control over unprofitable agricultural-based railway systems still remains a contentious fiscal responsibility issue even 100 years after the event occurred.[84]

Fiscal propriety through mandating usage of a monopoly public utility does not always result in efficient outcomes.[85] What may be convenient to government and good for taxpayers through mandating usage and applying user-pay principles, may be a bad outcome for other stakeholders. Apart from supply chain ineffectiveness and inefficiencies, there usually are opportunity costs of lost export competitiveness through higher costs and lack of product relevance. This will impact on domestic employment, economic growth, trade balances, and lower subsequent fiscal revenue. The intervention exercise may therefore become self-defeating even from a fiscal perspective.

Similarly, the adoption of a logistics and inventory management approach may preclude the achievement of other goals such as meeting customer needs, providing service satisfaction, and balancing stakeholder requirements. Bias towards a particular system approach may preclude value being created at each stage of the supply chain. This lack of value creation was the inherent problem for logistics and inventory management within the Lange socialist production model.[86] Supply chain efficiency is unlikely unless broader stakeholder objectives are met.

Deregulating and privatizing public transport and freight networks can be exceedingly difficult after many decades of regulation.[87] Many stakeholders feel threatened by change, management can preserve the existing organizational culture, whilst vested interests will actively resist change. This is particularly the case when transport costs are pooled[88] or cross-subsidized.[89] Existing monopolies will tend to shift only towards self-preservation policies,[90] and can prevent change because of fixed and sunk costs as well as entry barriers.[91] This often entrenches inefficiency in food supply chain infrastructure.

One continual problem for monopolies is the need to react to internationally driven change rather than domestically driven change. Demand for public rail-port complexes that were earlier perceived as efficient and suitable for monopoly control over bulk exports was quickly replaced during the 1980s by the need for decentralized flexible ports, particularly when many governments were deregulating and privatizing domestic transport. However, this occurred when governments usually were debt-laden and striving for fiscal budget balance, and had little willingness to fund large public infrastructural investments. This has resulted in port-user ownership and access problems, management inflexibility, and restricted terminal and freight logistics.[92]

There remains an urgent need for new private flexible port facilities,[93] especially those that facilitate containerization, but private incentives in large long-term port infrastructure development are often inadequate.[94] Such a large private investment requirement with uncertain financial returns remains the challenge for efficient global food supply chains, and is highly unlikely to occur with supply chain regulation or under the threat of re-regulation.

Efficiency and technology

Efficiency is the slave of technology. Product design, processes, organization, strategy, and management will mostly reflect prevailing technology, but with adoption time lags. The problem is that technological changes can quickly outdate systems and practices. Any early competitive advantages over competitors might be eroded quickly with technological change. As well, new technology does not always translate into success, especially when lead times to technological inventions and implementation can be extremely long.

It took the shipping industry 100 years after the introduction of agricultural bulk handling before the first Panamax ship was introduced in the 1960s, which could adequately handle large bulk cargoes. This may have solved the economies of scale problem, but Panamax-size bulk handling was dependent on extremely large bulk cargoes that only governments could afford, fund, and handle. This reduced the usefulness of the Panamax ship to the remnant years of the Cold War. By the late 1970s, increasing numbers of private end users were demanding either segregated containers or smaller shiploads to enable quality control, regular payment, and efficient inventory logistics handling. Efficiency criterion was just as mercurial as the underlying technological processes.

There is a timing element required in technological development. The paradox suggests that the long invention and implementation phase involved with aligning technological innovation and development to a perceived gap is unlikely to solve the problem, because the void is not static and is very dependent on dynamic externalities, whilst perceptions and priorities change with time. Technological breakthroughs are more likely to have an element of luck with the correct timing at the right place, rather than any planned outcomes through problem solving. The Panamax ship really came at the end of a historical era in agricultural commodity trade, and found more relevance in the advent of shipping bulk mineral resources.

There can be a huge divide in efficiency between different stages of the supply chain that can be attributable to different technological developments. One stage in the supply chain may potentially have efficiency but be prevented from benefiting because of inefficiencies at other stages of the supply chain. Bottlenecks and congestion can still plague otherwise efficient supply chains, particularly when technology lags high growth phases.

Although there is a need for adequate research, development and investment that efficiently align technological needs with problem solving, success largely depends on the luck of good timing. The uncertain question remains as to the role of government and taxpayers in funding such research investment, and whether there are enough incentives for private research and development. New technological development is crucial to maintain competitive advantage; however, there is always the dilemma over large investment, long lead times, the high incidence of technological failures, adoption and change problems, and the need for luck in implementation timing.

Efficiency and cluster industry formation

Private cluster industries can be another measure of supply chain efficiency through the amount of support complementariness to the existing stakeholders. Successful cluster industries usually arise under a suitable supply chain environment of deregulation, competition, price signal transparency, innovation, research and development, and expected niche market profitability.[95] Cluster industries and related investment are unlikely to occur if there are perceived obstacles or actual impediments to their success. This makes cluster industries a good barometer for supply chain efficiency, with few cluster industries existing under food supply chain regulation and intervention inefficiency.

However, the corollary requires close examination because mere deregulation is unlikely to quickly induce new clustering of industries. There can be a very long time lag to the successful development of cluster industries, and much depends on efficiency in macroeconomic policies including the removal of import–export barriers, and the introduction of microeconomic practices such as enforcing competition and preventing the re-emergence of monopolies.[96]

Supermarket chains are developing their own cluster industries, whilst previous small food manufacturers and service providers are finding new synergies with emerging supermarkets. Such developments are dramatically changing and challenging previous government perspectives and policies regarding domestic food chain regulation and intervention. Bureaucracy currently requires a completely different mindset compared to the singular homogenized food supply chains that so characterized much of the twentieth century.

Governments do have a major role in the development of cluster industries, but there is a very delicate balance between too much and too little control. Highly regulated supply chains are unlikely to encourage cluster industry development because of entry barriers, a lack of competition, imbalances between risk and reward, and few transparent price signals to base good decision making to enable profitability to occur. Too little government control can result in uncompetitive practices, the formation of monopolies, high instances of corruption, and a lack of infrastructure.

It is doubtful whether governments can successfully decide on a strategic location of agribusiness cluster industries based on theoretical optimal location theories. Agribusiness cluster industries develop on their own initiative, provided that the supply chain environment is conducive for their development.[97] Optimum cluster location should be based on maximizing the response to meeting identifiable needs within existing supply chains, and not based solely on geographical considerations or theoretically new supply chain concepts. After all, it was government policies that generally destroyed much cluster industry activity in the first instance. However, new road infrastructure by government can influence decision making in private cluster planning.

Clustering, intense rivalry and competition, low entry–exit barriers, and good positioning between customers and suppliers is what drives both industry and supply chain efficiency.[98] There is a need for competition to enable benchmarking against competitors and competing supply chains to maximize supply chain performances such as customer satisfaction and cost reduction.[99] However, benchmarking inevitably assumes static variables when dynamics prevail, and such competition comparisons may lead to targeting average performance levels rather than leadership goals.

Efficiency in government actions

Some government control is vital to maintain efficiency in food supply chains. The perennial problem is in deciding how much control,[100] especially in a dynamic environment that changes dramatically over time with many paradoxes, and how to prevent reintervention.[101] Even government promotion of particular food supply chains is fraught with problems of selective industry bias and potential health conflict issues. Research funding can be wasted on losers and be discriminative against winners.

Government policies are never going to satisfy all supply chain participants and their perceptions of efficiency criteria. What may be desirable and efficient for government may not be efficient for other stakeholders. Yet governments have a public responsibility to function efficiently,[102] not only for fiscal audits but also for service outcomes that are at minimum cost for maximum benefit, despite the difficulty in measuring such outcomes. Costs can easily outweigh benefits from public actions, which can be detrimental to supply chain efficiency and international competitiveness.

Paradoxes and trade-offs can frequently undermine government efficiency.[103] Enforcing competition and anti-monopoly laws within supply chains may decrease consumer prices, which then acts as a disincentive for farmers to produce or as an incentive for them to hoard. If farmers decrease production or withhold supply because of price disincentives, food shortages could eventuate, which drives food prices higher and benefits farmers. Cycles of profitability should therefore be part of the supply chain dynamics. Governments may have to ebb and flow in policy according to supply chain changes.[104] Alternatively, any bureaucratic rigidity could seriously impact on supply chain efficiency. Too little or too late can have a disproportionate impact on the supply chain, even when compared to overreaction.

Government institutional catering can have a disproportionate impact on food supply chains despite the seemingly public profile of private competition policies.[105] In most countries, the government is the single biggest buyer in the food supply chain. Public institutional buying can therefore crowd out private supply chains and distort prices through large tenders and supply-demand manipulation.[106] There is a requirement for some minimum amount of government institutional catering, however, there is also a need for growth constraints, change mechanisms, and efficiency targets,[107] otherwise institutional food buying could overwhelm private supply chains. Regulatory and institutional reform processes need to be permanently in place.[108]

Notes

1 Rama, Ismo and Sallyann Harvey, 2009, *Market Failure and the Role of Government in the Food Supply Chain*, Policy and Strategy Group, Victorian Department of Primary Industries, Melbourne, June.
2 Pfeffer, Jeffrey and Gerald R. Salancik, 2003, *The External Control of Organizations: a Resource Dependence Perspective*, Stanford Business Books, Stanford, CA.
3 Barr, Nicholas, 2004, *Economics of the Welfare State*, Oxford University Press, New York.
4 Desrochers, Pierre, 2010, *The Wealth and Poverty of Regions: Why Cities Matter*, Regional Studies, 44:8, 1097–1098.
5 Grossman, Stanford J. and Joseph E. Stiglitz, 1980, *On the Impossibility of Informationally Efficient Markets*, American Economic Review, 70, 393–408.
6 Lauck, Jon, 2000, *American Agriculture and the Problem of Monopoly: The Political Economy of Grain Belt Farming, 1953–1980*, University of Nebraska Press, Lincoln.
7 McColl, Jim, 1988, *Royal Commission into Grain Storage, Handling, and Transport*, Australian Government Publishing Service, Canberra.
8 Parfitt, Julian, Bathel, Mark, and Sarah Macnaughton, 2010, *Food Waste Within Food Supply Chains: Quantification and Potential for Change to 2050*, The Royal Society, London.
9 Fay, S.B., 1950, *Bismarck's Welfare State*, Current History, XVIII, January, 1–7.
10 Beck, Hermann, 1995, *The Origins of the Authoritarian Welfare Sate in Prussia: Conservatives, Bureaucracy, and the Social Question, 1815–70*, University of Michigan Press, Ann Arbor, MI.
11 Refer to Chapter 2.
12 Australian Competition and Consumer Commission, 2008, *Report of the ACCC Inquiry into the Competitiveness of Retail Prices for Standard Groceries, July 2008*, ACCC, Canberra, 5 August.
13 Mayer, Robert N., 1989, *The Consumer Movement: Guardians of the Marketplace*, Twayne, Boston.
14 von Mises, Ludwig, 1949, *Human Action: A Treatise on Economics*, Auburn, AL, Ch. 15, section 4.

15 Pfefferman, G. and G. Fields, 2003, *Paths Out of Poverty: Economic Mobility and the Private Sector in Developing Countries*, Kluwer and World Bank Publications, Washington, DC.
16 Landsman, Mark, 2005, *Dictatorship and Demand*, Harvard University Press, Cambridge, MA.
17 Montefiore, Simon Sebag, 2004, *Stalin: The Court of the Red Tsar*, Phoenix-Orion, London.
18 Atkinson, Robert D., 2006, *Supply-side Follies: Why Conservative Economics Fails, Liberal Economics Falters, and Innovation Economics is the Answer*, Rowman and Littlefield, Lanham, MD.
19 Creighton, Lucy Black, 1976, *Pretenders to the Throne: The Consumer Movement in the United States*, DC Heath, Lexington, MA.
20 Hoek, Remko I. van, 1998, *Measuring the Unmeasurable – Measuring and Improving Performance in the Supply Chain*, Supply Chain Management, 3:4, 187–191.
21 Hines, Peter, Lamming, Richard, Jones, Dan, Cousins Paul, and Nick Rich, 2000, *Strategic Performance Measurement Systems*, in M. Kald, and F. Nilsson, 2000, *Value Stream Management: Strategy and Excellence in the Supply Chain*, Prentice Hall, Harlow.
22 Hilmer, Frederick G., 1993, *National Competition Policy: Hilmer Report*, Australian Government Publishing Service, Canberra.
23 Hines, Peter, Holweg, Matthias, and Nick Rich, 2004, *Learning to Evolve: A Review of Contemporary Lean Thinking*, International Journal of Operations and Production Management, 24:10, 994–1011.
24 Latham, Gary P. and Glen Whyte, 1994, *The Futility of Utility Analysis*, Personal Psychology, 47:1, 31–46.
25 National Competition Policy Wheat Marketing Act Review Committee, 2000, *National Competition Policy Review of the Wheat Marketing Act 1989*, Canberra.
26 Griffith, Garry, 2000, *Competition in the Food Marketing Chain*, The Australian Journal of Agricultural and Resource Economics, 44:3, 333–367.
27 Yemtsov, Ruslan, 2008, *The Food Crisis: Global Perspectives and Impact on MENA, Fiscal and Poverty Impact*, Middle East and North Africa Poverty Reduction and Economic Management Department, The World Bank, Washington, DC.
28 Shaw, John D., 2007, *World Food Security: A History Since 1945*, Palgrave Macmillan, New York.
29 Stevenson, Lois, 2010, *Private Sector and Enterprise Development: Fostering Growth in the Middle East and North Africa*, Cheltenham, in conjunction with the International Development Research Centre.
30 Anderson, Ronald W., Ed., 1984, *The Industrial Organization of Futures Markets*, Lexington Books, Lexington, MA.
31 Irwin, Scott H., Garcia, Philip, Good, Darrel L., and Eugene L. Kunda, 2009, *Poor Convergence Performance of CBOT Corn, Soybeans, and Wheat Futures Contracts: Causes and Solutions*, Marketing and Outlook Research Report 2009–02, University of Illinois at Urbana-Champaign.
32 Carlton, Dennis W., 1984, *Futures Markets: Their Purpose, Their History, Their Growth, Their Success and Failures*, Journal of Futures Markets, 4:3, 237–271.
33 Permanent Subcommittee on Investigations, 2009, *Excessive Speculation in the Wheat Market*, US Senate Majority and Minority Staff Paper, Washington, DC, 24 June.
34 Working, Holbrook, 1954, *Whose Markets? Evidence on Some Aspects of Futures Trading*, Journal of Marketing, 19:1, 1–11.
35 Ross, David F., 2008, *The Intimate Supply Chain: Leveraging the Supply Chain to Manage the Customer Experience*, Taylor and Francis, Abingdon.
36 Burch, David and Geoffrey Lawrence, Eds., 2007, *Supermarkets and Agri-Food Supply Chains: Transformations in the Production and Consumption of Foods*, Edward Elgar, Cheltenham.
37 Abrahams, Caryn, 2009, *Transforming the Region: Supermarkets and the Local Food Economy*, African Affairs, 109:434, 115–134.
38 Readon, Thomas, Henson, Spencer, and Julio Berdeque, 2007, '*Proactive Fast-tracking' Diffusion of Supermarkets in Developing Countries: Implications for Market Institutions and Trade*, Journal of Economic Geography, 7:4, 399–431.

39 Issacs. Harold R., 1975, *Idols of the Tribe: Group Identity and Political Change*, Harper and Row, New York.
40 Karmarkar, Uday, 1989, *Getting Control of Just-in-Time*, Harvard Business Review, 67:5, 122–131.
41 Irwin, Scott H., Garcia, Philip, Good, Darrel L., and Eugene L. Kunda, 2009, *Poor Convergence Performance of CBOT Corn, Soybeans, and Wheat Futures Contracts: Causes and Solutions*, Marketing and Outlook Research Report 2009–02, University of Illinois at Urbana-Champaign.
42 Hobbs, Jill E. and Linda M. Young, 2001, *Vertical Linkages in Agri-Food Supply Chains in Canada and the United States*, Strategic Policy Branch, Agriculture and Agri-Food Canada, June.
43 Permanent Subcommittee on Investigations, 2009, *Excessive Speculation in the Wheat Market*, US Senate Majority and Minority Staff Paper, Washington, DC, 24 June.
44 Roy, Subroto, 1984, *Pricing, Planning, and Politics: A Study of Economic Distortions in India*, Institute of Economic Affairs, London.
45 Faruqee, Rashid, Coleman, Jonathan R., and Tom Scott, 1997, *Managing Price Risk in the Pakistan Wheat Market*, The World Bank Economic Review, 11:2, 263–292.
46 Aulerich, Nicole M., Fishe, Raymond P., and Jeffrey H. Harris, 2011, *Why Do Expiring Futures and Cash Prices Diverge for Grain Markets?* Journal of Futures Markets, 31:6, 503–533.
47 Wagner, Stephan M. and Christopher Bode, 2008, *An Empirical Examination of Supply Chain Performance Along Several Dimensions of Risk*, Journal of Business Logistics, 29:1, 307-325.
48 Irwin, Scott H., Garcia, Philip, Good, Darrel L., and Eugene L. Kunda, 2009, *Poor Convergence Performance of CBOT Corn, Soybeans, and Wheat Futures Contracts: Causes and Solutions*, Marketing and Outlook Research Report 2009–02, University of Illinois at Urbana-Champaign.
49 Permanent Subcommittee on Investigations, 2009, *Excessive Speculation in the Wheat Market*, US Senate Majority and Minority Staff Paper, Washington, DC, 24 June.
50 Gupta, Sanjeev, 2000, *Equity and Efficiency in the Reform of Price Subsidies: A Guide for Policymakers*, International Monetary Fund, Paris.
51 Möller, Kristian E. and Pekka Törrönen, 2003, *Business Suppliers' Value Creation Potential: A Capability-based Analysis*, Industrial Marketing Management, 32:2, 109–118.
52 Harland, Christine, Lamming, Richard, Jones, Dan, and Paul Cousins, 1999, *Developing the Concept of Supply Strategy*, International Journal of Operations and Production Management, 19:7, 650–673.
53 McColl, Jim, 1988, *Royal Commission into Grain Storage, Handling, and Transport*, Australian Government Publishing Service, Canberra.
54 Fawcett, Stanley E. and Gregory M. Magnan, 2002, *The Rhetoric and Reality of Supply Chain Integration*, International Journal of Physical Distribution and Logistics Management, 32:5, 339–361.
55 Walter, Achim, Ritter, Thomas, and Hans G. Gemünder, 2001, *Value Creation in Buyer–Seller Relationships: Theoretical Considerations and Empirical Results from a Supplier's Perspective*, Industrial Marketing Management, 30, 365–377.
56 Cohen, S. and J. Rousell, 2005, *Strategic Supply Chain Management: The Five Disciplines for Top Performance*, McGraw-Hill, New York.
57 Kingwell, Ross, 1994, *Effects of Tactical Responses and Risk Aversion on Farm Wheat Supply*, Review of Marketing and Agricultural Economics, 62:1, 29–42.
58 Johnson, Glenn L., 1952, *Handling Problems of Risk and Uncertainty in Farm Management Analysis*, Journal of Farm Economics, 34:5, 807–817.
59 Rosin, Christopher, Stock, Paul, and Hugh Campbell, 2011, *Food Systems Failure*, Earthscan, London.
60 Friend, John and Allen Hickling, 1987, *Planning under Pressure – The Strategic Choice Approach*, Pergamon Press, Oxford.
61 Refer to Chapter 7.

62 Gustavsson, Jenny, Cederberg, Christel, Sonesson, Ulf, van Otterdijk, Robert, and Alexandra Meybeck, 2011, *Global Food Losses and Food Waste: Extent, Causes and Prevention*, FAO, Rome.
63 Buzby, Jean C., Farah Wells, Hodan, Axtman, Bruce, and Jana Mickey, 2009, *Supermarket Loss Estimates for Fresh Fruit, Vegetables, Meat, Poultry, and Their Use in the ERS Loss-Adjusted Food Availability Data*, USDA Economic Information Bulletin, 44.
64 Bloom, Jonathan, 2010, *American Wasteland: How America Throws Away Nearly Half of its Food (and What We Can Do About It)*, Da Capo Press, Cambridge, MA.
65 Gopinath, Munisamy, 2008, *Global Productivity Distribution and Trade Liberalisation: Evidence from Processed Food Industries*, European Review of Agricultural Economics, 35:4, 439–460.
66 Baldwin, Cheryl, Ed., 2009, *Sustainability in the Food Industry*, IFT Press, Chicago, Ch. 2.
67 Smith, Nigel and Mansoor Ali, 2002, *Waste from Food: Review of Developing and Transitional Countries*, FAO, Rome.
68 Bala, B.K., Haque, M.A., Hossain, A., and S. Majumdar, 2010, *Post Harvest Loss and Technical Efficiency of Rice, Wheat, and Maize Production System: Assessment and Measures for Strengthening Food Security*, Bangladesh Agricultural University, Report 6/08.
69 Fleming, Peter and Stelios C. Zyglidopoulos, 2009, *Charting Corporate Corruption*, Edward Elgar, Cheltenham.
70 Klein, B. and R.G. Crawford, 1978, *Vertical Integration, Appropriate Rents, and the Competitive Contracting Process*, Journal of Law and Economics, 21, 297–326.
71 Wagner, Stephan M. and Christopher Bode, 2008, *An Empirical Examination of Supply Chain Performance Along Several Dimensions of Risk*, Journal of Business Logistics, 29:1, 307–325.
72 Blank, Steven C., Carter, Colin A., and Jeffrey McDonald, 1997, *Is the Market Failing Agricultural Producers Who Wish to Manage Risks?* Contemporary Economic Policy, 15:3, 103–112.
73 Fischer, Christian, Hartmann, Monika, Reynolds, Nikolai, Leat, Philip, Revoredo-Giha, Ce'sar, Henchion, Maeve, Albisu, Luis Miguel, and Azucena Gracia, 2009, *Factors Influencing Contractual Choice and Sustainable Relationships in European Agri-food Supply Chains*, European Review of Agricultural Economics, 36:4, 541–569.
74 Sheldon, I.M., 1996, *Contracting, Imperfect Information, and the Food System*, Review of Agricultural Economics, 18, 7–19.
75 Williams, John and William Schroder, 1999, *Agricultural Price Risk Management – The Principles of Commodity Trading*, Oxford University Press, Melbourne, Ch. 1.
76 Lee, Hau L., 2002, *Aligning Supply Chain Strategies with Product Uncertainties*, California Management Review, 44:3, 105–119.
77 Hobbs, Jill E. and Linda M. Young, 2001, *Vertical Linkages in Agri-Food Supply Chains in Canada and the United States*, Strategic Policy Branch, Agriculture and Agri-Food Canada, June.
78 Sachs, Jeffrey D., Warner, Andrew, Åslund, Anders, and Stanley Fischer, 1995, *Economic Reform and the Process of Global Integration*, Brookings Papers on Economic Activity, 1, 1–118.
79 Beuthe, Michel, Himanen, Veli, Reggiani, Aura, and Luca Zamparini, 2003, *Transport Developments and Innovations in an Evolving World*, Springer-Verlag, New York.
80 Hein, Treena, 2011, *As Grain Export Volumes Increase, Canada Struggles to Boost Infrastructure Efficiency*, Australian Farm Journal, 21:7, 6–8
81 Brennan, D. and R. Linder, 1991, *Investing in Grain Storage Facilities under Fluctuating Production*, Australian Journal of Agricultural Economics, 35:2, 159–178.
82 Harris, Robert G., 1977, *Economics of Traffic Density in the Rail Freight Industry*, Bell Journal of Economics, 8:2, 556–564.
83 McColl, Jim, 1988, *Royal Commission into Grain Storage, Handling, and Transport*, Australian Government Publishing Service, Canberra.
84 Brennan, D.C., 1995, *Deregulation and Efficiency in Grain Transport*, Australian Journal of Agricultural Economics, 39:2, 129–141.

85 Friedlaender, Ann F. and Richard H. Spady, 1980, *Freight Transport Regulation*, MIT Press, Cambridge, MA.
86 Lange, Oskar R., 1936, *On the Economic Theory of Socialism*, Part 1, Review of Economic Studies, 4:1, 53–71.
87 Bromley, Daniel W., 1993, *Creating Market Economies from Command Economies*, Economic Issues, Department of Agricultural Economics, University of Wisconsin-Madison, No. 121.
88 Quiggin, John, 1994, *Cost Pooling in the Australian Grain Handling and Transport Industries*, American Journal of Agricultural Economics, 76:2, 262–269.
89 Faulhaber, G.R., 1975, *Cross-Subsidisation: Pricing in Public Enterprises*, American Economic Review, 65, 966–977.
90 Sharkey, William W., 1982, *The Theory of Natural Monopoly*, Cambridge University Press, Cambridge.
91 Baumol, William J. and Robert D. Willig, 1981, *Fixed Costs, Sunk Costs, Entry Barriers, and Sustainability of Monopoly*, The Quarterly Journal of Economics, 96:3, 405–431.
92 Hall, Peter V. and Jacobs Wouter, 2010, *Shifting Proximities: The Maritime Ports Sector in an Era of Global Supply Chains*, Regional Studies, 44:9, 1103–1115.
93 Frankel, Ernst G., 1985, *Bulk Shipping and Terminal Logistics*, The World Bank, Washington, DC.
94 United Nations, 2004, *Commercial Development of Regional Ports as Logistics Centres*, New York.
95 Porter, Michael E., 1998, *Clusters and the New Economics of Competition*, Harvard Business Review, 77, 1–7.
96 Baumol, William J. and Andrew E. Burke, 2001, *Competition Policy in Dynamic Markets*, International Journal of Industrial Organization, 19:5, 613–634.
97 Porter, Michael E., 2003, *The Economic Performance of Regions*, Regional Studies, 37: 6&7, 549–578.
98 Porter, Michael E., 1985, *Competitive Advantage – Creating and Sustaining Superior Performance*, The Free Press – Macmillan, New York.
99 Ryder, Ray and Andrew Fearne, 2003, *Procurement Best Practice in the Food Industry: Supplier Clustering as a Source of Strategic Competitive Advantage*, Supply Chain Management: An International Journal, 8:1, 12–16.
100 Barro, Robert J., 1996, *Getting It Right: Markets and Choices in a Free Society*, MIT Press, Cambridge, MA.
101 Vogel, Stephen, 1996, *Freer Markets, More Rules: Regulatory Reform in Advanced Industrial Countries*, Cornell University Press, Ithaca, NY.
102 Head, Brian and Elaine McCoy, 1991, *Deregulation or Better Regulation: Issues for the Public Sector*, Macmillan, Melbourne.
103 Okun, Arthur M., 1975, *Equality and Efficiency: The Big Tradeoff*, Brookings Institution, Washington, DC.
104 Ramadanadham, Venkata Vemuri, *Privatization and After: Monitoring and Regulation*, Routledge, London.
105 Vietor, Richard H.K., 1996, *Contrived Competition: Regulation and Deregulation in America*, Harvard University Press, Cambridge, MA.
106 World Bank, 1995, *Bureaucrats in Business: The Economics and Politics of Government Ownership*, Oxford University Press, New York.
107 North, Douglass, 1990, *Institutions, Institutional Change and Economic Performance*, Cambridge University Press, Cambridge.
108 Breyer, Stephen, 1982, *Regulation and its Reform*, Harvard University Press, Cambridge, MA.

Further readings on supply chain efficiency

Australian Productivity Commission, 1999, *Impact of Competition Policy Reforms on Rural and Regional Australia*, Canberra.

Baumol, William J. and Andrew E. Burke, 2001, *Competition Policy in Dynamic Markets*, International Journal of Industrial Organization, 19:5, 613–634.

Bloch, Harry and Peter Kenyon, 2001, *Creating an Internationally Competitive Economy*, Palgrave-Macmillan, Sydney.
Cassen, Robert and Vijay Joshi, Eds., 1995, *India: The Future of Economic Reform*, Oxford University Press, New Delhi.
Colclough, Catherine M. and James Manor, Eds., 1995, *States or Markets? Neo-Liberalism and the Development Policy Debate*, Oxford University Press, Oxford.
Friend, John and Allen Hickling, 1987, *Planning under Pressure – The Strategic Choice Approach*, Pergamon Press, Oxford.
Griffith, Garry R., 2000, *Competition in the Food Marketing Chain*, Australian Journal of Agricultural and Resource Economics, 44, 333–367.
Hines, Tony, 2006, *Supply Chain Strategies: Customer-Driven and Customer Focused*, Elsevier, Boston, MA.
Holden, Paul and Sarath Rajapatirana, 1995, *Unshackling the Private Sector: A Latin American Story*, The World Bank, Washington, DC.
Howard, Philip K., 1994, *The Death of Common Sense: How Law is Suffocating America*, Warner Books, New York.
Johnson, Glenn L., 1957, *Methodology for Studying Decision Making*, Journal of Farm Economics, 39:5, 1215–1226.
Kahneman, Daniel and Amos Tversky, 1973, *On the Psychology of Prediction*, Psychological Review, 80, 237–251.
Knight, Frank H., 1922, *Risk, Uncertainty and Profit*, The Quarterly Journal of Economics, 36:4, 682–690.
Nelson, Richard R. and Sidney G. Winter, 1982, *An Evolutionary Theory of Economic Change*, Belknap Press, Cambridge, MA.
North, Douglass, 1994, *The Evolution of Efficient Markets in History*, in John A. James and Mark Thomas, 1994, *Capitalism in Context: Essays on Economic Development and Cultural Change in Honor of R.M. Harwell*, University of Chicago Press, Chicago.
Overholt, William H., 1993, *The Rise of China: How Economic Reform is Creating a New Superpower*, W.W. Norton and Co., New York.
Shapira, Zur, 1993, *Ambiguity and Risk Taking in Organizations*, Journal of Risk and Uncertainty, 7, 89–94.
Watson, Alistair S., 1999, *Grain marketing and National Competition Policy: reform or reaction?* Australian Journal of Agricultural and Resource Economics, 43:4, 429–455.
Wilson, William W., Carlson, Donald C.E. and Bruce L. Dahl, 2004, *Logistics and Supply Chain Strategies*, Agribusiness, 20:4, 449–464.

GLOSSARY OF TERMS

Arbitrage. This is riskless simultaneous buying and selling (or selling and buying) of a product with the same specifications to take advantage of price discrepancies between markets. It forms one of the main reasons for trade.

Asset fixity. This refers to the inability of farm assets to shift between different farm enterprises because of restrictions relating to land, machinery and equipment, labour, and capital.

Backwardation. As a traditional merchant term, it referred to the discount or percentage that a seller had to accept by delaying the delivery of physical inventory or stock to a buyer. It refers to a forward market whereby either the end user pays a premium for spot delivery that is above the price paid for future deliveries, or else pays a discount for forward deliveries. It is characteristic of a shortage market. The forward month prices are at a discount to the spot month price.

Barème. It is a publicly listed price schedule. There could be a list of prices for a range of products, or it could refer to a premium and discount schedule for the one product type. It was common under French price controls.

Basis. Traditionally it refers to the basis of how price is determined. Quite often for storable agricultural products that have a futures market from which to benchmark local prices, basis is the difference between the local price and the futures price. Spot basis is the difference between the spot local price and the spot month futures price, whereas forward basis is the difference between the local forward price and the forward month futures price.

Basis risk. This is the risk of major differences between local prices and a benchmark price over time (the risk of market prices not being parallel). In futures and options hedging, basis risk is extremely important in determining hedge (offsetting) effectiveness. Initial basis may not equal final basis.

Black market. It can refer to any private market or supply chain that is deemed illegal under government law.

Bounties. These refer to government (taxpayer) subsidies to domestic manufacturers such as fertilizer companies to keep farm input prices lower than what they otherwise would have been. They are usually implemented when there are high tariffs or import protection.

Buffer funds. These are centralized funds accumulated through a levy or voluntary deposit when prices are high, form which payments can be made to farmers when prices are low.

Buffer stock. This is commodity either purchased by government through a stock reserve policy (buffer against food scarcity) or else held collectively by farmers through a statutory selling authority (buffer against low prices – *see* reserve price schemes).

Buy call options. The buyer of call options is given the right but not the obligation to take up the underlying buy futures contracts, whenever prices rise. There is a cost, referred to as the premium or price of the option.

Buy put options. The buyer of put options is given the right but not the obligation to take up the underlying sell futures contracts, whenever prices fall. There is a cost, referred to as the premium or price of the option.

CAP. Common Agricultural Policy of the European Union.

Carry. This refers to the reimbursement by the end user to a merchant on a monthly basis for the cost of storing the specified product over a period of time. It has a time value component. Carry occurs in a contango surplus forward market, whereby forward monthly prices are stepped upwards. Carry does not refer to creating place value through transport logistics.

cif. Commission (cost), insurance, and freight. It is destination-port pricing.

Contango. As a traditional merchant term, it referred to the charge or percentage paid by the buyer to a seller to delay the delivery of physical inventory. This occurs in surplus markets when the end user pays a storage charge (carry) to accumulating merchants to enable just-in-time deliveries to occur. The forward month prices are at a premium to the spot month price.

Contract default. This refers to when either party to a contract cannot meet contractual obligations. If physical delivery cannot occur by the seller, then the contract is either cash settled or an agreed upon alternate physical exchange occurs. If cash payment cannot be made by the buyer, then bankruptcy proceedings are implemented by the seller.

Cornering. It refers to purchasing and hoarding sufficient quantity of commodity in an attempt to 'squeeze' the market and force prices higher. Cornering is a specific type of hoarding.

Countertrade. International trade whereby the buyer makes payment that is not in the currency or method preferred by the seller. Barter may be one form of countertrade.

Dumping. This refers to a strategy of selling at a significantly lower price in an overseas market than in the supplier's home market. It also has more formal legal definitions under the World Trade Organization.

Economies of scale. Unit cost decreases with quantity produced, but only up to a critical point and then diseconomies of scale occur.

Elasticity of demand. This measures the responsiveness of demand (by an end user or consumer) to a change in price. The greater the responsiveness, the higher is the elasticity.

Elasticity of supply. This measures the responsiveness of supply (by a farmer or speculative merchant) to a change in price. The greater the responsiveness, the higher is the elasticity.

Fiscal policy. Government policy associated with budgetary income and expenditure.

fob. Free-on-board, fully stowed on-board ship. It is origin-port pricing.

Futures market. It is a central exchange where standardized forward contracts are traded and either cash-settled or delivered-against into the underlying 'trade' market.

Hermetic. Airtight conditions in storage that prevent living organisms.

Hoarding. This is a very deliberate attempt either by a farmer, merchant, or commodity pool manager to withhold commodity from the supply chain for the purpose of forcing prices higher. It is highly speculative and does not involve product already committed to the 'trade' markets or institutional catering. Any government stocks in excess of institutional catering needs can be also considered hoarding.

Home consumption price. This occurs when the domestic price is controlled, either higher or lower than the prevailing international price.

JIT. Just-in-time inventory delivery.

Merchant. This is a supply chain intermediary who takes ownership title over the product, is characterized by buying and selling functions, and is a risk taker except for price. Domestic merchants usually integrate with 'trade' markets with specific product, whereas international merchants might be more involved with freight forwarding, global networks, and funding the supply chain. Any merchant who undertakes price risk becomes a physical speculator.

Monetary policy. Government policy associated with the supply and demand of money, interest rates, inflation and currency.

Money laundering. Unofficial monetary transactions and money flows that are designed to avoid official scrutiny.

Monopoly. Product or service offered by one seller but with many buyers. It derives from the Greek 'monos' meaning alone or single, and 'poleein' meaning to sell.

Monopoly rent. Excessive profit gained through monopoly advantage.

Monopsony. Product or service offered by several sellers but with only one buyer.

Oligopoly. Product or service offered by a few (limited) sellers but with many buyers.

Pool price. This is collective price averaging in a comingled commodity pool over a fixed selling period, which facilitates equal payments to farmers per unit sold.

Pooling. This refers to the commingling of commodity by a collective of farmers, generally with the aim of averaging price from a spread of sales over a certain time period.

Price discrimination. This occurs under controlled price differentiation when there are separate markets with different price elasticities of demand, and the quantity supplied to each market can be controlled and different prices applied.

Price stabilization schemes. These usually are an attempt to collectively manage price through domestic supply and demand manipulations.

Quotas. These are predetermined quantitative limits on either production or acreage.

Reserve price scheme. This establishes a collective minimum price and operates through the use of a buffer stock that withholds domestic supply.

Spot sale. It refers to an immediate transaction at the current spot price when the seller is willing and able to sell and deliver, and when the buyer is willing and able to pay.

Supply diversion. This is a collective means of transferring supply between markets with the specific purpose of supporting the prevailing price. Dumping is one method of supply diversion.

'Trade' market. This is a supply chain agreement between an end user and a merchant for the latter to accumulate, store, and deliver the specified product when required by the end user. As a reward for creating time value, the end user pays a monthly carry charge to the merchant. Because of this carry characteristic, 'trade' markets usually only exist in surplus markets, and can disintegrate when product shortages occur.

USD Index. This is an index of the US dollar against a basket of currencies of the major trading countries that undertake trade or currency exchange with the US.

Vesting powers. In agricultural supply chains, it refers to the authority of government or statutory organization to assume ownership of agricultural output before it is actually produced or finally harvested. It can refer to 'in-paddock' ownership.

INDEX